Food Phytochemicals for Cancer Prevention II

ACS SYMPOSIUM SERIES **547**

Food Phytochemicals for Cancer Prevention II

Teas, Spices, and Herbs

Chi-Tang Ho, EDITOR
Rutgers, The State University of New Jersey

Toshihiko Osawa, EDITOR
Nagoya University

Mou-Tuan Huang, EDITOR
Rutgers, The State University of New Jersey

Robert T. Rosen, EDITOR
Rutgers, The State University of New Jersey

Developed from a symposium sponsored
by the Division of Agricultural and Food Chemistry
at the 204th National Meeting
of the American Chemical Society,
Washington, D.C.,
August 23–28, 1992

American Chemical Society, Washington, DC 1994

Library of Congress Cataloging-in-Publication Data

Food phytochemicals for cancer prevention II: teas, spices, and herbs /
Chi-Tang Ho, editor ... [et al.].

    p.   cm.—(ACS symposium series, ISSN 0097–6156; 547)

"Developed from a symposium sponsored by the Division of
Agricultural and Food Chemistry at the 204th National Meeting of the
American Chemical Society, Washington, D.C., August 23–28, 1992."

Includes bibliographical references and index.

ISBN 0–8412–2769–1

1. Cancer—Chemoprevention—Congresses.   2. Cancer—Nutritional
aspects—Congresses. 3. Herbs—Therapeutic use—Congresses.

    I. Ho, Chi-Tang, 1944–   . II. American Chemical Society. Division
of Agricultural and Food Chemistry. III. Series.

RC268.15.F66   1994
616.99′405—dc20   93–33774
                                                               CIP

The paper used in this publication meets the minimum requirements of American National
Standard for Information Sciences—Permanence of Paper for Printed Library Materials, ANSI
Z39.48–1984.   ∞

PRINTED IN THE UNITED STATES OF AMERICA

Second printing 1995

Foreword

THE ACS SYMPOSIUM SERIES was first published in 1974 to provide a mechanism for publishing symposia quickly in book form. The purpose of this series is to publish comprehensive books developed from symposia, which are usually "snapshots in time" of the current research being done on a topic, plus some review material on the topic. For this reason, it is necessary that the papers be published as quickly as possible.

Before a symposium-based book is put under contract, the proposed table of contents is reviewed for appropriateness to the topic and for comprehensiveness of the collection. Some papers are excluded at this point, and others are added to round out the scope of the volume. In addition, a draft of each paper is peer-reviewed prior to final acceptance or rejection. This anonymous review process is supervised by the organizer(s) of the symposium, who become the editor(s) of the book. The authors then revise their papers according to the recommendations of both the reviewers and the editors, prepare camera-ready copy, and submit the final papers to the editors, who check that all necessary revisions have been made.

As a rule, only original research papers and original review papers are included in the volumes. Verbatim reproductions of previously published papers are not accepted.

M. Joan Comstock
Series Editor

Contents

PERSPECTIVES

PHYTOCHEMICALS FROM TEA

ANTIOXIDANTS

LICORICE, GINSENG, AND OTHER
MEDICINAL PLANTS

INDEXES

Preface

EFFORTS IN CANCER CHEMOTHERAPY have intensified over the past several decades, but many cancers still remain difficult to cure; cancer prevention could become an increasingly useful strategy in our fight against cancer. Human epidemiology and animal studies have indicated that cancer risk may be modified by changes in dietary habits or dietary components. Humans ingest large numbers of naturally occurring antimutagens and anticarcinogens in food. These antimutagens and anticarcinogens may inhibit one or more stages of the carcinogenic process and prevent or delay the formation of cancer. Recent studies indicate that compounds with antioxidant or antiinflammatory properties, as well as certain phytochemicals, can inhibit tumor initiation, promotion, and progression in experimental animal models. Epidemiological studies indicate that dietary factors play an important role in the development of human cancer. Attempts to identify naturally occurring dietary anticarcinogens may lead to new strategies for cancer prevention.

The two volumes of *Food Phytochemicals for Cancer Prevention* present recent research data and review lectures by numerous prestigious experts. Contributors from academic institutions, government, and industry were carefully chosen to provide different insights and areas of expertise in these fields. Volume I covers many phytochemicals in fruits and vegetables, and their chemical and biological properties as well as their effects on health. Special emphasis is on isolation, purification, and identification of novel phytochemicals from fruits and vegetables. Biological, biochemical, pharmacological, and molecular modulation of tumor development in experimental animal models, and possibly humans, is also included. Volume II explores the chemical, biological and molecular properties of some phytochemicals in teas, spices, oriental herbs, and food coloring agents, as well as their effects on modulation of the carcinogenic process. This book provides valuable information and useful research tools for chemists, biochemists, pharmacologists, oncologists, and molecular biologists, as well as researchers in the field of food science.

Acknowledgments

We are indebted to the contributing authors for their creativity, promptness, and cooperation in the development of this book. We also sincerely appreciate the patience and understanding given to us by our wives, Mary

Ho, Mari Osawa, Chiu Hwa Huang, and Sharon Rosen. Without their support, this work would not have materialized. We thank Thomas Ferraro for his excellent reviews and suggestions and for preparing the manuscripts as camera-ready copy.

We acknowledge the financial support of the following sponsors: Campbell Soup Company; Kalsec, Inc.; Merck Sharp & Dohme Research Laboratories; The Quaker Oats Company; Schering-Plough Research Institute; Takasago USA; Tea Council of the USA; Thomas J. Lipton Company; and the Division of Agricultural and Food Chemistry of the American Chemical Society.

CHI-TANG HO
Department of Food Science
Cook College
Rutgers, The State University
    of New Jersey
New Brunswick, NJ 08903

TOSHIHIKO OSAWA
Department of Food Science
    and Technology
Nagoya University
Chikusa, Nagoya 464–01, Japan

Received August 20, 1993

MOU-TUAN HUANG
Laboratory for Cancer Research
Department of Chemical Biology
    and Pharmacognosy
College of Pharmacy
Rutgers, The State University
    of New Jersey
Piscataway, NJ 08855–0789

ROBERT T. ROSEN
Center for Advanced Food
    Technology, Cook College
Rutgers, The State University
    of New Jersey
New Brunswick, NJ 08903

PERSPECTIVES

Chapter 1

Phytochemicals in Teas and Rosemary and Their Cancer-Preventive Properties

Chi-Tang Ho[1], Thomas Ferraro[2], Qinyun Chen[1], Robert T. Rosen[3], and Mou-Tuan Huang[2]

[1]Department of Food Science, Cook College, Rutgers, The State University of New Jersey, New Brunswick, NJ 08903
[2]Laboratory for Cancer Research, College of Pharmacy, Rutgers, The State University of New Jersey, Piscataway, NJ 08855–0789
[3]Center for Advanced Food Technology, Cook College, Rutgers, The State University of New Jersey, New Brunswick, NJ 08903

Our laboratories, among many others, have been long interested in the isolation and identification of food phytochemicals that have antioxidant and anticarcinogenic activity. The extensive work in extraction, isolation, analysis, and identification resulted in detection of several phytochemicals from leaves of teas and rosemary that have high antioxidant and antiinflammatory activities. Studies of the effects of a fraction of green tea polyphenols, an extract of leaves of rosemary and the pure phytochemicals on the carcinogenic process in short-term animal studies (biochemical markers) and long-term animal tumor studies indicate that they have potent inhibitory effects on biochemical marker changes associated with tumor initiation and promotion, and anticarcinogenic activity in several animal models. In this chapter, methods for the isolation and identification of phytochemicals in teas and rosemary and their inhibitory effects on carcinogenic processes are discussed and reviewed.

A multistage model of carcinogenesis was first postulated over 50 years ago by the sequential topical application of chemical agents possessing initiating and promoting activities to mouse skin (*1*). Since then, a similiar multistage carcinogenic process has been found in many different tissues or organs in different animal models. The multistage model of chemical carcinogenesis divides carcinogenesis into at least 3 stages — initiation, promotion and progression (*2*).

Chemoprevention of cancer by phytochemicals aims to block one or more steps in the process of the carcinogenesis. The steps of carcinogenesis and possible points of inhibition by food phytochemicals are shown in Figure 1.

A number of short-term animal studies, which use biochemical markers, and long-term animal tumor studies *in vivo* are available for phytochemicals with the potential to inhibit or to delay the process of carcinogenesis. The short-term models are based on changes in tissue morphology, inflammation, cellular proliferation rate, oncogene activation and expression, and the activities of key enzymes or enzyme systems, that is, parameters that are considered key biochemical markers

0097–6156/94/0547–0002$06.00/0

for pre-neoplastic change to tissues, or for alteration of the rates of metabolic activation or disposal of carcinogens. Long-term tumor models are based on the modulation of carcinogen-induced tumorigenesis in animals, resulting in a lengthening of the latency period to tumor appearance, a reduction in tumor incidence or type, or a decrease in tumor size and in tumor volume.

Certain naturally occurring and synthetic substances are capable of interfering with the carcinogenic process. In the past decade, both laboratory animal and epidemiological studies have indicated that ascorbic acid, α-tocopherol, β-carotene, retinoic acid, and tamoxifen, and certain steroid antiinflammatory agents can effectively inhibit carcinogenesis. In this chapter, we discuss some phytochemicals that have been isolated from tea and rosemary and their anti-carcinogenic activity.

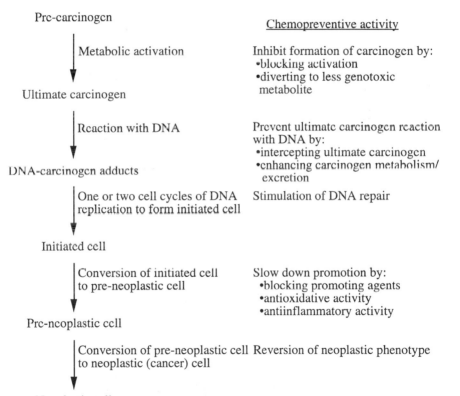

Figure 1. Multistage carcinogenesis model and possible points of intervention by chemopreventive agents.

Tea Chemistry and Its Role in Cancer Chemoprevention

Chemistry of Teas. Tea was first cultivated in China and then in Japan. With the opening of ocean routes to the East by European traders, during the fifteenth to

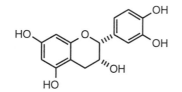

I. (-) Epicatechin

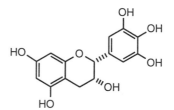

II. (-) Epigallocatechin

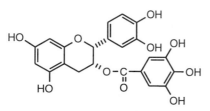

III. (-) Epicatechin-3-gallate

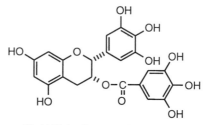

IV. (-) Epigallocatechin-3-gallate

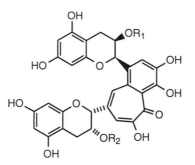

<u>Theaflavins</u>
V. Theaflavin (TF1) $R_1=R_2=H$
VI. Theaflavin monogallate A (TF2A) R_1=galloyl R_2=H
VII. Theaflavin monogallate B (TF2B) R_1=H R_2=galloyl
VIII. Theaflavin digallate (TF3) $R_1=R_2$=galloyl

Figure 2. Structures of several constituents of tea.

seventeenth centuries, commercial cultivation gradually expanded to Indonesia and then to the Indian subcontinent, including Sri Lanka (*3*). Tea is now one of the most widely consumed beverages worldwide. Black and green teas are the two main types, defined by their respective manufacturing techniques. Green tea is consumed mostly in Asian countries such as China and Japan, while black tea is more popular in North America and Europe. Oolong tea is an intermediate variant between green and black tea. Its production is confined to some regions of China including Taiwan (*4*). Most commercially prepared tea is obtained from the leaf of the plant *Camellia sinensis*, and an estimated 2.5 million tons of dried tea were manufactured in 1990 (*4*).

The term green tea refers to the product manufactured from fresh *Camellia sinensis* leaves in which significant oxidation of the major leaf polyphenols known as catechins is prevented. The principal catechins present in green tea are (-)-epicatechin (EC) (Figure 2, **I**), (-)-epigallocatechin (EGC) (**II**), (-)-epicatechin-3-gallate (ECG) (**III**) and (-)-epigallocatechin-3-gallate (EGCG) (**IV**). During the production of black tea leaves, there is extensive enzymatic oxidation of the leaf polyphenols to dark products such as theaflavins and thearubigens. The major theaflavins in black tea are theaflavin (TF1) (**V**), theaflavin monogallate A (TF2A) (**VI**), theaflavin monogallate B (TF2B) (**VII**) and theaflavin digallate (TF3) (**VIII**). The structure and chemistry of thearubigins are not well characterized. Oolong tea is partially oxidized and retains a considerable amount of the original catechins

The green tea polyphenol fraction is prepared in our laboratory as follows. One hundred grams of green tea leaves are extracted 3 times with 300 ml methanol at 50°C for 3 h, and the samples are filtered after each extraction. Solvent is removed from the combined extract with a vacuum rotary evaporator. The residue is dissolved in 500 ml water (50°C) and extracted 3 times with 200 ml hexane to remove pigments and 3 times with 200 ml chloroform to remove some of the caffeine. The aqueous phase is extracted 3 times with 180 ml ethyl acetate, and the ethyl acetate is evaporated under reduced pressure. The residue is redissolved in 300 ml water and lyophilized to obtain 8–9 g of green tea polyphenol fraction (*5*). The composition of the green tea polyphenol fraction is shown in Figure 3.

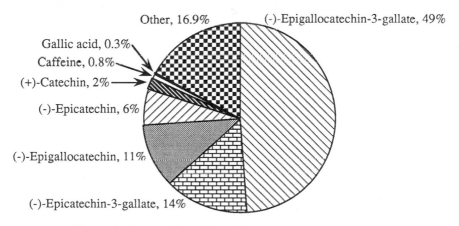

Figure 3. Composition of the green tea polyphenol fraction.

A detailed procedure for the preparation of black tea polyphenols (crude theaflavins) from black tea is given by Hara (Vol. II, Chapter 3).

Cancer Chemoprevention by Tea and Its Constituents. Recent studies in our laboratory and others have found that green and black teas and some polyphenolic compounds isolated from them have anticarcinogenic activity in many laboratory animal models *in vivo*. The broad effects of green tea as a cancer chemopreventive agent in many different animal models are of considerable interest.

The oral administration of green tea inhibits the formation of chemically-induced tumors in several models. Oral administration of green tea inhibits *N*-nitrosodiethylamine (NDEA)-induced forestomach and lung tumors in mice (*6*), 4-(methyl-*N*-nitrosamino)-1-(3-pyridyl)-1-butanone (NNK)-induced lung tumors in mice (*6-8*), and *N*-nitrosomethylbenzylamine (NMBzA)-induced esophageal tumors in rats (*9,10*). Orally administered black tea infusion also inhibits NMBzA-induced esophageal tumors in rats (*9,11*) and NNK-induced lung tumors in mice (*10*).

Several components of green tea can have similar effects. Oral administration of EGCG inhibits *N*-ethyl-*N'*-nitro-*N*-nitrosoguanidine (ENNG)-induced duodenal tumors (*12*), NNK-induced lung tumors (*8*) and spontaneous hepatomas (*13*) in mice. In addition to these studies, the oral administration of a green tea polyphenol fraction or a green tea infusion inhibits 7,12-dimethylbenz[*a*]anthracene (DMBA)- and ultraviolet B (UVB) light-induced tumorigenesis in mouse skin (*14–16*).

Green tea and its components also inhibit tumor promotion in several models. Oral administration of a green tea polyphenol fraction or a green tea infusion inhibits 12-*O*-tetradecanoylphorbol-13-acetate (TPA)-induced tumorigenesis in SKH-1 mouse skin (*14-16*). Topical application of a green tea polyphenol fraction or its constituent, EGCG, to mice inhibits skin tumor promotion induced by TPA (*5*), teleocidin, okadaic acid (*17,18*), and mirex (*19*). The perioral administration of EGCG inhibits metastasis of B16 meanoma cell lines, such as B16-F10 and BL6, in both experimental and spontaneous systems (*20*)

Mechanisms of Cancer Chemoprevention by Tea. Teas and their constituents have a wide range of biological activities that may contribute to chemopreventive effects. Table I summarizes some possible sites of inhibitory action.

Green tea can prevent the formation of carcinogens from precursor substances. Different varieties of Chinese and Japanese green tea have been shown to prevent the formation of nitrosoamine *in vitro* and *in vivo* (*9,21*), and to inhibit nitrosamine-induced formation of lesions and papillomas in the esophageal mucosa of the rat (*11*). In addition, several green tea components are able to scavenge carcinogenic electrophiles. Green tea polyphenols or EGCG accelerated the disappearance of the ultimate mutagen and/or carcinogen, benzo[*a*]pyrene diol epoxide-2 in solution (*14*).

Table II shows some carcinogens and tumor promoters that are commonly used as research tools in the laboratory to study the effects of green tea and its constituents on the process of carcinogenesis *in vitro* and *in vivo*. The indirect chemical carcinogens — benzo[*a*]pyrene (BP), DMBA, 3-methylcholanthracene (3MC), NDEA, NNK, NMBzA — all require monooxygenase-catalyzed conversion to their electrophilic ultimate carcinogens. Green tea and its constituents may inhibit monooxygenase activity or may serve as scavengers for the carcinogenic electrophile (*22,23*). Green tea polyphenols also inhibit mutations induced by the direct carcinogen *N*-methyl nitrosourea (*14*).

Green tea and its constituents also modulate some enzyme activities that relate to cell proliferation (Table III).

Table I. Some Possible Mechanisms of Inhibitory Action by Green Tea and Its Constituents on Carcinogenic Processes *InVitro* and *In Vivo*

Inhibition of Tumor Initiation

Inhibition of formation of carcinogen by prevention of formation of NMBzA from methylbenzylamine (MBzA) and sodium nitrite (*9*).

Inhibition of metabolic activationby inhibition of cytochrome P450 activities in skin and liver of rats (*14,22*).

Protection against ultimate carcinogen reaction with DNA: green tea polyphenols (a) function as nucleophiles, intercepting electrophilic ultimate carcinogens and (b) induce phase II enzymes that speed ultimate carcinogen removal (*5,14,23,24*).

Stimulation of cellular DNA repair by enhancing DNA repair enzyme activity (*25*).

Inhibition of Tumor Promotion

Inhibition of TPA-induced production of hydrogen peroxide and other active oxygen species in cell culture and mouse skin (*2,5,26,27*).

Inhibition of TPA- and carcinogen-induced formation of oxidized DNA bases, e.g. 8-hydroxydeoxyguanine, 5-hydroxymethyl uracil and thymine glycol (*8,28*).

Inhibition of TPA-induced hyperplasia in mouse skin (*5,29*).

Inhibition of TPA-induced ornithine decarboxylase activity in mouse epidermis (*5,7*).

Enhancement of immunological function; the derivatives of catechin polyphenols with a galloyl group [(-)-epicatechin-3-gallate, (-)-epigallocatechin-3-gallate and theaflavin digallate] enhance spontaneous proliferation of B cells (Hu *et al.*, Vol. II, Chapter 6).

Rosemary Chemistry and Its Role in Cancer Chemoprevention

Antioxidants are added to fats and oils or foods containing fats to prevent the formation of various off-flavors and other objectionable compounds that result from the oxidation of lipids. Butylated hydroxyanisole (BHA) and butylated hydroxytoluene (BHT), the most widely used synthetic antioxidants, have (besides

Table II. Several Physical and Chemical Carcinogens and Tumor Promoters That are Broadly Protected Against by Green Tea and Its Constituents

I. Indirect chemical carcinogens
    A. Benzo[a]pyrene (BP)
    B. 7,12-Dimethylbenz[a]anthracene (DMBA)
    C. 3-Methylcholanthracene (3MC)
    D. 4-(Methylnitrosamino)-1-(3-pyridyl)-1-butanone (NNK)
    E. N-Nitrosodiethylamine (NDEA)
    F. Azoxymethane (AOM)
    G. N-Nitrosomethylbenzylamine (NMBzA)

II. Direct chemical carcinogens
    A. N-Ethyl-N'-nitro-N-nitrosoguanidine (ENNG)
    B. N-Methyl-N'-nitro-N-nitrosoguanidine (MNNG)
    C. N-Methyl nitrosourea (MNU)

III. Physical carcinogens
    A. Ultraviolet B (UVB) light
    B. Ionizing radiation

IV. Tumor promoters
    A. 12-O-Tetradecanoylphorbol-13-acetate (TPA)
    B. Teleocidin
    C. Okadaic acid
    D. Mirex

Table III. Some Enzyme Activities That are Associated with Cellular Proliferation and Carcinogenesis and are Modulated by Green Tea and Its Constituents *In Vitro* and *In Vivo*

1.	Protein kinase C	(*17*)
2.	DNA polymerase	Nakane *et al.*, Vol. II
3.	RNA polymerase	Nakane *et al.*, Vol. II
4.	Lipoxygenase	(*29*)
5.	Cyclooxygenase	(*29*)
6.	Ornithine decarboxylase	(*5,56*)
7.	Cytochrome p-450	(*22*)
8.	Induction of phase II enzyme activity (glutathione S-transferase, glutathione peroxidase, quinone reductase, catalase)	(*24*)
9.	Salivary α-amylase	Honda *et al.*, Vol. II
10.	Intestinal sucrase	Honda *et al.*, Vol. II
11.	Intestinal maltase	Honda *et al.*, Vol. II

their high stability, low cost, and other practical advantages) unsurpassed efficacy in a variety of food systems (*30*). Their use in food, however, has been decreasing because of their suspected action as promoters of carcinogenesis, as well as the general consumer rejection of synthetic food additives (*30*). The search for and development of antioxidants of natural origin are, therefore, highly desirable. Such new antioxidants might also play a role in combatting carcinogenesis as well as the aging process.

Rosemary (*Rosmarinus officinalis* L.) and sage (*Salvia officinalis*) leaves are commonly used as spices and flavoring agents. The dried leaf of the plant *Rosmarinus officinalis* L. is one of the most widely used spices in food processing because it has a desirable flavor and because it prossess high antioxidant activity (*31-33*). Rosemary is used in food processing for the preparation of poultry, lamb, veal, shellfish, sausages, and salads as well as soups and breadings (*32,33*). Rosemary is also used as a spice in potato chips and french fries.

The use of the extract from rosemary or sage leaves as an antioxidant was first reported by Rac and Ostric-Matijasevic in 1955 (*34*). Over the years, several reports have appeared on the preparation of rosemary and sage extracts for retarding lipid oxidation (*35-38*). Many antioxidants in rosemary are non-volatile, and the antioxidant activity of rosemary in lard is comparable with that of BHA and BHT (*31,33*).

Several phenolic diterpenes with antioxidant activities have been isolated from rosemary leaves. Carnosic acid (Figure 4, **IX**), carnosol (**X**), rosmanol (**XI**) and rosmaridiphenol (**XII**) have the same vicinal diphenol feature (*31,33,39-42*).

Characterization and Measurement of Antioxidative Components in the Extracts of Rosemary and Sage. Crude and refined extracts of rosemary and sage are commercially available. Manufacturing procedures generally involve two steps (*13*). In the first step, the essential oils of rosemary and sage are removed by steam distillation. The residue containing the active antioxidant principles is extracted with different solvents — for example, methanol, ethanol, acetone or hexane.

The content of active antioxidant substances in rosemary or sage extracts can be analyzed using HPLC with either an electrochemical detector (*44*) or a mass detector (*45*). With the recent advances in liquid chromatography/mass spectrometry (LC/MS), particularly particle beam (PB)- or electron-ionization (EI)-LC/MS, the antioxidant components in rosemary and sage can be conveniently identified. PB-LC/MS features an interface for mass spectrometry which uses a heated thermospray probe for nebulization and vaporization of the solvent and sample. The majority of the solvent vapor is removed in a heated countercurrent flow gas diffusion separator. Any remaining vapor or carrier gas is removed in a two-stage momentum separator, leaving only the analyte particles. This system ensures the proper pressure needed for the EI source. EI-MS provides the fragmentation patterns necessary for the interpretation of the mass spectra. Figures 5 and 6 show the mass spectra of carnosic acid and carnosol obtained by PB-LC/MS. The fragmentation pattern of carnosol shown in Figure 6 is identical to the spectrum obtained by the traditional EI-MS (*31*).

Table IV lists the content of carnosic acid, carnosol and ursolic acid (**XIII**) in the hexane, acetone and methanol extracts of rosemary and sage. Carnosic acid and carnosol are well-known active antioxidant principals of rosemary. A recent report by Schwarz and Ternes (*44*) indicated that the antioxidant activity of rosemary extracts depended directly on the concentration of carnosic acid and carnosol. Ursolic acid is not an active antioxidant (31), but it markedly inhibited 12-O-tetradecanoylphorbol-13-acetate (TPA) induced inflammation and tumor promotion in mouse skin (46).

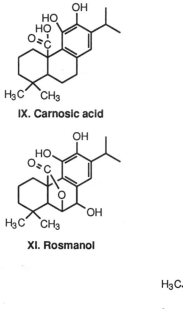

IX. Carnosic acid

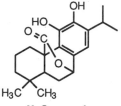

X. Carnosol

XI. Rosmanol

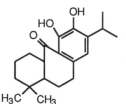

XII. Rosmaridiphenol

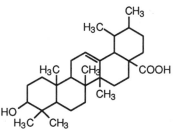

XIII. Ursolic acid

Figure 4. Structures of several constituents of rosemary.

Conversion of Carnosic Acid and Carnosol. Brieskorn and Dömling first observed that carnosic acid changed to carnosol in methanol (47). Table V shows the patterns of degradation of carnosic acid and carnosol in either methanol or a mixture of methanol and water. In both methanol and methanol:water, the amount of carnosic acid decreased while carnosol was formed, which in turn decomposed to form the γ-lactone, rosmanol. The stability of carnosic acid decreased in the presence of water. The degradation of carnosol in either methanol or the mixture of methanol and water led to the formation of rosmanol (48). The results were similar to those found the study by Schwarz et al. (49) and was consistent with the proposed mechanism shown in Figure 7.

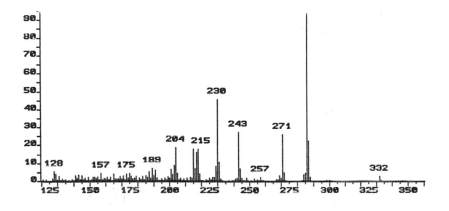

Figure 5. The mass spectrum of carnosic acid (**IX**) obtained by PB-LC/MS.

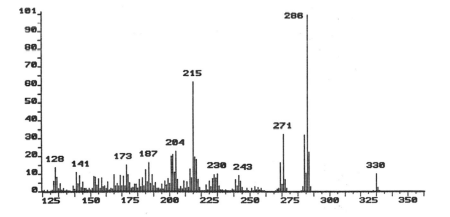

Figure 6. The mass spectrum of carnosol (**X**) obtained by PB-LC/MS.

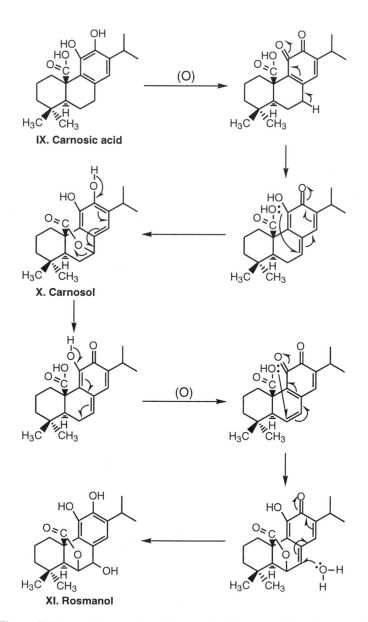

Figure 7. Proposed conversion of carnosic acid, carnosol, and rosmanol.

Table IV. Content of Major Antioxidant Components
in Rosemary and Sage Extracts

Extraction solvent	Carnosic acid[a]	Carnosol[a]	Ursolic acid[a]
Rosemary			
Hexane	100.3	16.4	13.7
Acetone	58.4	36.5	88.5
Methanol	trace	24.1	76.6
Sage			
Hexane	61.7	23.2	16.2
Acetone	79.6	37.3	14.9
Methanol	trace	6.2	95.3

[a]mg/g extract

Table V. Conversion of Carnosic Acid and Carnosol[a]

Starting compound	Solvent	Carnosic acid	Carnosol	Rosmanol
Carnosic acid	MeOH	22.3%	34.8%	trace
Carnosic acid	MeOH:H$_2$O (5:1)	13.1%	51.6%	3.6%
Carnosol	MeOH		82.0%	2.9%
Carnosol	MeOH:H$_2$O (5:1)		74.0%	9.3%

[a]Stored at room temperature for 30 days.

Cancer Chemoprevention by Rosemary

Preparation of Rosemary Extract. A rosemary extract was prepared according to the previously reported method of Wu *et al.* (*31*) as recently described by Huang *et al.* (*50*). Fifty g of ground leaves of the plant *Rosmarinus officinalis* L. were extracted twice with 250 ml methanol at 60°C for 2 hours, and the samples were filtered after each extraction. The combined extract was bleached with 100 g activated carbon at 60°C and then filtered to yield a light-brown filtrate. The methanol filtrate was concentrated to a final volume of 50 ml with a vacuum rotatory evaporator and then filtered to remove the precipitate. Seventy-five ml water were then added to the filtrate. The precipitate formed after the addition of water was filtered and dried to yield a methanol extract of leaves of *Rosmarinus officinalis* L. (rosemary extract). Preparation of carnosol and ursolic acid are described previously (*50*).

Inhibition of Tumor Initiation. Topical application of rosemary extract 5 minutes prior to treatment with [3H]BP to the backs of CD-1 mice inhibited [3H]BP metabolite-DNA adduct formation in mouse epidermis. The formation of [3H]BP-DNA adducts in mouse epidermis is proportional to the dose of [3H]BP and reaches a maximum at 12 to 15 hours after the topical application of 20 nmol [^{3}H]BP (*51*). Topical application of 20 nmol [^{3}H]BP to vehicle-treated control mice resulted in the covalent binding of 0.74 pmol [^{3}H]BP metabolites per mg epidermal DNA at 15 hr after the dose. Topical application of 1.2 or 3.6 mg rosemary extract 5 min before the application of 20 nmol [^{3}H]BP resulted in the formation of 0.52 and 0.34 pmol (30 and 54% inhibition) of covalently bound [^{3}H]BP metabolites per mg epidermal DNA, respectively, at 15 hr after the dose (*46,52*).

Rosemary markedly inhibited skin tumor initiation by BP in CD-1 mice. Female CD-1 mice (30 per group) were treated topically with 200 µl acetone or with rosemary extract in acetone 5 min before each application of 20 nmol BP in 200 µl acetone once weekly for 10 weeks. One week after the last application of BP, the mice were treated with TPA (15 nmol) twice weekly. After 13 and 19 weeks of promotion with TPA, the mice developed 3.4 and 6.9 tumors per mouse, respectively, and the tumor incidence was 87 and 100%, respectively. In a parallel group of animals, 3.6 mg rosemary extract was topically applied to the backs of mice 5 min prior to the application of BP. After 13 and 19 weeks of TPA application, there were 1.2 and 3.0 tumors per mouse, respectively, and the tumor incidence was 48 and 73%, respectively (*52*).

Mice that were initiated with 200 nmol DMBA and promoted with 5 nmol TPA twice weekly for 19 weeks developed an average of 17 skin tumors per mouse, and 83% of the mice had tumors. In another group of similarly initiated mice, topical application of 0.4, 1.2 or 3.6 mg rosemary extract together with 5 nmol TPA twice weekly for 19 weeks inhibited TPA-induced tumor formation (tumors per mouse) by 60, 68 or 99%, respectively, and the percent of mice with tumors was inhibited by 57, 37 or 92%, respectively. Additional groups of mice were initiated with DMBA and then treated with acetone alone or with 3.6 mg of rosemary twice a week for 19 weeks. None of these animals developed tumors indicating that rosemary is not a tumor promoter. In addition, mice were treated topically with 3.6 mg of rosemary and then promoted with 5 nmol TPA twice a week for 19 weeks; none of these animals developed tumors, indicating rosemary is not a tumor initiator on mouse skin (*50*).

Inhibition of Tumor Promotion. Topical application of rosemary extract together with 5 nmol TPA to the backs of mice inhibited TPA-induced increases in epidermal ornithine decarboxylase activity in a dose-dependent manner. Topical application of 0.4 or 3.6 mg rosemary extract together with 5 nmol TPA to the backs of CD-1 mice inhibited TPA-induced increases in epidermal ornithine decarboxylase activity by 66 or 100%, respectively (*52,53*).

Rosemary also inhibited TPA-induced skin inflammation and hyperplasia (*50*). Topical application of TPA (1 nmol) twice a day for 4 days to the dorsal surface of CD-1 mice resulted in a 4- to 5-fold increase in the number of epidermal cell layers and in the epidermal thickness. Inflammatory cell infiltration in the dermis and intercellular edema in the epidermis were also observed. Topical application of 3.6 mg rosemary extract together with 1 nmol TPA twice a day for 4 days inhibited these effects of TPA (*46,50,52*).

Components of Rosemary. Carnosol and ursolic acid are major constituents of rosemary. We isolated carnosol and ursolic acid from rosemary extract and studied the biological effects of these compounds. Both carnosol and ursolic acid inhibited

TPA-induced ear edema, ornithine decarboxylase activity and tumor promotion in mouse skin.

Topical application of 1, 3 or 10 μmol carnosol together with 5 nmol TPA twice weekly for 20 weeks to the backs of CD-1 mice previously initiated with DMBA inhibited the formation of TPA-induced skin tumors per mouse by 38, 63 or 88%, respectively (*46,50,52*).

Topical application of 0.1 or 2 μmol ursolic acid together with 5 nmol TPA twice weekly for 20 weeks inhibited skin tumors per mouse by 52 or 61%, respectively (*46,50,52*). Other laboratories have also shown ursolic acid to inhibit TPA-induced tumor promotion in mouse skin (*54*).

Dietary Rosemary. Singletary and Nelshoppen reported that administration of 1% crude rosemary extract in the diet to female Sprague-Dawley rats for 3 weeks before a single oral dose of DMBA reduced mammary gland tumor incidence by 47% at 16 weeks after DMBA treatment (*55*). Singletary and Nelshoppen also reported that 0.5% and 1% rosemary extract in the diet inhibited the *in vivo* binding of DMBA to mammary epithelial cell DNA and the formation of two major DNA adducts (*55*).

Recent studies in our laboratory found that rosemary extract (2% of the diet) inhibited BP-induced forestomach and lung tumors in A/J mice as well as azoxymethane-induced colon tumors in CF-1 mice. Dietary rosemary extract also markedly inhibited the toxicity of intraperitoneally administered DMBA in Sencar mice (*46,50,52*).

See Table VI for a summary of the anticarcinogenic activities of rosemary and its constituents.

Summary and Conclusions

The broad effects of green tea as a cancer chemopreventive agent in many different animal models are of considerable interest. Green tea infusion and tea catechin polyphenolic compounds are non-specific and broad-spectra anticarcinogens. Tea catechin polyphenols are able to inhibit every stage of the carcinogenic process, the development of a wide variety of tumors in different sites and organs in several experimental animal models, a wide variety of enzyme activities that are associated with cell proliferation, and a variety of carcinogens and tumor promoters (including chemical and physical carcinogens). The major active constituents in green tea are (-)-epigallocatechin-3-gallate, (-)-epigallocatechin, (-)-epicatechin-3-gallate and (-)-epicatechin.

Leaves of the plant *Rosmarinus officinalis* L. (rosemary) are commonly used as a spice. Topical application of a methanol extract of leaves of rosemary to the backs of mice inhibts formation of [³H]BP-DNA adducts in mouse epidermis, skin tumor initiation by BP and DMBA, and TPA-induced inflammation, ornithine decarboxylase activity, hyperplasia, and tumor promotion in mouse epidermis. Dietary rosemary inhibits the formation of [³H]DMBA-DNA adducts in mammary epithelial cells of rat and DMBA-induced mammary gland tumorigenesis in rats. Carnosol and ursolic acid are major constituents of rosemary which also inhibit TPA-induced inflammation, ornithine decarboxylase and tumor promotion in mouse skin.

**Table VI. Some Inhibitory Effects of Rosemary on Short-term Biochemical
Marker Studies and Long-term *In Vivo* Tumor Studies in Rodents**

Inhibitory site		Reference
Initiation		
A. Biochemical markers:	Topical application of rosemary inhibits the formation of [^{3}H]BP-DNA adducts in mouse epidermis.	*(46)*
	Dietary rosemary inhibits the formation of [^{3}H]DMBA-DNA adduct in rat mammary gland epithelial cells.	*(55)*
B. Tumorigenesis:	Topical application of rosemary inhibits skin tumor initiation by BP and DMBA in CD-1 mice.	*(46)*
	Dietary rosemary inhibits DMBA-induced mammary gland tumorigenesis in SD rats.	*(55)*
Promotion		
A. Biochemical markers:	Rosemary and its constituents carnosol and ursolic acid inhibit TPA-induced epidermal ornithine decarboxylase activity.	*(46)*
	Rosemary and its constituents carnosol and ursolic acid inhibit TPA-induced mouse ear edema.	*(46)*
	Rosemary inhibits TPA-induced hyperplasia.	*(46)*
B. Tumorigenesis:	Topical application of rosemary and its constituents carnosol and ursolic acid inhibit TPA-induced tumor promotion in mouse epidermis.	*(46,53,54)*
	Dietary rosemary inhibits BP-induced forestomach and lung tumorigenesis.	
	Dietary rosemary inhibits azoxymethane-induced colon tumorigenesis.	

Literature Cited

1. Berenblum, I. *Arch. Pathol.* **1944**, *38*, 233–271.
2. Bertram, J. S.; Kolonel, L. N.; Meyskens, F. L., Jr. *Cancer Res.* **1987**, *47*(11), 3012–3031.
3. Forrest, D. *The World Trade: A Survey of the Production, Distribution and Consumption of Tea*; Woodhead-Faulkner, Ltd.: Cambridge, UK, 1985.
4. Graham, H. N. *Preventive Med.* **1992**, *21*, 334–350.
5. Huang, M.-T.; Ho, C.-T.; Wang, Z.-Y.; Ferraro, T.; Finnegan-Olive, T.; Lou, Y.-R.; Mitchell, J. M.; Laskin, J. D.; Newmark, H.; Yang, C. S., et al. *Carcinogenesis* **1992**, *13*(6), 947–954.
6. Wang, Z. Y.; Hong, J.-Y.; Huang, M.-T.; Reuhl, K. R.; Conney, A. H.; Yang, C. S. *Cancer Res.* **1992**, *52*(7), 1943–1947.
7. Agarwal, R.; Katiyar, S. K.; Zaidi, S. A.; Mukhtar, H. *Cancer Res.* **1992**, *52*, 3582–3588.
8. Chung, F.-L.; Xu, Y.; Ho, C.-T.; Desai, D.; Han, C. In *Phenolic Compounds in Food and Their Effects on Health II: Antioxidants and Cancer Prevention*; Huang, M.-T., Ho, C.-T.; Lee, C. Y., Ed.; ACS Symposium Series No. 507; American Chemical Society: Washington, D.C., 1992; pp 300–307 (Chapter 23).
9. Xu, Y.; Han, C. *Biomed. Environ. Sci.* **1990**, *3*(4), 406–412.
10. Wang, Z. Y.; Hong, J.-Y.; Huang, M.-T.; Conney, A. H.; Yang, C. S. In *Phenolic Compounds in Food and Their Effects on Health II: Antioxidants and Cancer Prevention*; Huang, M.-T., Ho, C.-T.; Lee, C. Y., Ed.; ACS Symposium Series No. 507; American Chemical Society: Washington, D.C., 1992; pp 292–299 (Chapter 22).
11. Han, C.; Xu, Y. *Biomed. Environ. Sci.* **1990**, *3*(1), 35–42.
12. Fujita, Y.; Yamane, T.; Tanaka, M.; Kuwata, K.; Okuzumi, J.; Takahashi, T.; Fujiki, H.; Okuda, T. *Jpn. J. Cancer Res.* **1989**, *80*(6), 503-5.
13. Nishida, H.; Omori, M.; Fukutomi, Y.; Ninomiya, M.; Nishiwaki, S.; Moriwaki, H.; Muto, Y. *Abstracts of Thirteen International Symposium on Cancer: The Sapporo Cancer Seminar, Current Strategies of Cancer Chemopreventionm, Sapporo, Japan, July 6-9, 1993* **1993**, , 48.
14. Wang, Z. Y.; Cheng, S. J.; Zhou, Z. C.; Athar, M.; Khan, W. A.; Bickers, D. R.; Mukhtar, H. *Mutat. Res.* **1989**, *223*, 273–285.
15. Wang, Z. Y.; Agarwal, R.; Bickers, D. R.; Mukhtar, H. *Carcinogenesis* **1991**, *12*(8), 1527–1530.
16. Wang, Z. Y.; Huang, M.-T.; Ferraro, T.; Wong, C.-Q.; Lou, Y.-R.; Reuhl, K.; Iatropoulos, M.; Yang, C. S.; Conney, A. H. *Cancer Res.* **1992**, *52*(5), 1162–1170.
17. Yoshizawa, S.; Horiuchi, T.; Fujiki, H.; Yoshida, T.; Okuda, T.; Sugimura, T. *Phytother. Res.* **1987**, *1*, 44-47.
18. Yoshizawa, S.; Horiuchi, T.; Suganuma, M.; Nishiwaki, S.; Yatsunami, J.; Okabe, S.; Okuda, T.; Muto, Y.; Frenkel, K.; Troll, W., et al. In *Phenolic Compounds in Food and Their Effects on Health II: Antioxidant & Cancer Prevention*; Huang, M.-T., Ho, C.-T.; Lee, C. Y., Ed.; ACS Symposium Series No. 507; American Chemical Society: Washington, D.C., 1992; pp 316–325 (Chapter 25).
19. Huang, M.-T.; Lou, Y.-R.; Ho, C.-T.; Conney, A. H. (unpublished results)

20. Taniguchi, S.; Fujiki, H.; Kobayashi, H.; Go, H.; Miyado, K.; Sadano, H.; Shimokawa, R. *Cancer lett.* **1992**, *65*, 51-54.
21. Nakamura, M.; Kawabata, T. *J. Food Sci.* **1981**, *46*, 306–307.
22. Wang, Z. Y.; Das, M.; Bickers, D. R.; Mukhtar, H. *Drug Meta. Dispos.* **1988**, *16*, 98–103.
23. Khan, W. A.; Wang, Z. Y.; Athar, M.; Bickers, D. R.; Mukhtar, H. *Cancer Lett.* **1988**, *42*, 7–12.
24. Khan, S. G.; Katiyar, S. K.; Agarwal, R.; Mukhtar, H. *Cancer Res.* **1992**, *52*, 4050–4052.
25. Shimoi, K.; Nakamura, Y.; Tomita, I.; Hara, Y.; Kada, T. *Mutat. Res.* **1986**, *173*, 239–244.
26. Zhao, B. L.; Li, X. J.; Cheng, S. J.; Xin, W. J. *Cell. Biophys.* **1989**, *14*(2), 175–185.
27. Osawa, T.; Namiki, M.; Kawakishi, S. In *Antimutagenesis and Anticarcinogenesis Mechanisms II*; Kuroda, Y., Shankel, D. M.; Waters, M. D., Eds.; Basic Life Sciences; Plenum Press: New York, 1990, Vol. 52; pp 139–153.
28. Xu, A.; Ho, C.-T.; Amin, S. G.; Han, C.; Chung, F.-L. *Cancer Res.* **1992**, *52*, 3875-3879.
29. Katiyar, S. K.; Agarwal, R.; Wood, G. S.; Mukhtar, H. *Cancer Res.* **1992**, *52*, 6890–6897.
30. Namiki, M. *CRC Crit. Rev. Food Sci. Nutr.* **1990**, *29*, 273–300.
31. Wu, J. W.; Lee, M.-H.; Ho, C.-T.; Chang, S. S. *J. Am. Oil Chem. Soc.* **1982**, *59*, 339-345.
32. Fisher, C. In *Phenolic Compounds in Food and Their Effects on Health I: Analysis, Occurence, and Chemistry*; Ho, C.-T., Lee, C. Y.; Huang, M.-T., Ed.; ACS Symposium Series No. 506; American Chemical Society: Washington, D.C., 1992; pp 118–129 (Chapter 9).
33. Nakatani, N. In *Phenolic Compounds in Food and Their Effects on Health II: Antioxidants and Cancer Prevention*; Huang, M.-T., Ho, C.-T.; Lee, C. Y., Ed.; ACS Symposium Series No. 507; American Chemical Society: Washington, D.C., 1992; pp 72–86 (Chapter 6).
34. Rac, M.; Ostric-Matijasevic, B. *Rev. Fr. Corps Gras* **1955**, *2*, 796–800.
35. Berner, D. L.; Jacobson, G. A., U.S. Patent 3,732,111, 1973.
36. Chang, S. S.; Ostric-Matijasevic, B.; Hsieh, O. A.-L.; Huang, C.-L. *J. Food Sci.* **1977**, *42*, 1102–1106.
37. Bracco, U.; Loliger, J.; Viret, J.-L. *J. Am. Oil Chem. Soc.* **1981**, *58*, 686–690.
38. Tateo, M.; Fellin, M.; Bianchi, A.; Bianchi, L. *Perfumer Flavorist* **1988**, *13*(6), 48–54.
39. Nakatani, N.; Inatani, R. *Agric. Biol. Chem.* **1981**, *45*, 2385–2386.
40. Inatani, N.; Nakatani, N.; Fuwa, H.; Seto, H. *Agri. Biol. Chem.* **1982**, *46*, 1661–1666.
41. Houlihan, C. M.; Ho, C.-T.; Chang, S. S. *J. Am. Oil Chem. Soc.* **1984**, *61*, 1036–1039.
42. Nakatani, N.; Inatani, R. *Agri. Biol. Chem.* **1984**, *48*(8), 2081–2085.
43. Loliger, J. In *Free Radicals and Food Additives*; Okezie, I.; Halliwell, B., Eds.; Tatlor & Francis: New York, 1991; pp 121–145 (Chapter 6).
44. Schwarz, K.; Ternes, W. *Z. Lebensm. Unters. Forsch.* **1992**, *195*, 99–103.
45. Chen, Q.; Shi, H.; Ho, C.-T. *J. Am. Oil Soc.* **1992**, *69*, 999–1002.

46. Huang, M.-T.; Ho, C.-T.; Ferraro, T.; Wang, Z. Y.; Stauber, K.; Georigiadis, C.; Laskin, J. D.; Conney, A. H. *Proc. Am. Assoc. Cancer Res.* **1992**, *33*, 165 (abstr. 990).

47. Brieskorn, C. H.; Domling, H. *J. Arch. Pharm.* **1969**, *302*, 641–649.

48. Chen, Q.; Ho, C.-T. (unpublished results)

49. Schwarz, K.; Ternes, W.; Schmauderer, E. *Z. Lebensm. Unters. Forsch.* **1992**, *195*, 104–107.

50. Huang, M.-T.; Ho, C.-T.; Wang, Z. Y.; Ferraro, T.; Lou, Y.-R.; Stauber, K.; Ma, W.; Georgiadis, C.; Laskin, J. D.; Conney, A. H. submitted to *Cancer Res.* **1993**

51. Smart, R. C.; Huang, M.-T.; Chang, R. L.; Sayer, J. M.; Jerina, D. M.; Wood, A. W.; Conney, A. H. *Carcinogenesis* **1986**, *7*(10), 1669-1675.

52. Huang, M.-T.; Ho, C.-T.; Wang, Z. Y.; Stauber, K.; Georgiadis, C.; Laskin, J. D.; Conney, A. H. In *Food and Cancer Prevention: Chemical and Biological Aspects*; Waldron, K., Johnson, I. T.; Fenwick, G. R., Ed.; Royal Society of Chemistry: Cambridge, 1993; (in press).

53. Huang, M.-T.; Ho, C.-T.; Cheng, S.-J.; Laskin, J. D.; Stauber, K.; Georgiadis, C.; Conney, A. H. *Proc. Am. Assoc. Cancer Res.* **1989**, *30*, 180 (abstr. 714).

54. Tokuda, H.; Ohigashi, H.; Koshimizu, K.; Ito, Y. *Cancer Lett.* **1986**, *33*, 279-285.

55. Singletary, K. W.; Nelshoppen, J. M. *Cancer Lett.* **1991**, *60*, 169-175.

56. Bhimani, R.; Frenkel, K. *Proc. Am. Assoc. Cancer Res.* **1991**, *32*, 126 (abstr. 756).

RECEIVED September 10, 1993

Chapter 2

Inactivation of Oxygen Radicals by Dietary Phenolic Compounds in Anticarcinogenesis

Michael G. Simic[1] and Slobodan V. Jovanovic[2]

[1]Department of Radiation Oncology, School of Medicine, University of Pennsylvania, 195 John Morgan Building, 37th and Hamilton Walk, Philadelphia, PA 19194–6072
[2]Gamma Laboratory, Institute Vinca, Belgrade, Yugoslavia

Hydroxylated and polyhydroxylated aromatic and heterocyclic compounds may have antioxidant and anticarcinogenic properties. Whether an antioxidant is an anticarcinogen may depend on its efficacy as an oxygen radical (peroxyl, alkoxyl, superoxide, hydroxyl) inactivator and inhibitor. Location, concentration *in situ*, reaction kinetics (rate constants), energetics (redox potentials), and products (intermediates and final) contribute to the efficacy of an antioxidant. The kinetics and energetics of an antioxidant are governed by the type and position of the substituents. The rate constant for the reaction of an antioxidant and an oxy radical depends on the type of radical. In general, the reactivity of a radical decreases in the following order: hydroxyl > alkoxyl > peroxyl > superoxide. One-electron oxidation potential of an antioxidant at pH 7, E_7, can be calculated from the Hammett correlation using Brown substituent constants.

Free radical damage to biosystems is one of the major processes that contributes to degenerative diseases (cancer and cardiovascular) and aging (*1*). Even before that relationship was evident, it was demonstrated that certain antioxidants (BHA and BHT) are inhibitors of chemical carcinogenesis (*2*). On the basis of the measured rate constants for the reaction of these anticarcinogenic antioxidants (mainly substituted phenols) with diverse free radicals and their efficacy as anticarcinogens, it was suggested that the major component of carcinogenesis may be based on the damage by oxy radicals and that the major role of anticarcinogens is inhibition of oxidative processes (*3*).

In continuing pioneering work on anticarcinogens (*2*), attention was shifted to dietary antioxidants (*4,5*). Demonstrating an active role of dietary antioxidants in the inhibition of degenerative diseases (*5*) was an important milestone in recognizing the role of diets and dietary antioxidants in cancer prevention (*6*).

Proposed molecular mechanisms of oxidative carcinogenesis (*7–9*) indicated intrinsic complexities despite major advances in the field. Detailed free

0097–6156/94/0547–0020$06.00/0

radical mechanisms of direct (ionizing radiations, metabolism) and indirect (UV light, initiators, promoters, progressors) inducers of free radicals *in vivo*, and their quantitative contribution to carcinogenesis, are still lacking (*10,11*). Despite these uncertainties, it is clear that dietary antioxidants may inhibit endogenous, metabolically driven, oxidative DNA damage (*12*), as well as mutation (*13*) and tumor formation (*14*) by exogenous agents.

Generation of Oxy Radicals in Biosystems

Numerous chemical and biochemical sources of free radicals *in vivo* have been reviewed recently (*1,10,15*). Different types of radicals and their origins are summarized in Table I.

Table I. Classes of Oxy Radicals and Related Species

Species	Name	Origin
$^3O_2$	Triplet oxygen	Stable atmospheric form
$^1O_2$	Singlet oxygen	$^3O_2$, peroxidation, photosensitization
RH	Parent molecule	Biocomponents or subunits
R•	Free radical	RH (redox, abstraction, addition)
ROO•	Peroxyl radical	R• + O_2, oxidation of ROOH
H_2L	Linoleic acid, PUFA	Fats, membranes, lipoproteins
HL•	Bisallylic radical	H_2L (abstraction, addition)
HLOO•	Linoleic acid peroxyl radical	HL• + O_2, HLOOH (oxidation)
•O_2H	Hydroperoxyl radical	•O_2^- + H^+, O_2 + H
•O_2^-	Superoxide radical	O_2 + e^-
•OH	Hydroxyl radical	H_2O_2 + e^-, H_2O (radiation)
H_2O_2	Hydrogen peroxide	•O_2^-, biogeneration
RO•	Alkoxyl radical	ROOH + e^-, ROOR
HLO•	Linoleic acid alkoxyl radical	HLOOH + e^-
ROOH	Hydroperoxide	ROO•, $^1O_2$
ROOR	Peroxide	Peroxidation processes
⫧C—C⫨ (epoxide)	Epoxide	Unsaturated ROO•
C_6H_5O•	Phenoxyl radical	Phenol (oxidation)
ArO•	Aroxyl radical	Hydroxylated aromatics (oxidation)

The essence of free radical generation *in vivo* is the ubiquitous presence of oxygen in biosystems and its reactivity with abundant electron donors,

$$O_2 + e^- \rightarrow •O_2^- \tag{1}$$

The electron may come from reactive ferrous moieties such as Fe^{2+}, heme(II) and certain other Fe(II) complexes,

$$O_2 + Fe(II) \rightarrow •O_2^- + Fe(III) \tag{2}$$

A major and continuous contributor to reaction (1) is the mitochondrial electron transport chain (*16–18*) which "leaks" electrons via ubiquinone semiquinone radicals, $\bullet UQ^-$, or reduced ferredoxin. Under certain conditions, such as ischemia-reperfusion, xanthine/xanthine oxidase is an important source of superoxide radical (*19*).

Once the superoxide radical is generated, it recombines in an SOD-catalyzed reaction,

$$\bullet O_2^- + \bullet O_2^- + 2H^+ \rightarrow H_2O_2 + O_2 \tag{3}$$

Reaction (3) is expected to be the predominant reaction in cells because the reactivity of $\bullet O_2^-$ is very limited, excluding the reaction with cytochrome c (*19*).

Another significant source of oxy radicals is endogenously generated peroxides (*20*), i.e., hydrogen peroxide (H_2O_2) and hydroperoxides (ROOH). These peroxy compounds can be readily reduced by the well-known Haber-Weiss reaction (2) in which ferrous iron and some of its complexes act as electron donors,

$$H_2O_2 + Fe(II) \rightarrow \bullet OH + OH^- + Fe(III) \tag{4}$$

$$ROOH + Fe(II) \rightarrow RO\bullet + OH^- + Fe(III) \tag{5}$$

Reactions (4) and (5) are relatively fast ($k \sim 1 \times 10^4 \ M^{-1} \ s^{-1}$) and do not require a high concentration of Fe(II) in a biosystem.

Although it is an antioxidant, ascorbate may promote generation of oxy radicals by recycling ferrous moieties,

$$Fe(III) + AH^- \rightarrow Fe(II) + \bullet A^- + H^+ \tag{6}$$

Oxy radicals may also be generated directly or indirectly by exogenous agents and processes. The most thoroughly studied agents are ionizing radiations (*21*). Interaction of ionizing radiations with biosystems leads to ionization of biocomponents, with yields roughly proportional to the mass fraction of each component.

Water is the major component of most biosystems and is split into hydroxyl radicals in two steps. The first step,

$$H_2O \xrightarrow{\nu} H_2O\bullet^+ + e^- \tag{7}$$

is followed by rapid deprotonation ($\tau < 1$ ps), and formation of hydroxyl radicals,

$$H_2O\bullet^+ + H_2O \rightarrow \bullet OH + H_3O^+ \tag{8}$$

Reactivities of Oxy Radicals

Predominant free radical species in biosystems under oxic conditions are the four oxy radicals: hydroxyl, $\bullet OH$; alkoxyl, RO$\bullet$; peroxyl, ROO$\bullet$; and superoxide, $\bullet O_2^-$.

The reactivities of oxy radicals with biocomponents may vary from highly reactive ($k \sim 10^{10} \ M^{-1} \ s^{-1}$) to nonreactive (*15*). The rate constants (reactivity) decrease in the following order: hydroxyl > alkoxyl > peroxyl > superoxide (Table II).

Table II. Reaction Rate Constants for Oxy Radicals with Biochemical Substrates in Solutions at Room Temperature at pH ~ 7

Substrate	$k(R\bullet + S)$, $M^{-1} s^{-1}$				
	ROO•	•O_2H^a	•O_2^-	RO•	•OH
Stearic acid	10^{-4}–10^{-3}	low	low	2.3×10^6	10^9–10^{10}
Oleic acid	0.1–1	low	low	3.3×10^6	10^9–10^{10}
Linoleic acid	~ 60	1.2×10^3	low	8.8×10^6	9.0×10^9
Linolenic acid	~ 120	1.7×10^3	low	1.3×10^7	7.3×10^9
Arachidonic acid	~ 180	3.0×10^3	low	2.0×10^7	~ 10^{10}
Aldehydes	2.7×10^3	50	n.m.	n.m.	~ 10^9
GSH	< 10^6	1.8×10^5	< 15	n.m.	1.4×10^{10}
BHT	10^4	2.4×10^3	n.m.	4×10^7	~ 10^{10}
BHA	2.6×10^6	n.m.	n.m.	n.m.	~ 10^{10}
QH_2	1.2×10^5	10^4 - 10^5	n.m.	n.m.	~ 10^{10}
α-Tocopherol	5.7×10^6	2.6×10^5	~ $10^{4\ b}$	n.m.	~ 10^{10}
Ascorbate	2.2×10^6	n.m.	5.0×10^4	n.m.	1×10^{10}
Thymine	n.o.	n.o.	n.o.	n.m.	7×10^9
Guanine	n.o.	n.o.	n.o.	n.m.	9×10^9
Deoxyribose	very slow	n.o.	n.o.	~ 10^7	2×10^9

SOURCE: Adapted from ref. 15.
[a] At pH<4
[b] Ref. 22
n.m. not measured, n.o. not observed

Reacting with most biocomponents at close to diffusion controlled rates, hydroxyl radicals are not selective, reacting indiscriminately within a short range. For example, if iron in reaction (4) is complexed by a biomolecule such as DNA, the generated •OH will react with the DNA subunits closest to complexed Fe.

Alkoxyl radicals are 2 to 3 orders of magnitude less reactive than hydroxyl radicals and, consequently, are more selective. Nevertheless, their efficacy of biodamage is not increased due to their rapid decay (k ~ 10^6–10^7 s^{-1}) via β-scission (23). If generated close to DNA via reaction (5), they would abstract an H from sugar and induce a DNA strand break (21).

Peroxyl radicals are much less reactive than hydroxyl and alkoxyl radicals. Hence, they are considerably more selective and have larger radii of action than any other oxy radicals (except superoxide, which is relatively unreactive). Because of their relative unreactivity with DNA and proteins, their damage is mainly limited to fatty acids in membranes and lipoproteins. Oxidation of lipids often proceeds via chain reactions propagated by peroxyl radicals (15,24).

Reactions of superoxide radicals in biosystems proceed either via protonated conjugates •O_2H (pK$_a$ = 4.75), which are basically peroxyl radicals, or via electron transfer reactions. Reduction of Fe(III) in cytochrome c by superoxide radicals is a well-known example (k ~ 10^5 M^{-1} s^{-1}) of a redox reaction.

Reactivities of Hydroxylated Aromatics with Oxy Radicals

Hydroxyl radical is very reactive with aromatic and heterocyclic substrates, with $k \sim 10^5$ M^{-1} s^{-1} (25). The reaction proceeds via addition to benzene and other aromatic and heterocyclic rings, specifically, by addition to the double bonds within the ring (21),

$$ArH + \bullet OH \rightarrow \bullet ArH{-}OH \tag{9}$$

Mono and polyhydroxylated derivatives react similarly,

$$ArOH + \bullet OH \rightarrow \bullet Ar(OH)_2 \tag{10}$$

Both adducts in reactions (9) and (10) react with O_2 ($k \sim 10^8$ M^{-1} s^{-1}). There is a major difference, however, in the stability of the two adducts. The first one is relatively stable, except for its reaction with oxygen and other radicals. The dihydroxy and polyhydroxy intermediates are unstable and eliminate OH^-, yielding a phenoxyl radical. Hence, in the absence of oxygen,

$$\bullet Ar(OH)_2 \rightarrow Ar{-}O\bullet + OH^- + H^+ \tag{11}$$
$$k \sim 10^3{-}10^4 \ s^{-1}$$

whereas in the presence of oxygen (in contrast to phenoxyl radicals, most of which are unreactive with oxygen) these intermediates react with oxygen,

$$\bullet Ar(OH)_2 + O_2 \rightleftharpoons \bullet OOAr(OH) \tag{12}$$

Reaction (12) is reversible to a limited extent because of the weak resonance character of $\bullet Ar(OH)_2$ (10).

Peroxyl radicals are considerably less reactive with phenolics than corresponding $\bullet OH$ reactivities. The reaction mechanism is different because the site of attack by peroxyl is the hydroxyl group of the phenol. Oxy radicals, in principle, could abstract an H atom from ArO-H bonds,

$$ArOH + ROO\bullet \rightarrow ROOH + ArO\bullet \tag{13}$$

There is growing evidence, however, that the interaction of peroxyl radicals and phenolics is an electron transfer (redox) process (26–28). Reaction (13) therefore takes place in two steps,

$$ArOH + ROO\bullet \rightarrow ROO^- + ArOH^+ \tag{14}$$

followed by rapid deprotonation of the charged intermediate,

$$ArOH^+ + H_2O \rightarrow ArO\bullet + H_3O^+ \tag{15}$$

The reactions of phenolics with alkoxyl and superoxide radicals are expected to be similar to those with peroxyl radicals. The reactivities of alkoxyls are higher than those of peroxyl radicals, whereas those of superoxide much lower (see Table II).

The rate constants, in general, are higher for deprotonated phenol,

$$ArOH \rightleftharpoons ArO^- + H^+ \tag{16}$$

and for deprotonated polyhydroxy derivatives, because the deprotonated forms are better electron donors than the protonated forms (27).

Similarly to parent phenols,

$$HO-Ar-OH \rightleftharpoons HO-Ar-O^- + H^+ \tag{17}$$

aroxyl radicals also have acid-base properties,

$$\bullet O-Ar-OH \rightleftharpoons \bullet O-Ar-O^- + H^+ \tag{18}$$

In general, the pK_a of a radical is lower than that of the corresponding phenol, because of the strong electron-withdrawing character of the $O\bullet$ atom.

Redox Potentials

Relative energetics of electron-transfer processes is represented by redox potentials, E. Because the pH of most biosystems is close to 7, redox potential values at pH 7 are more relevant to biological conditions, than standard redox potentials (27).

If a phenol donates an electron to an electron acceptor (e.g., peroxyl and aroxyl radicals), then we refer to the oxidation potential of the phenol. $E(ArO^-/ArO\bullet)$ or $E(ArOH/ArO\bullet)$ are used to define the process and distinguish the state of protonation of the phenol under experimental conditions. For the reverse process, $E(ArO\bullet/ArO^-)$ is considered to be the reduction potential of the aroxyl radical. By definition the oxidation potential of the parent compound, $E(ArO /ArO\bullet)$, is equal numerically to the reduction potential of the radical, $E(ArO\bullet/ArO^-)$, but with the opposite sign. Throughout this paper only the reduction potential values of the aroxyls are given, but for simplicity only their parent compounds are listed.

In general, one-electron redox potentials are derived from the equilibrium between a known redox standard, S^- (or SH), an antioxidant, A^- (or AH), and their free radical intermediates,

$$\bullet A + S^- \underset{k_r}{\overset{k_f}{\rightleftharpoons}} A^- + \bullet S \tag{19}$$

by use of k_f and k_r or from absorptions (concentrations) of intermediates at the equilibrium,

$$K = \frac{k_f}{k_r} = \frac{[A^-] [\bullet S]}{[\bullet A] [S^-]} \tag{20}$$

and the Nernst equation,

$$\Delta E = \frac{RT}{nF} \ln K \tag{21}$$

At 20°C, the difference between the redox potentials is

$$\Delta E = E(\bullet A/A^-) - E(\bullet S/S^-) = 0.059 \log K \qquad (22)$$

The reduction potentials of some well-known antioxidants are shown in Table III.

Table III. One-electron Reduction Potentials (vs. NHE) of Selected Free Radicals ($\bullet$R) of Biological and Nutritional Interest in Aqueous Solutions at pH = 7 and 20°C Obtained by Pulse Radiolysis

$\bullet$R or RH	E_7 ($\bullet$R/RH), V[a]	$\bullet$R or RH	E_7 ($\bullet$R/RH), V[a]
$\bullet$OH	2.31	Uric acid	0.69
Indole	1.15	Serotonin	0.64
$CH_3OO\bullet$	1.06	Sesamol	0.62
Trp	1.05	α-Tocopherol	0.48
GSH	1.00	Hydroquinone	0.46
Phenol	0.97	Ascorbate	0.28
Tyr	0.89		

see ref. 26, 29, 30, 32–35.
[a] In some cases parent compound RH is deprotonated, hence E_7 ($\bullet$R/R$^-$).

Reduction potentials determine the direction of reaction (19). If the reduction potential of $\bullet$S/S$^-$ or $\bullet$S/SH is lower than for $\bullet$A/A$^-$ or $\bullet$A/AH at the pH of interest, then k_f will be higher than k_r and the reaction proceeds in the forward direction. For example, ascorbate ($E_7 = 0.28$ V) rapidly reduces (k = 1.5×10^6 M^{-1} s^{-1}) α-tocopherol radical ($E_7 = 0.48$ V). Conversely, when the reduction potential of $\bullet$S/S$^-$ is higher than that of $\bullet$A/A$^-$, then $k_r > k_f$ and reaction (19) proceeds in the reverse direction. Hence, phenol ($E_7 = 0.97$ V) or tyrosine ($E_7 = 0.89$ V) cannot reduce the α-tocopherol radical, whereas phenol derivatives with $E_7 < 0.48$ V can.

Calculation of Redox Potentials

One-electron redox potentials of phenols are determined by charge separation and electronic configuration of the intermediates (26,29,30). This is affected by the nature and position of the substituents and expressed by the Hammett correlation. For a single substituent, the redox potential at pH 7 may be calculated as follows (26):

$$E_7 = 0.95 + 0.31 \, \sigma^+ \qquad (23)$$

where 0.95 is the extrapolated E_7 value (V) for phenol, and σ^+ is the Brown substituent constant (31). The following relationship, however, was discovered for polysubstituted phenols (26):

$$E_7 = 0.95 + 0.31 \, \Sigma \, \sigma^+ \qquad (24)$$

The substituted phenols, the measured values, and the Hammett correlation for E_7 are shown in Table IV and Figure 1.

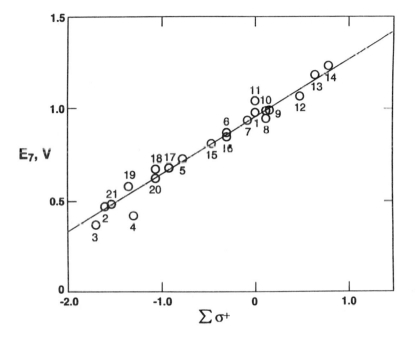

Figure 1. The Hammett plot of the reduction potentials at pH 7 (E_7) of various substituted phenoxyl radicals (see Table IV) in aqueous solution at 20°C vs. $\Sigma\sigma^+$. The line is described by $E_7 = 0.95 + 0.31\ \Sigma\sigma^+$; the correlation coefficient = 0.987 (Reproduced from reference 26. Copyright 1991 American Chemical Society.)

Starting from $E_7 = 0.95$ V, the extrapolated value for phenol (the measured value $E_7 = 0.97$ V is only slightly higher), the reduction potentials of substituted phenols can be higher or lower, depending on whether the substituents have electron-withdrawing or electron-donating properties, respectively. For example, a nitro group or halogen atoms are electrophilic and increase E_7, making the hydroxyl group a poorer electron donor. On the other hand, methyl, hydroxy, and methoxy groups are good electron donors and reduce the E_7 value, making those derivatives better electron donors. The σ^+ values for a few substituents in ortho), meta, and para positions to the hydroxy groups of phenol are shown in Table V. The corresponding decreases of measured reduction potentials (26) relative to $E_7 = 0.97$ V for phenol are shown in Figure 2.

A similar correlation between E_7 and σ^+ was observed for heterocyclic aromatic compounds, i.e., indole derivatives (30).

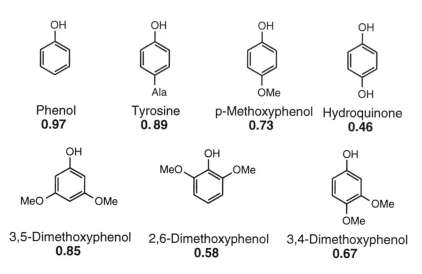

Figure 2. Structures and reduction potentials (E_7, V) of phenoxyl radicals originating from selected substituded phenols (26).

Antioxidants and Protoantioxidants

Antioxidants. By definition an antioxidant is a compound capable of inhibiting oxygen-mediated oxidation of diverse substrates, from simple molecules to polymers and complex biosystems. There are two types of antioxidants. The first type inhibits formation of free radicals, e.g., reactions (2), (4), and (5), which may initiate oxidation. In most cases they are chelators of metal ions. The second type inhibits free-radical chain-propagation reactions (15,24),

$$H_2L + {}^\bullet X \rightarrow HL^\bullet + XH \tag{25}$$

$$HL^\bullet + O_2 \rightarrow HLOO^\bullet \tag{26}$$

**Table IV. Acid-Base Properties of Substituted Phenols (pK$_a$),
Reduction Potentials of Their Phenoxyl Radicals at pH = 7 (E$_7$) and 20°C,
and Brown Substituent Constants (σ^+)**

Substituent in the radical at pH 7	Parent pK$_a$[a]	E$_7$, V	σ^+ [c]
1. H	10.0	0.97	0
2. 4-O$^-$	9.91; 12.0	0.46	-1.6
3. 4-(CH$_3$)$_3$N	8.0	0.36	-1.70
4. 4-NH$_2$	8.5; 10.3	0.41	-1.31
5. 4-OCH$_3$	10.2	0.73	-0.78
6. 4-CH$_3$	10.19	0.87	-0.31
7. 4-F	9.95	0.93	-0.08
8. 4-Cl	9.43	0.94	0.11
9. 4-Br	9.34	0.96	0.15
10. 4-I	9.31	0.09	0.13
11. 4-CO$_2^-$	9.39; 4.58	1.04	0.0
12. 4-COCH$_3$	8.05	1.06	0.50
13. 4-CN	7.95	1.17	0.66
14. 4-NO$_2$	7.14	1.23	0.79
15. 3-O$^-$	9.8; 11.3	0.81	-0.47
16. 3,5-(CH$_3$O)$_2$	9.345	0.85	-0.28
17. 3,4-(CH$_3$O)$_2$	9.80	0.67	-0.92
18. 3,4,5-(CH$_3$O)$_3$	8.95	0.66	-1.06
19. 2,6-(CH$_3$O)$_2$	9.63	0.58	-1.34
20. (Sesamol)	9.70	0.62	-1.06
21. (α-Tocopherol)	11.9	0.48	-1.52

SOURCE: Adapted from ref. 26.

**Table V. Brown Substituent Constants (σ^+) for
Some Ortho, Meta, and Para Substituted Phenols**

Substituent	Ortho	Meta	Para
O$^-$	-1.13[a]	-0.47	-1.6
OH	—	-0.04	-0.92
OCH$_3$	-0.67	-0.14[b]	-0.79
CH$_3$	-0.27	-0.07	-0.33
C$_2$H$_5$	-0.24	-0.06	-0.30

SOURCE: See ref. 31.
[a] Calculated from E$_7$ for dihydroxy derivatives.
[b] From ref. 26.

$$H_2L + HLOO\bullet \rightarrow HL\bullet + HLOOH \qquad (27)$$

where $\bullet X$ stands for superoxide, hydroxyl, or alkoxyl radical; H_2L stands for un-saturated or polyunsaturated fatty acid. Reaction (27) proceeds via H abstraction at bisallylic position due to lower bond strength (BS) of the bisallylic C-H bond [BS(HL-H) = 75 kcal/mol] than BS of other C-H bonds in the molecule, and a relatively low value for hydroperoxides [BS(HLOO-H) = 88-92 kcal/mol (32)]. The chain-breaking antioxidant must therefore quantitatively inhibit reaction (25) or (27).

The reactivities of hydroxyl and alkoxyl radicals are about 10^9–10^{10} M^{-1} s^{-1} and 10^6–10^7 M^{-1} s^{-1}, respectively. Because the concentration of biomaterials is very high and because $k = 10^{10}$ M^{-1} s^{-1} is the upper limit of any antioxidant reaction with free radicals, inhibition of $\bullet OH$ radical reactions is impossible. Inhibition of $RO\bullet$ reactions, however, could be substantial, if not complete. Consequently, the reaction of peroxyl radical — e.g., reaction (27) — is the reaction that an antioxidant must inhibit. In contrast to $\bullet OH$ and $RO\bullet$, the reaction rate constants, k, and the energetics, E, favor inactivation of $ROO\bullet$ by phenols and polyphenols vs. $ROO\bullet$ driven propagation, as discussed below.

The reduction potential of $CH_3OO\bullet$ is $E_7 = 1.05 \pm 0.05$ V (32). Natural phenols have a reduction potential lower than phenol, $E_7 = 0.97$ V, and should consequently react with methyl peroxyl radical. The reduction potential of other peroxyl radicals may be lower by 0.1 V due to the inductive effect of R in $ROO\bullet$. Even those peroxyl radicals should react with most substituted phenols.

The kinetics of reaction (27) vs. the reaction

$$HLOO\bullet + AH \rightarrow HLOO^- + \bullet A + H^+ \qquad (28)$$

also favors the antioxidant, AH, because $k_{27} \leq 240$ M^{-1} s^{-1}, whereas $k_{28} \leq 10^7$ M^{-1} s^{-1}. The large difference between k_{27} and k_{28} is due to the difference in the reaction mechanisms (15). Reaction rate constants are much higher, in general, for electron transfer processes than H-atom abstraction. Consequently, even at $AH/H_2L \sim 0.01$, an antioxidant concentration achievable in practice, the antioxidant could inhibit chain-propagation reactions. It is clear from these examples that E_7, reactivity (k), and concentration (overall or local) are important parameters for efficient antioxidant activity in biosystems.

Protoantioxidants. Numerous phytochemicals have $E_7 > 1$ V and, therefore, are inefficient antioxidants. Metabolic alteration of these compounds may result in metabolites with sufficiently lower reduction potentials to qualify as antioxidants. The idea originated (33) from experiments associated with the measurements of chain propagation length, redox potentials, kinetics of free radical reactions, and biochemical hydroxylation processes. For example, E_7 equals 1.05 V for Trp and tryptamine (34) (decarboxylation on the side chain of Trp does not affect redox potential). Hence, these two compounds are without antioxidant properties. Enzymatic hydroxylation of Trp *in vivo* generates 5-hydroxytryptophan and serotonin with $E_7 = 0.64$ V. Consequently, Trp is by our definition a physiological protoantioxidant that is converted *in vivo* into active and efficient physiological and neurophysiological antioxidants (35).

The concept of a protoantioxidant may be of crucial importance in categorization of potential anticarcinogenic phytochemicals. For example, the anticarcinogenic properties of indole derivatives in cruciferous vegetables (36) may be induced by physiological conversion of indole-derived protoantioxidants.

Conclusions

The induction of mutation and tumors by chemical mutagens and carcinogens is amenable to modulation by antioxidants to a greater or lesser extent (*36*). A few important questions arise.
- What are the mechanisms of action of antioxidants as anticarcinogens in chemical carcinogenesis?
- What is the role of antioxidants in signal transduction?
- Are antioxidants activating defense enzymes (*36,37*), inactivating oxidative processes (*3*), or both?
- What is the extent of oxidative processes induced by promoters and progressors in multistage carcinogenesis (*38*)?
- What is the difference between inhibition of endogenous vs. chemical carcinogenesis?

   It is unfortunate that the current status of our mechanistic understanding of molecular carcinogenesis, including oxidative processes (oxidative carcinogenesis), does not provide definitive answers to these fundamental questions. It is hoped that further advances in free radical mechanisms in chemical and endogenous carcinogenesis will contribute to development of comprehensive mechanistic anticarcinogenesis and its successful application to cancer prevention.

Acknowledgments

Supported by PHS grant number CA44982-07 awarded by the National Cancer Institute.

Literature Cited

1. Davies, K.J.A. *Oxidative Damage and Repair*; Pergamon Press, New York, 1991.
2. Wattenberg, L. W. *Adv. Cancer Res.* **1978**, *26*, 197–226.
3. Simic, M.; Hunter, E. In *Radioprotection and Anticarcinogenesis*; Nygaard, O; Simic, M. Eds.; Academic Press 1983; pp 449–460.
4. Doll, R.; Peto, R. *J. Natl. Cancer Inst.* **1981**, *66*, 1191–1308.
5. Ames, B. N. *Science*, **1983**, *221*, 1256–1264.
6. Visek, W. J., Editor *Cancer Research* **1992**, *52*, 2019S–2126S.
7. Cerutti, P. A. *Science*, **1985**, *227*, 375–381.
8. Ames, B. N. *Environ. Mol. Mutagen.* **1989**, *14*, 66–77.
9. Cerutti, P. A.; Trump, B. F. *Cancer Cells* **1991** *3*, 1–7.
10. Simic, M. G. *J. Environ. Sci. Health* **1991**, *C9*, 113–153.
11. Frenkel, K. *Environ. Health. Perspect.* **1989**, *81*, 45–54.
12. Simic, M. G.; Bergtold, D. S. *Mutat. Res.* **1991**, *250*, 17–24.
13. Hartman, P. E.; Shankel, D. M. *Environ. Mol. Mutagen.* **1990**, *15*, 145–182.
14. Wattenberg, L. W. *Cancer Res.* **1992**, *52*, 2085s–2091s.
15. Simic, M. G.; Jovanovic, S. V.; Niki, E. *Am. Chem. Soc. Symp. Ser.* **1992**, *500*, 14–32.
16. Boveris, A.; Oshino, N.; Chance, B. *Biochem. J.* **1972**, *128*, 617–630.
17. Loschen, G. A.; Azzi, A.; Richter, C.; Flohe, L. *FEBS Letters*, **1974**, *41*, 68–72.
18. Nohl, H; Jordan, W. *Biochem. Biophys. Res. Comm.* **1986**, *138*, 533–539.
19. McCord, J. M.; Fridovich, I. *J. Biol. Chem.* **1968**, *243*, 5753–5759; ibid. **1969**, *244*, 6049–6055.

20. Chance, B.; Sies, H.; Boveris, A. *Physiol. Rev.* **1979**, *59*, 527–605.
21. von Sonntag, C. *The Chemical Basis of Radiation Biology*; Taylor and Francis: New York, NY, 1987.
22. Niki, E. private communication.
23. Neta, P.; Dizdaroglu, M.; Simic, M.G. *Israel J. Chem.* **1983**, *24*, 25–28.
24. Al-Sheikhly, M.; Simic, M. G. *J. Phys. Chem.* **1988**, *93*, 3103–3106.
25. Buxton, G. V.; Greenstock, C. L.; Helman, W. P.; Ross, A. B. *Phys. Chem. Ref. Data* **1988**, *17*, 513.
26. Jovanovic, S. V.; Tosic, M.; Simic, M. G. *J. Phys. Chem.* **1991**, *95*, 10824–10827.
27. Simic, M. G. In *Oxidative Damage and Repair*; Davies, K. J. A., Ed.; Pergamon Press: New York, 1991; 47–56.
28. Jovanovic, S. V. in *Oxidative Damage and Repair*, Davies, K. J. A., Ed.; Pergamon Press: New York, 1991; 93–97.
29. Lind, J.; Shen, X.; Eriksen, T.E.; Merenyi, G. *J. Am. Chem. Soc.* 1990, **112**, 479.
30. Jovanovic, S. V.; Steenken, S. *J. Phys. Chem.* **1992**, *96*, 6674–6679.
31. Hansch, C.; Leo, A. *Substituent Constants for Correlation Analysis in Chemistry and Biology*; John Wiley & Sons: New York, 1979.
32. Jovanovic, S. V.; Jankovic, I.; Josimovic, L. *J. Am. Chem. Soc.* **1992**, in press.
33. Simic, M. G.; Jovanovic, S. V. In *Antimutagenesis and Anticarcinogenesis Mechanisms II*; Kuroda, Y.; Shankel, D. M.; Waters, M. D., Eds.; Plenum: New York, 1990; pp 127–137.
34. Jovanovic, S. V.; Steenken, S.; Simic, M. G. *J. Phys. Chem.* **1990**, *94*, 3583–3588.
35. Jovanovic, S.V; Simic, M. *Life Chemistry Reports* **1985**, *3*, 124–130.
36. Zhang, Y.; Talalay, P.; Chio, C-G.; Posner, G.H. *Proc. Natl. Acad. Sci. USA* **1992**, *89*, 2399–3403.
37. Weinstein, I. B. *Cancer Res.* **1991**, *51*, 5080s–5085s.
38. Harris, C. C. *Cancer Res.* **1991**, *51*, 5023s–5044s.

RECEIVED May 17, 1993

PHYTOCHEMICALS FROM TEA

Chapter 3

Prophylactic Functions of Tea Polyphenols

Yukihiko Hara

Food Research Laboratories, Mitsui Norin Company Ltd., Fujieda 426, Japan

Since ancient times tea has been known to be effective in keeping "body and soul" in good condition. Among the soluble components of tea, the polyphenolic fraction constitutes the major part. With a view to linking the traditionally acclaimed health benefits of tea to these polyphenolic compounds we have fractionated and purified tea polyphenols and subjected them to a variety of experiments, not only in the laboratory, but also in animals and humans. As a result, tea polyphenols have proven to be multi-functionally effective in preventing rancidity and putrefaction of foods, infection of the alimentary canal, respiratory diseases, occurrence of cardiovascular disease and development of malignant tumors. In this review, methods of separating tea polyphenols from green tea or black tea and some of their functions revealed so far will be described in brief.

Method of Extracting Tea Polyphenols

Catechins. The polyphenolic fraction is mostly composed of catechins, which are the major components of fresh tea leaves as well as the soluble matter in green tea. Crude catechin powder is obtainable by treating tea leaves with water and organic solvents, as shown in Figure 1. The total catechin content of this crude product is a little more than 90%. This crude catechin powder is then submitted to liquid chromatography to isolate five kinds of catechin compounds (1). The composition of the crude catechin powder is shown in Table I.

Theaflavins. In the process of black tea manufacture, catechins are mostly oxidized to form pigments such as theaflavins or thearubigins (2). In the case of tropical black tea, these fractions exist in a ratio of 10-15% thearubigins, 1-2% theaflavins and 5-10% catechins on a dried tea basis. The amount of theaflavins is closely related to the commercial value of black tea.

To separate theaflavins, black tea is extracted with water, then washed with solvents to remove impurities (3). The extraction process is shown in Figure 2 and

0097–6156/94/0547–0034$06.00/0

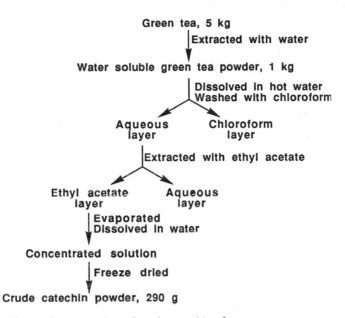

Figure 1. Preparation of crude catechins from green tea.

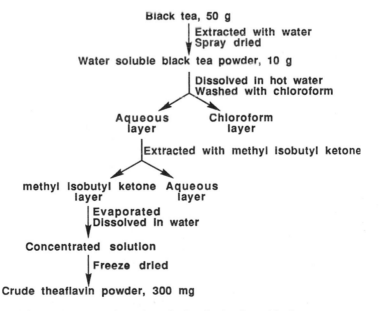

Figure 2. Preparation of crude theaflavins from black tea.

the composition of individual theaflavins is shown in Table II. Structural formulae of catechins and theaflavins are shown in Figure 3.

Table I. Composition of crude catechins in green tea

Catechins	Absolute %	Relative %
(+)-Gallocatechin (GC)	1.44	1.6
(-)-Epigallocatechin (EGC)	17.57	19.3
(-)-Epicatechin (EC)	5.81	6.4
(-)-Epigallocatechin gallate (EGCg)	53.90	59.1
(-)-Epicatechin gallate (ECg)	<u>12.51</u>	<u>13.7</u>
	91.23	100.0

Table II. Composition of crude theaflavins in black tea

Theaflavins	Absolute %	Relative %
Theaflavin (TF1)	13.35	15.8
Theaflavin monogallate A (TF2A)	18.92	22.4
Theaflavin monogallate B (TF2B)	18.64	22.1
Theaflavin digallate (TF3)	<u>33.46</u>	<u>39.7</u>
	84.37	100.0

Antioxidative Action

Edible oils. The antioxidative potency of crude catechin powder and individual catechins was tested in experiments on lard using Active Oxygen Method (AOM) (*1*). As shown in Figure 4, crude catechins reduced the formation of peroxides far more effectively than *dl*-α-tocopherol or BHA. In the same way, the antioxidative potency of each catechin was measured as shown in Figure 5. The antioxidative activity increased on a molarity basis in the order: EC<ECg<EGC<EGCg. From these experiments, we found that both crude catechin powder and EGCg are 20 times more potent than *dl*-α-tocopherol and nearly 4 times more potent than BHA on a weight basis. Other noteworthy characteristics of the antioxidative activity of tea catechins are: (1) Catechins work synergistically with tocopherol and ascorbic acid as well as with organic acids such as citric acids, malic acid, tartaric acid; (2) Catechins work as antioxidants in salad oil whereas tocopherol and BHA are ineffective; (3) Catechins show antioxidative activity against photooxidation (*4*); (4) Catechins show antioxidative activity in oil which is emulsified into water.

Erythrocyte membrane lipids. While theaflavins are ineffective against the oxidation of edible oils, they show fairly potent antioxidative activity in the rabbit red blood cell membrane ghost system *in vitro* (Shiraki M. *et al.*, unpublished data). According to the method by Ames *et al.* (*5*), peroxidation of rabbit red blood cell

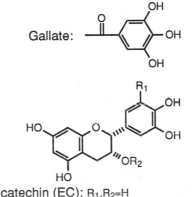

Gallate:

(-)-Epicatechin (EC): R₁,R₂=H
(-)-Epigallocatechin (EGC): R₁=OH, R₂=H
(-)-Epicatechin gallate (ECg): R₁=H, R₂=Gallate
(-)-Epigallocatechin gallate (EGCg): R₁=OH, R₂=Gallate

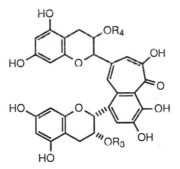

Theaflavin (TF1): R₃,R₄=H
Theaflavin monogallate A (TF2A): R₃=H, R₄=Gallate
Theaflavin monogallate B (TF2B): R₃=Gallate, R₄=H
Theaflavin digallate (TF3): R₃,R₄=Gallate

Figure 3. Structures of tea polyphenols.

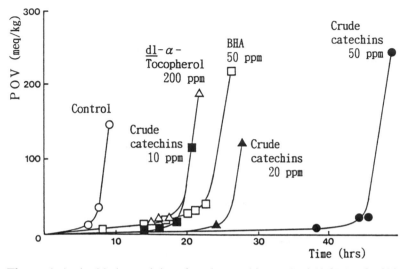

Figure 4. Antioxidative activity of crude catechins on lard (AOM at 97.8°C).

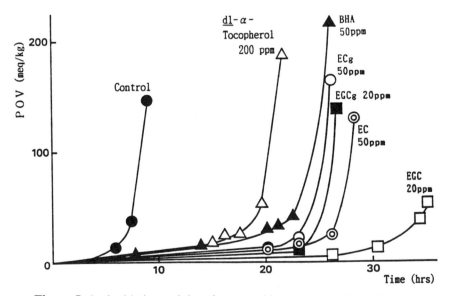

Figure 5. Antioxidative activity of tea catechins on lard (AOM at 97.8°C).

membrane was induced by *t*-butylhydroperoxide with the addition of test compounds. After a coloring reaction with thiobarbituric acid, the formation of the TBA-reacting substance was measured. As shown in Figure 6, all four kinds of theaflavins exhibited much stronger antioxidative effects than α-tocopherol or propyl gallate. Among theaflavins, TF3 showed the most potent antioxidative activity, inhibiting about 80% of the peroxidation.

Plasma. The antioxidative action of catechins in the plasma was also observed *in vivo* in experiments where tea catechins were fed to rats (Nanjo, F., Hara Y., unpublished data). Two kinds of oil rich diets for rats were prepared. One contained 30% palm oil and the other contained 30% perilla oil. Palm oil is rich in saturated fatty acids while perilla oil is rich in unsaturated fatty acids. One percent crude catechin powder was added to each experimental group. After 4 weeks of feeding, the rats were sacrificed and their plasma was collected to determine the degree of peroxidation (indicated as TBA value) and the content of α-tocopherol.

As shown in Figure 7, TBA values in the palm oil group were lower (nearly normal) and showed no difference between the control and catechin diet. In the perilla oil group, however, TBA values were much higher and there was significant suppression in the catechin-fed group. These results indicate that there are more peroxides in the plasma when eating an unsaturated fat diet than when eating a saturated fat diet. Further, according to these indications tea catechins were supposed to have suppressed peroxidation in the unsaturated groups. Figure 8 shows a much higher content of α-tocopherol in the palm oil groups than in perilla oil groups. In the perilla oil groups, more residual α-tocopherol remained in the catechin-fed group than in the control. These results indicate less reduction of α-tocopherol in the saturated fat group than in the unsaturated fat group, and more α-tocopherol remained by catechin-feeding compared to that of the controls.

Radioprotective Action

The C57BL/6 strain of mouse is prone to develop lymphomas after irradiation. In order to investigate the effect of tea catechins on the suppression of thymic lymphomas caused by repeated γ-radiation, three hundred female mice of this strain were raised for two years (most of their life span) (6). Half of them were fed a commercial diet while the rest were fed on the same diet with the addition of 0.2% EGCg (EGCg was tentatively complexed with active aluminum hydroxide to alleviate pungency). Mice from each group were γ-irradiated at 0.5, 1.0, 1.5 or 2.0 Gy once a week from 5 to 8 weeks of age. In this two year feeding study, the influence of EGCg on survival rates (Figure 9) and on the incidence of tumors (Figure 10) was observed. As shown in the figures, a significant decrease in the incidence of tumors was observed in the experimental group of mice receiving 1.0 Gy/week irradiation and fed with EGCg complex. From these results, it is assumed that daily intake of EGCg could protect against radiation damage in humans.

Antimutagenic Action

Spontaneous mutations in *Bacillus subtilis* NIG 1125. A mutator strain of *Bacillus subtilis* NIG 1125 has a temperature sensitive mutation in the structural gene of DNA-polymerase (7). Because of this characteristic, even at 30°C with a normal growth rate, very frequent spontaneous reverse mutations at *his* and *met* loci occur.

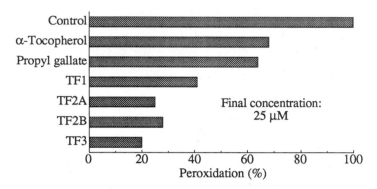

Figure 6. Antioxidative activity of theaflavins as determined by the rabbit erythrocyte ghost system.

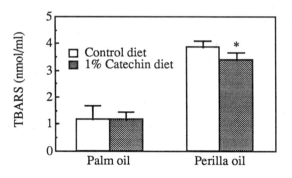

Figure 7. Effects of tea catechins in the diet (four weeks in a 30% oil diet) on plasma peroxides. *Significantly different from control (p<0.05).

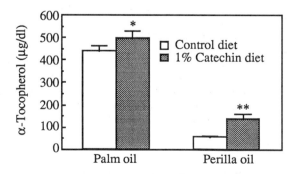

Figure 8. Effects of tea catechins in the diet (four weeks in a 30% oil diet) on plasma α-tocopherol. *Significantly different from control (p<0.05). **Significantly different from control (p<0.001).

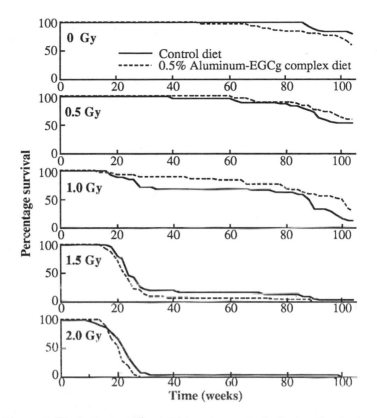

Figure 9. The influence of catechin on the survival of γ-irradiated mice.

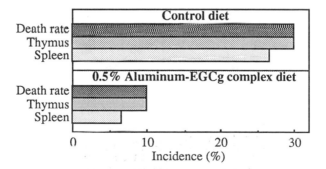

Figure 10. Incidence of malignant lymphomas 40 weeks after 1.0 Gy γ-irradiation once per week for 4 weeks.

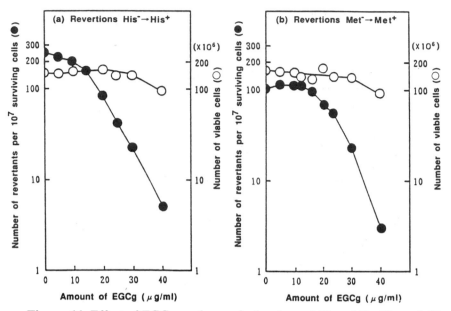

Figure 11. Effect of EGCg on the survival and mutability of *Bacillus subtilis* NIG 1125 (*his met*).

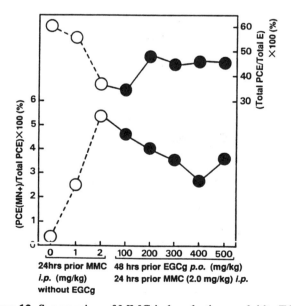

Figure 12. Suppression of MMC-induced micronuclei by EGCg.

Agents which lower these frequent mutations without affecting cellular growth can be considered to be bio-antimutagens working at the level of DNA replication.

EGCg was mixed in histidine deficient agar medium. Bacteria were then spread on the agar plate and cultured for 48 hrs. As shown in Figure 11, the frequency of spontaneous reversions with regard to histidine requirement significantly decreased without there being any significant growth suppression. Similar results were also obtained for reversions with regard to the methionine requirement. Other catechins did not show this decrease of reversions whereas all four theaflavins showed similar curves to EGCg. From these results, it was considered that EGCg and theaflavins have bio-antimutagenic properties.

Suppression of MMC-induced micronuclei in mouse bone marrow cells. The influence of orally dosed tea catechins on Mitomycin C (MMC)-induced chromosome aberrations was investigated (8). EGCg was orally administered to groups of mice in doses of 100-500 mg/kg. After 24 hrs, 2 mg/kg MMC was injected intraperitoneally into each mouse. As controls, 0, 1 mg or 2 mg/kg MMC was injected into mice not administered EGCg. Another 24 hrs after MMC injection, all mice were sacrificed and their femur bone marrow cells were fixed on slides according to conventional procedure. Under the microscope, the number of micronucleated polychromatic erythrocytes cells [PCE(MN+)] was counted. As shown in Figure 12, the percentages of PCE(MN+) increased with the dose of MMC. On the other hand, pre-administered EGCg dose-dependently suppressed the number of PCE(MN+). As shown in the figure, 500 mg/kg EGCg (with 2 mg/kg of MMC) might have worked toxically. We have also confirmed that EGCg administration 48 hrs before, 2 hrs before, or 2 hrs after MMC injection was not effective in suppressing the number of polychromatic micronuclei in this system.

Antitumor Action

Based on the data that tea catechins have antioxidative or antimutagenic property, a range of experiments was conducted with regard to the effect of catechin feeding on tumorigenesis in mice.

Implanted tumor cells. Mice (ddY) fed diets containing crude catechin powder for 9 months were inoculated subcutaneously with sarcoma 180 cells. After 19 days, tumors were resected (9). As shown in Figure 13, the growth of tumors was suppressed in catechin-fed groups in a dose-dependant fashion.

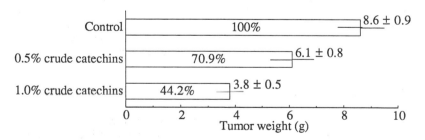

Figure 13. Effect of crude catechins on sarcoma 180 tumors implanted in ddY mice.

Tumorigenesis by a chemical carcinogen. The carcinogen 3-methylcholanthrene (3MC) was administered subcutaneously to mice (*10*). From around the 70th day onward, solid tumors developed under the skin of the mice. As shown in Figure 14, tumorigenesis by 3MC was delayed in the catechin fed group. (In this experiment, crude catechin powder was tentatively complexed with active aluminum hydroxide to alleviate pungency. The actual content of catechins in the diet was 0.2%. These complexed catechins will separate in the stomach.)

Spontaneous breast cancer. Female mice of a certain strain (C3H/HeN) spontaneously develop breast cancer after 3–4 deliveries (*11*). As shown in Figure 15, the age at which the catechin feeding is commenced determines the time it takes before breast cancer develops. With feeding from the 180th day after birth, there was no delay of tumor development. Feeding from the 60th day after birth, however, delayed breast cancer development. Tumorigenesis was even more markedly delayed when catechins were fed from birth on. These results imply certain roles of tea catechins in cancer prevention.

Anti-hypercholesterolemic Action

In order to prevent cardiovascular diseases such as heart attack or brain stroke, it is very important to keep the cholesterol level in the blood within the normal range. The following experiments demonstrate that tea catechins suppress an excessive rise of cholesterol concentration (*12*).

In the first experiment, rats were fed a high fat, high cholesterol diet for 4 weeks to increase the cholesterol in the blood. As shown in Figure 16, addition of 1% crude catechin powder to this diet suppressed the rise of total cholesterol concentration. In the second experiment, rats were fed a high fat, high cholesterol diet to which a small amount of cholic acid was added to facilitate lipid absorption (*13*). After 4 weeks, all rats were sacrificed to examine their cholesterol levels. The results in Figure 17 show that the total cholesterol concentration of the test group rose to more than twice the level of that of the control group. Particularly, the LDL-cholesterol increased as much as 15 times and the HDL-cholesterol decreased to less than half of the control values. But the addition of EGCg to the above diet moderated the increase of LDL-cholesterol and the decrease of HDL-cholesterol. This tendency was dose dependent, in the 1% EGCg group, the LDL-cholesterol level was only 5 times that of the control group.

In another series of experiments we have proved that 1% catechin in a fat inducing diet will reduce the excess accumulation of body and liver fat (data not shown). These results suggest that dietary hyperlipidemia in humans could be controlled by the ingestion of tea catechins. In yet another experiment we confirmed that the addition of tea catechins up to 2% in a normal diet will not affect any of the body organs nor dietary intake, hence there is no fear there being any of adverse effects in a balanced diet.

Anti-hypertensive Action

Most of the hypertension in humans occurs as essential hypertension, where angiotensin I converting enzyme (ACE) converts angiotensin I to vasoconstrictive angiotensin II. Tea polyphenols were confirmed to inhibit the activity of ACE (Table III).

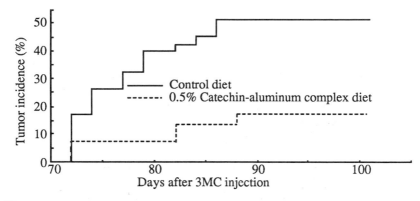

Figure 14. Effect of catechin-aluminum complex on 3-methylcholanthrene (3MC)-induced tumorigenesis.

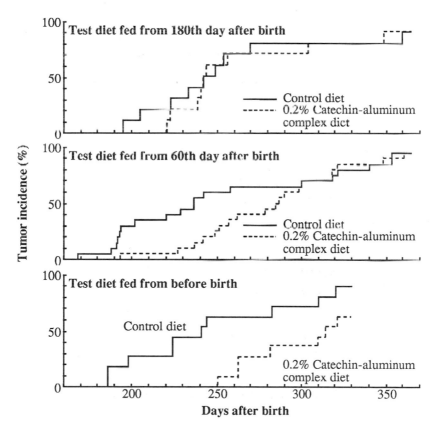

Figure 15. Inhibition of spontaneous mammary tumor incidence in C3H/HeN mice fed catechin-aluminum complex in the diet.

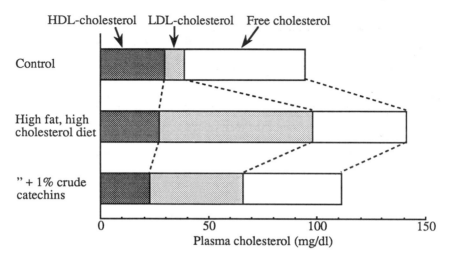

Figure 16. Hypocholesteremic effect of crude catechins.

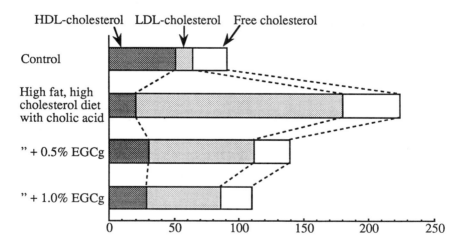

Figure 17. Hypocholesteremic effect of EGCg.

**Table III. Inhibitory effects of
tea polyphenols against ACE**

Tea polyphenols	Inhibition (IC_{50})
(-)-Epicatechin gallate	1400 μM
(-)-Epigallocatechin gallate	90 μM
Theaflavin	400 μM
Theaflavin monogallate A	115 μM
Theaflavin monogallate B	110 μM
Theaflavin digallate	35 μM

Accordingly, the effect of crude catechin powder on the blood pressure of spontaneously hypertensive rats (SHR) was studied. The SHR were divided into two groups. The first group received a normal diet, while the second group was given a diet containing 0.5% crude catechin powder from one week after weaning. Although the blood pressure in the normal diet group already exceeded 200 mm Hg at 10 weeks of age, significant suppression was noted in 1% catechin group (Figure 18). When the diet of both groups was switched at 16 weeks of age, the blood pressures, in time, also changed accordingly, as shown in the figure. We also confirmed the effect of crude catechin powder, using the death rate of stroke-prone SHR (SHRSP) as an index (Figure 19). As shown in the figure, addition of 0.5% crude catechin powder to the normal diet delayed the outbreak of stroke and extended the life span more than 15% as compared to the control group. These results imply that tea may be effective in preventing hypertension in humans.

Antibacterial Action

For a long time tea has been thought to be effective in mitigating diarrhea caused by bacterial infection. Tea polyphenols were separated and the antibacterial property of each compound against various bacteria was studied. Many varieties of food borne pathogenic bacteria were found to be susceptible to tea polyphenols, even in a concentration far lower than would be consumed in everyday drinking of tea (Table IV). We have also found that tea polyphenols show marked bactericidal effects against spores and vegetative cells of most fatal *Clostridium botulinum*. In connection with the above, improvement of the condition of the large intestine was noted (*14*) since tea polyphenols are bactericidal against pathogenic bacteria and ineffective against acid forming good bacteria. Tea polyphenols are also effective against various kinds of phytopathogenic bacteria (*15*).

Another interesting feature of tea polyphenols is that they inhibit not only the growth of cariogenic *Streptococcus mutans* but also the extracellular glucosyltransferase of the same bacteria (*16*), thus they may prevent the formation of dental plaque and caries at a far lower concentration than an everyday cup of tea.

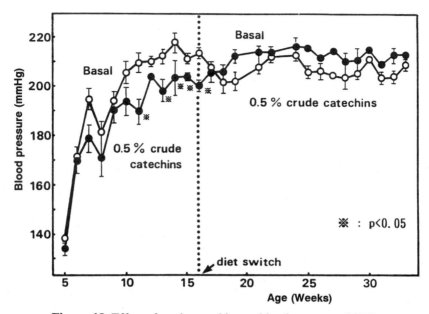

Figure 18. Effect of crude catechins on blood pressure of SHR.

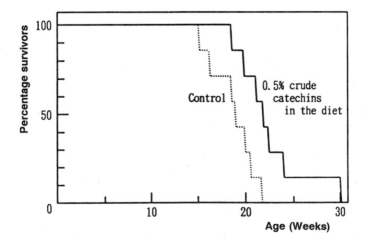

Figure 19. Effect of crude catechins on the life span of SHRSP.

Table IV. Minimum inhibitory concentrations of tea polyphenols against food borne pathogenic bacteria

Bacteria	MIC (ppm)									
	Crude catechins	EC	ECg	EGC	EGCg	Crude theaflavins	TF1	TF2A	TF2B	TF3
Staphylococcus aureus IAM 1011	450	>800	800	150	250	500	500	400	150	200
Vibrio fluvialis JCM 3752	200	800	300	300	200	400	500	500	400	500
V. parahaemolyticus IFO 12711	200	800	500	300	200	300	300	500	400	500
V. metschnikovii IAM 1039	500	—	—	500	1000	300	600	800	700	800
Clostridium perfringens JCM 3816	400	—	400	1000	300	200	500	600	500	400
C. botulinum A, B mix.	<100	—	200	300	<100	200	150	250	150	200
Bacillus cereus JCM 2152	600	—	600	—	600	500	600	1000	1000	800
Plesiomonas shigelloides IID No.3	100	700	100	200	100	100	200	200	200	100
Aeromonas sobria JCM 2139	400	—	700	400	300	300	600	600	400	500
A. hydrophila subsp. hydrophyla JCM 1072	—	—	—	—	—	700	—	1000	1000	1000
Escherichia coli IID 952	—	—	—	—	—	—	—	—	—	—
E. coli IID 954	—	—	—	—	—	—	—	—	—	—
Salmonella enteritidis IFO 3313	—	—	—	—	—	—	—	—	—	—
S. typhimurium IFO 12529	—	—	—	—	—	—	—	—	—	—
Campylobacter jejuni JCM 2013	—	—	—	—	—	—	—	—	—	—
C. coli JCM 2529	—	—	—	—	—	—	—	—	—	—
Yersinia enterocolitica IID 981	—	—	—	—	—	—	—	—	—	—

—: >1000

Literature Cited

1. Matsuzaki, T.; Hara, Y. *Nippon Nogeikagaku Kaishi* **1985**, *59*, 129–134.
2. Roberts, E.A.H.; Smith, R.F. *J. Sci. Fd. Agric.* **1963** *14*, 689–700.
3. Hara, Y.; Matsuzaki, T.; Suzuki, T. *Nippon Nogeikaku Kaishi* **1987**, *61*, 803–808.
4. Hara, Y. *Advances in Food Science and Technology* Nippon Shokuhin Kogyo Gakkai; Korin: Tokyo, **1990**, *4*, 21–39.
5. Ames, B.N.; Cathcart, R.; Schwiers, E.; Hochstein, P. *Proc. Natl. Acad. Sci. U.S.A.* **1981**, *78*, 6858–6862.
6. Hara, Y.; Tono-oka, F.; Tezuka, H. *Proc. Jpn. Radiation Res. Soc. (34th Annual Meeting, Tokyo)* **1991**, 143.
7. Kada, T.; Kaneko, K.; Matsuzaki, S.; Matsuzaki, T.; Hara, Y. *Mutation Res.* **1985**, *150*, 127–132.
8. Hara, Y. Ph.D. Thesis, Tokyo University, Tokyo, Japan, 1990, 150.
9. Hara, Y.; Matsuzaki, S.; Nakamura, K. *Nippon Eiyo Shokuryo Gakkaishi* **1985**, *42*, 39–45.
10. Asai, H.; Hara, Y.; Nakamura, K.; Hosaka, H.; Kokue, S. *Proc. Jpn. Cancer Assoc. (45th Annual Meeting, Sapporo)* **1986**, 387.
11. Asai, H.; Hara, Y.; Nakamura, K.; Hosaka, H.; Kokue, S. *Proc. Jpn. Cancer Assoc. (47th Annual Meeting, Tokyo)* **1988**, 139.
12. Muramatsu, K.; Fukuyo, M.; Hara, Y. *J. Nutr. Sci. Vitaminol.* **1986**, *32*, 613–622.
13. Fukuyo, M.; Hara, Y.; Muramatsu, K. *Nippon Eiyo Shokuryo Gakkaishi* **1986**, *39*, 495–500.
14. Terada, A.; Hara, H.; Nakajyo, S.; Ichikawa, H.; Hara, Y.; Fukai, K.; Kabayashi, Y.; Mitsuoka, T. *Mibrobial Ecology in Health and Disease* **1993**, *6*, 3–9.
15. Fukai, K.; Ishigami, T.; Hara, Y. *Agric. Biol. Chem.* **1991**, *55*, 1895–1897.
16. Hattori, M.; Kusumoto, I. T.; Namba, T.; Ishigami, T.; Hara, Y. *Chem. Pharm. Bull.* **1990**, *38*, 717–720.

RECEIVED April 14, 1993

Chapter 4

Preventive Effect of Green Tea Polyphenols on Colon Carcinogenesis

Mujo Kim[1], Nobuyuki Hagiwara[1], S. J. Smith[1], Takehiko Yamamoto[2], Tetsuro Yamane[3], and Toshio Takahashi[3]

[1]Central Research Laboratories, Taiyo Kagaku Company Ltd., 1–3 Takaramachi, Yokkaichi, Mie 510, Japan
[2]Department of Biotechnology, Faculty of Engineering, Fukuyama University, Fukuyama, Hiroshima 729–02, Japan
[3]First Department of Surgery, Kyoto Prefectural University of Medicine, Kawaramachi-Hirokoji, Kamigyo-ku, Kyoto 602, Japan

We have recently reported that one of the main constituents in green tea polyphenols (GTP), (-)-epigallocatechin gallate, inhibited the promotion stage of mouse duodenal carcinogenesis induced by N-ethyl-N-nitro-N'-nitrosoguanidine. In addition, GTP was found to inhibit the growth of intestinal clostridia, both *in vitro* and *in vivo*, which are considered to participate in the production of carcinogens. This evidence prompted us to study the influence of GTP on colon carcinogenesis. Male Fisher rats were injected s.c. with azoxymethane (7.4 mg/kg) once a week for 10 weeks. One week after treatment, they received 0.01 or 0.1% GTP dissolved in drinking water from weeks 11 to 26. Autopsy on week 26 showed that tumor incidence and average tumor yield in rats receiving GTP were significantly lower than those of rats without GTP. Since GTP was given one week after the azoxymethane treatment, it may inhibit the promotion stage of AOM-induced colon carcinogenesis.

Green tea polyphenols have been reported to have anti-mutagenic (*1*) and anti-tumor activities (*2*). Recently, (-)-epigallocatechin gallate (EGCg), one of its main polyphenolic constituents, has been shown to be an anti-tumor promoter in a two-stage carcinogenesis experiment on mouse skin (*3*). Although the mechanisms underlying these activities remain to be investigated, the above findings on green tea attract much interest from the standpoint of cancer chemoprevention.

Previously, we reported that EGCg inhibited the promotion stage of duodenal carcinogenesis induced by N-ethyl-N'-nitro-N-nitrosoguanidine (*4*), suggesting it has a preventive effect on carcinogenesis in the alimentary tract. In addition, green tea polyphenols were found to inhibit the growth of intestinal clostridia *in vitro* (*5*) and *in vivo* (*6*) which are considered to participate in the biotransformation of a variety of ingested or endogenously formed compounds to carcinogenic products such as N-nitroso compounds and aromatic steroids. This evidence prompted us to study the influence of green tea polyphenols on colon carcinogenesis. This study deals with the inhibitory effect of green tea polyphenol fraction (GTP) on colon carcinogenesis induced by azoxymethane (AOM) in rats.

0097–6156/94/0547–0051$06.00/0

Materials and Methods

Animals and Diets. Eight-week-old Fisher rats were purchased from Japan SLC Inc. (Shizuoka, Japan) and given free access to drinking water and standard laboratory chow MF (Oriental Yeast Co., Tokyo, Japan).

Chemicals. AOM was purchased from Sigma Chemical Co. (St. Louis, MO, U.S.A.), and kept at -80°C before use. GTP used in this study was SUNPHENON, a product of Taiyo Kagaku Co., Ltd. (Yokkaichi, Japan). It contains 75% polyphenolic compounds, which consisted of six catechin derivatives: (+)-catechin, (-)-epicatechin, (-)-epicatechin gallate, (-)-gallocatechin, (-)-epigallocatechin and (-)-epigallocatechin gallate.

Experimental Protocol. AOM was dissolved in 0.9% NaCl solution, and injected s.c. at a dosage of 7.4 mg/kg body weight once a week for the first 10 weeks (7). One week after the last AOM treatment, the treated rats were divided into three groups: AOM-control (26 rats), AOM-GTP1 (26 rats) and AOM-GTP2 (25 rats). The AOM-control group received tap water throughout the experiment. The AOM-GTP1 and AOM-GTP2 groups received 0.01 and 0.1% GTP, respectively, dissolved in tap water as drinking water from week 11 to 26. In addition, three groups of 10 rats each without AOM treatment — *i.e.*, a control group (tap water throughout the experiment), and GTP1 and GTP2 (0.01% and 0.1% GTP, respectively, in tap water) — were prepared as the counterparts of the above AOM-treated groups. Rats in these groups were also injected s.c. for the first 10 weeks, but with 0.9% NaCl solution containing no AOM.

Histological Examination. All rats were sacrificed on week 26. Their esophagi, stomachs, small intestines and large intestines were removed and dissected longitudinally. The location, shape, size and number of tumors was recorded. All tumors were then fixed with 10% formalin for histological examination.

Statistical Analysis. The significance of differences in tumor incidence was analyzed using the χ^2 test; the remaining data were analyzed by Student's t-test.

Results

Body weights. During the experiment, body weights of rats treated with AOM were generally lower than those without the AOM treatment. In most cases, however, the differences were not statistically significant. Statistical differences of body weight were observed between the following groups: AOM-treated and untreated on week 9 ($p < 0.05$), and AOM-control and AOM-GTP2 on weeks 13, 17 and 21 ($p < 0.05$, $p < 0.005$ and $p < 0.05$, respectively). At autopsy on week 26, the body weights among the six groups were not significantly different (Figure 1).

Colon Tumors

Histology. Histologically, the colon tumors consisted of atypical glands, and basophilically stained cells with large number of mitotic bodies were often observed. The nuclei were large, and were located at different sites in the cytoplasm. Tumor cells often invaded submucosal space, proper muscle layer and subserosal space.

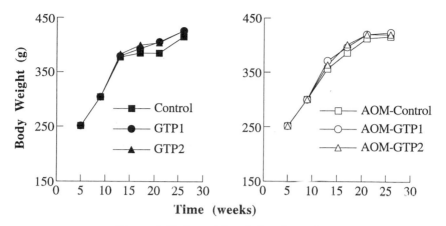

Figure 1. Body weights during the study.

Inhibition by GTP. Rats without the AOM treatment (*i.e.*, control, GTP1 and GTP2) had no colon tumors.

The inhibitory effect of GTP on AOM-induced colon tumors is summarized in Table I. The percent of rats bearing tumors in the AOM-control group was 77%. On the other hand, the percent in the AOM-GTP1 and AOM-GTP2 groups was 38% (p<0.005) and 48% (p<0.05), respectively. The difference between the AOM-GTP1 and the AOM-GTP2 groups was not statistically significant. The average number of tumors per rat in the AOM control group was 1.5, whereas for the AOM-GTP1 and AOM-GTP2 groups it was 0.6 (p<0.01) and 0.7 (p<0.01), respectively. The mean tumor diameter among the three AOM-treated groups was not significantly different. The average distance of colon tumors from the anus was 10.5 ± 5.5 cm (mean ± S.D.) for AOM-control, 7.1 ± 3.9 cm for AOM-GTP1 and 10.4 ± 5.4 cm for AOM-GTP2, and were not significantly different.

Table I. Inhibitory Effect of the Green Tea Polyphenol Fraction on AOM-Induced Colon Carcinogenesis

Treatment Group	Tumor Incidence[a]	Total Number of Tumors	Tumors per Rat[b]	Diameter (mm)[c]
AOM-Control	17/22 (77)	33	1.5 ± 0.2	5.6 ± 3.4
AOM-GTP1	8/21 (38)**	13	0.6 ± 0.2 ***	5.4 ± 2.8
AOM-GTP2	10/21 (48)*	14	0.7 ± 0.2 ***	6.4 ± 3.0

[a] Numbers in parentheses are the percent of rats bearing tumors.
[b] Mean ± S.E.   [c] Mean ± S.D.
  * Significantly different from the AOM-control group (p<0.05).
 ** Significantly different from the AOM-control group (p<0.005).
*** Significantly different from the AOM-control group (p<0.001).

Discussion

Recent studies have indicated a rapid increase in the incidence of large intestinal cancer in Japan. One of the main causative factors seems to be the general Westernization of Japanese eating habits, making diets higher in fat and lower in fiber. Green tea, which has been traditionally taken during and/or after meals, is still popular as a drink in Japan. Our recent studies indicate potential inhibitory activity of green tea polyphenols against carcinogenesis in the alimentary tract. In the present study, we focused on the effect of GTP on large intestinal cancer from the standpoint of cancer prevention through the components of daily diet.

Based on the above results, it can be concluded that GTP inhibited AOM-induced colon carcinogenesis. Although the mechanisms underlying the activity remain to be investigated, GTP might have inhibited the promotion stage of the colon carcinogenesis since it was administered to rats one week after the last AOM treatment. The effect of oral administration of GTP on colon mucosal ornithine decarboxylase activity and other biomarkers of colon carcinogenesis are now under investigation.

The GTP used in this study contained 75% polyphenolic compounds, including EGCg, which has been shown to be an anti-tumor promoter (*3,4*). These polyphenols are very similar in their structures and biological activities (*8*). Although the inhibition of AOM-induced colon carcinogenesis could also be contributed by components other than the polyphenols, this circumstantial evidence leads us to conclude that the polyphenols are the main inhibitory factors.

It is widely accepted that changes in intestinal microflora may be related to colon carcinogenesis. Fecal profiles of intestinal microflora in patients with colon carcinoma and patients with nonhereditary large bowel polyps are clinically different (*9*). Recently, it was suggested that tumor growth or malignant transformation was related to the increase of *Clostridium spp.* except *C. perfringnes* (*10*). Since green tea polyphenols have selective inhibitory activity against the growth of intestinal clostridia *in vitro* (*5*) and *in vivo* (*6*), it is quite tempting to suggest that GTP inhibited the AOM-induced colon carcinogenesis partially through its inhibitory effect on the intestinal microflora.

In this experiment, GTP was administered to rats in the drinking water at concentrations of 0.01 and 0.1%. In terms of average body weight of rats and their consumption of drinking water, the daily GTP intake by these rats was comparable to that consumed by average Japanese green tea drinkers. A recent epidemiological study showed a lower risk of gastric cancer among those with a high consumption of green tea (10 or more cups per day) (*11*). These findings, coupled with those previously reported (*4*), led to a postulation that green tea polyphenols are useful in preventing carcinogenesis in the alimentary tract.

The GTP used in this study is now one of the major food additives used in Japan, and its ability of suppressing dental caries is taken advantage of in confectioneries such as chocolates, candies and gums. From the standpoint of cancer prevention through the components of daily diet, the present results could be a step towards clinical trials of the prevention of colon cancer by green tea polyphenols.

Literature Cited

1. Okuda, T.; Mori, K.; Hayatsu, H. *Chem. Pharm. Bull.* **1984**, *32*, 3755–3758.

2. Hara, Y.; Matsuzaki, S.; Nakamura, K. *J. Jpn. Soc. Nutr. Food Sci.* **1989**, *42*, 39–45.
3. Yoshizawa, S.; Horiuchi, T.; Fujiki, H.; Yoshida, T.; Okuda, T.; Sugimura, T. *Phytother. Res.* **1987**, *1*, 44–47.
4. Fujita, Y.; Yamane, T.; Tanaka, M.; Kuwata, K.; Okusumi, J.; Takahashi, T.; Fujiki, H.; Okuda, T. *Jpn. J. Cancer Res.* **1989**, *80*, 503–505.
5. Ahn, Y. K.; Kim, M.; Yamamoto, T.; Mitsuoka, T. *Agric. Biol. Chem.* **1991**, *55*, 1425–1426.
6. Okubo, T.; Ishihara, N.; Oura, A.; Serit, M.; Kim, M.; Yamamoto, T.; Mitsuoka, T. *Biosci. Biotech. Biochem.* **1992**, *56*, 588.
7. Ward, J.; Yamamoto, S.; Brown, C. *J. Natl Cancer Inst.* **1973**, *51*, 1029–1039.
8. Hagiwara, N.; Ahn, Y.; Ishida, M.; Tateishi, M.; Sakanaka, S.; Kim, M.; Yamamoto, T. *Jpn. J. Pharmacognosy* **1991**, *45*, 199.
9. Mastromarino, A. J.; Reddy, B.S.; Wynder, E. L. *Cancer Res.* **1978**, *38*, 4458–4462.
10. Kubota, Y. *Jpn. J. Gastroenterol.* **1990**, *87*, 771–779.
11. Kono, S.; Ikeda, M.; Tokudome, S.; Kuratsune, M. *Jpn. J. Cancer Res.* **1988**, *79*, 1067–1074.

RECEIVED May 10, 1993

Chapter 5

Tea Polyphenols as a Novel Class of Inhibitors for Human Immunodeficiency Virus Reverse Transcriptase

Hideo Nakane[1], Yukihiko Hara[2], and Katsuhiko Ono[1]

[1]Laboratory of Viral Oncology, Aichi Cancer Center Research Institute, Chikusa-ku, Nagoya 464, Japan
[2]Food Research Laboratories, Mitsui Norin Company Ltd., Fujieda 426, Japan

Tea polyphenols (i.e., green tea catechins and black tea theaflavins) are strong inhibitors of human immunodeficiency virus (HIV)-reverse transcriptase. The galloyl moiety is important for their inhibitory effect because it is essential in catechins for inhibition and enhances the inhibitory potency of theaflavins. Tea polyphenols had considerable inhibitory activity against cellular DNA and RNA polymerases but were less effective than against HIV-reverse transcriptase. The mechanism of inhibition of DNA polymerases by the tea polyphenols was, in most cases, competitive with respect to the template•primer and noncompetitive to the nucleotide substrate. The inhibition of cellular polymerases by green tea catechins seems to cause their cytotoxicity to cultured cells, and might explain the epidemiological finding in Japan that the mortality of digestive tract cancer is significantly lower in areas where green tea ingestion is high.

Human immunodeficiency virus (HIV) is a causative agent of acquired immune deficiency syndrome (AIDS). Since HIV has proved to be a retrovirus, various chemotherapeutic approaches toward AIDS have been conducted using anti-retrovirals with HIV as a target. One of the appropriate molecular targets is HIV-associated reverse transcriptase because this enzyme is unique to retroviruses and is requisite for retrovirus infection. Various dideoxynucleoside analogs including azidothymidine (AZT) (*1,2*) and dideoxycytidine (DDC) (*3*) have been shown to inhibit proviral DNA synthesis by HIV-associated reverse transcriptase, protecting the host cells from HIV-induced cytopathogenicity. Various side effects, however, such as anemia and leukopenia for AZT (*4*) and peripheral neuropathy for DDC (*5*), have been demonstrated during the course of the treatment of AIDS patients. In addition, the appearance of AZT-resistant virus strains (*6*) has become a serious problem. The development of new kinds of anti-HIV agents is, therefore, an urgent task in the field.

0097–6156/94/0547–0056$06.00/0
© 1994 American Chemical Society

To find novel anti-HIV substances, we have been looking for natural products inhibitory to HIV-reverse transcriptase. We found that Sho-Saiko-To, a Chinese traditional drug, inhibited HIV-reverse transcriptase (*7*). We identified one effective substance as 5,6,7-trihydroxyflavone (baicalein) from *Scutellaria baicalensis*, a constituent of this herbal medicine (*8*). Since baicalein is a flavonoid, we extended the survey to other flavonoids and their related compounds, and found that (-)-epicatechin gallate (ECg) and (-)-epigallocatechin gallate (EGCg), two major components of Japanese green tea, were strong inhibitors of HIV-reverse transcriptase.

Inhibitory Effect of Green Tea Catechins on Reverse Transcriptase and Cellular DNA and RNA Polymerases

Green tea contains (-)-epicatechin (EC), (-) epigallocatechin (EGC) and their gallic acid esters at position 3 of their structures (Figure 1). We examined the inhibitory effect of these green tea catechins on HIV-1 reverse transcriptase with reaction conditions described previously (*8*). As shown in Figure 2, the 50% inhibition concentration (IC_{50}) of ECg and EGCg was 10–20 ng/ml. Neither EC, EGC nor gallic acid, however, was inhibitory to the enzyme. These results indicate that a galloyl moiety in the catechin structure is essential for an inhibitory effect.

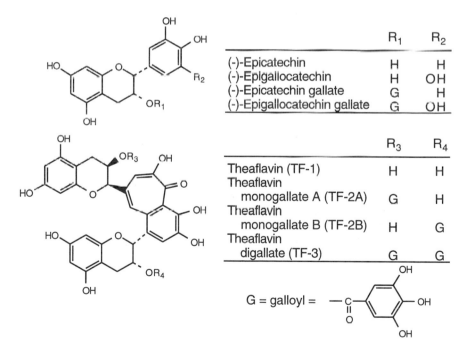

	R_1	R_2
(-)-Epicatechin	H	H
(-)-Epigallocatechin	H	OH
(-)-Epicatechin gallate	G	H
(-)-Epigallocatechin gallate	G	OH

	R_3	R_4
Theaflavin (TF-1)	H	H
Theaflavin monogallate A (TF-2A)	G	H
Theaflavin monogallate B (TF-2B)	H	G
Theaflavin digallate (TF-3)	G	G

G = galloyl =

Figure 1. Structures of green tea catechins and black tea theaflavins.

The inhibitory effects of green tea catechins on other polymerases were also evaluated with the reaction conditions described previously (*9*) (Figure 3). As

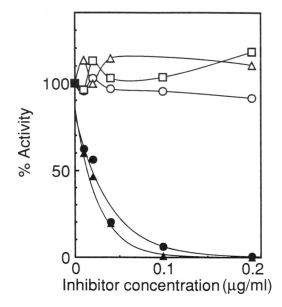

Figure 2. Effects of catechins on the activity of HIV-reverse transcriptase. Reverse transcriptase activity was measured by determining the incorporation of ^{3}H-TMP (400 cpm/pmol) with $(rA)_n \cdot (dT)_{12-18}$ as the template•primer in the presence of various concentrations of catechins or gallic acid as indicated in the figure. The symbols used are as follows: (-)-epicatechin gallate (●), (-)-epigallocatechin gallate (▲), (-)-epicatechin (○), (-)-epigallocatechin (Δ), gallic acid (□). The 100% value (pmol) was 16.2.

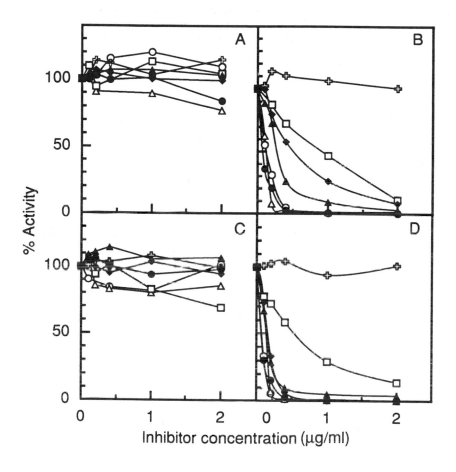

Figure 3. Effects of catechins on the activities of various DNA and RNA polymerases. The activities of Rauscher murine leukemia virus (R-MLV) reverse transcriptase and various DNA and RNA polymerases were measured in the presence of various concentrations of (-)-epicatechin (A), (-)-epicatechin gallate (B), (-)-epigallocatechin (C) and (-)-epigallocatechin gallate (D) as indicated in the figure. Figure symbols and 100% values (pmol) are: R-MLV reverse transcriptase, ●, 59.6; human DNA polymerase α, ○, 22.7; human DNA polymerase β, Δ, 8.9; human DNA polymerase γ, ☐, 1.45; calf thymus terminal deoxynucleotidyltransferase, ✢, 231.3; *E. coli* DNA polymerase I, ■, 65.0; RNA polymerase, ▲, 20.8. The specific radioactivities of ^{3}H-dNTP and ^{3}H-GTP were 6000 (☐), 1000 (○), 400 (●, Δ, ✢, ■), and 16 (▲).

shown in Figures 3A and 3C, neither EC nor EGC was inhibitory to other polymerases. On the other hand, ECg and EGCg inhibited various DNA and RNA polymerases though the necessary concentration for the inhibition was about 10 times higher than those for HIV-1 reverse transcriptase (Figures 3B and 3D). Only terminal deoxynucleotidyltransferase was insensitive to inhibition by these compounds.

The inhibition of cellular polymerases by the test compounds may explain their cytotoxicity to cultured cells. In fact, both ECg and EGCg were highly toxic to cultured MT4 lymphocytes. Because of this cytotoxicity, anti-HIV effects of these compounds cannot be evaluated using this cell culture system. Cellular DNA polymerases, however, especially DNA polymerase α, are regarded as the targets of some anticancer drugs because DNA polymerases play central roles in DNA replication which is indispensable for the proliferation of cancer cells. Therefore, the inhibitory effects of green tea catechins on cellular polymerases might explain, at least in part, the epidemiological finding in Japan that the mortality ratio of digestive tract cancer is significantly lower in areas where green tea ingestion is high (10).

Analysis of the Mode of Inhibition by ECg and EGCg and Determination of the Kinetic Constants. The mechanism of inhibition by ECg and EGCg was analyzed by changing the concentrations of either the template•primer or the triphosphate substrate in the presence of various concentrations of the inhibitors. These results are summarized in Table I. Very small inhibition constants of ECg and EGCg for HIV-1 reverse transcriptase indicate that these green tea catechins are the strong inhibitors of this enzyme. For most of the polymerases, the K_i values of EGCg are slightly smaller than those of ECg, indicating that the former is a stronger inhibitor than the latter. This difference between the two compounds in inhibitory potential is attributable to the presence or absence of the 5'-hydroxyl group of the B-ring.

As to the inhibition mechanism, both ECg and EGCg were competitive with respect to the template•primer and noncompetitive with the triphosphate substrate for most DNA polymerases including reverse transcriptase. Only HIV-reverse transcriptase exhibited a mixed-type inhibition with respect to the template•primer. On the other hand, RNA polymerase exhibited a completely different inhibition mechanism; the mechanism changed from noncompetitive-type to competitive-type by increasing the template•primer concentration, whereas the inhibition was purely competitive-type with respect to the nucleotide substrate GTP (Table I).

Inhibitory Effect of Black Tea Theaflavins

Black tea contains some novel polyphenols called theaflavins as the oxidation products of green tea catechins. Theaflavin has two binding sites of a galloyl moiety, and therefore there are three types of gallic acid esters — monogallate-A (TF-2A), monogallate-B (TF-2B) and digallate (TF-3) (Figure 1). These theaflavins were prepared from black tea as described previously (11). We found that theaflavins also were strong inhibitors of HIV-reverse transcriptase. The IC_{50} values determined from dose-response curves are summarized in Table II. Theaflavins inhibited not only reverse transcriptases but also all cellular polymerases tested except TdT. Although unesterified theaflavin (TF-1) was less inhibitory than the other three ester types, TF-1 itself showed an obvious inhibitory effect. Galloyl

Table I. Characterization of Inhibition of DNA and RNA Polymerases
by (-)-Epicatechin Gallate and (-)-Epigallocatechin Gallate

Polymerase	Variable substrate	K_m	(-)-Epicatechin gallate		(-)-Epigallocatechin gallate	
			Mode	K_i (nM)	Mode	K_i (nM)
Reverse transcriptase						
HIV	$(rA)_n \cdot (dT)_{12\text{-}18}$ (1:1)[a] dTTP	0.8 µg/ml 3.0 µM	Mix[c] NC[d]	7.2	Mix NC	2.8
R-MLV[b]	$(rA)_n \cdot (dT)_{12\text{-}18}$ (1:1) dTTP	0.4 µg/ml 15.5 µM	C[e] NC	47.5	C NC	34.9
Eukaryotic						
DNA polymerase α	activated DNA dTTP	162 µg/ml 0.9 µM	C NC	181	C NC	116
DNA polymerase β	$(rA)_n \cdot (dT)_{12\text{-}18}$ (1:2) dTTP	23.2 µg/ml 61.3 µM	C NC	23.7	C NC	71.1
DNA polymerase γ	$(rA)_n \cdot (dT)_{12\text{-}18}$ (10:1) dTTP	69.6 µg/ml 1.7 µM	C NC	298	C NC	286
E. coli						
RNA polymerase	$(dC)_n$ GTP	2.6 µg/ml 450 µM	NC→C[f] C	323	NC→C C	176

[a] Numbers in parentheses are the base ratios of the template to primer.
[b] Raucher murine leukemia virus
[c] Mix: Mixed type
[d] NC: Noncompetitive type
[e] C: Competitive type
[f] NC→C : Mode of inhibition changed from noncompetitive to competitive type by increasing the concentration of $(dC)_n$.

Table II. Inhibitory Effects of Theaflavins on the Activities of Various DNA and RNA Polymerases (IC$_{50}$ Values in μg/ml)

DNA or RNA Polymerase	TF-1	TF-2A	TF-2B	TF-3
HIV-1 reverse transcriptase	0.5	0.1	0.1	0.1
Mo-MLV[a] reverse transcriptase	0.7	0.05	0.04	0.04
DNA polymerase α	1.3	0.7	0.6	0.6
DNA polymerase ß	0.8	0.1	0.2	0.1
DNA polymerase γ	0.4	0.3	0.3	0.2
Terminal deoxynucleotidyltransferase	>10	>10	>10	>10
RNA polymerase	0.6	0.2	0.1	0.2

[a]Mo-MLV: Moloney murine leukemia virus

Table III. Characterization of Inhibition of DNA and RNA Polymerases by Theaflavin Digallate

DNA or RNA Polymerase	Variable Substrate	K_m	Mode[a]	K_i (μM)
Reverse transcriptase				
HIV-1	$(rA)_n \cdot (dT)_{12-18}$ (1:1)	0.6 μg/ml	C	0.02
	dTTP	4.4 μM	NC	
Mo-MLV	$(rA)_n \cdot (dT)_{12-18}$ (1:1)	5.1 μg/ml	C	0.01
	dTTP	8.8 μM	NC	
Eukaryotic				
DNA polymerase α	activated DNA	162 μg/ml	C	0.15
	dTTP	0.7 μM	NC	
DNA polymerase β	$(rA)_n \cdot (dT)_{12-18}$ (1:2)	33 μg/ml	C	0.33
	dTTP	57 μM	NC	
DNA polymerase γ	$(rA)_n \cdot (dT)_{12-18}$ (10:1)	70 μg/ml	C	0.14
	dTTP	1.7 μM	NC	
E. coli				
RNA polymerase	$(dC)_n$	2.6 μg/ml	NC→C	
	GTP	450 μM	C	0.32

[a] The abbreviations used are the same as those in Table I.

moiety is, therefore, not essential for, but enhances the inhibitory potency of theaflavin. Furthermore, little difference in inhibitory effect was observed among the three ester types. This indicates that the presence of the galloyl moiety, irrespective of the number or the binding position, is important for the enhancement of inhibitory potency.

Analysis of the Mode of Inhibition by Theaflavin Digallate. Since TF-2A, TF-2B and TF-3 showed a similar degree of inhibitory effect and all these compounds possess the same theaflavin structure, we chose TF-3 as the representative for kinetic analysis of the inhibition mechanism. The results are summarized in Table III. The kinetics of inhibition were almost the same as those of the green tea catechins. The only difference is that TF-3 exhibited a purely competitive-type inhibition in case of HIV-reverse transcriptase with respect to the template•primer. This difference in inhibition mechanism may be due to the difference in the shape or size of the molecule between catechins and theaflavins.

Like green tea catechins, theaflavins demonstrated strong cytotoxicity when anti-HIV activity was evaluated with cultured cell systems. Therefore, these tea polyphenols in themselves are not anti-HIV drugs. One possible explanation for the cytotoxicity is that tea polyphenols bind to and cannot pass through the cell membrane because of their strong affinity for membrane proteins. In order to reduce cytotoxicity and to enhance the anti-HIV effect, we are conducting some structural modifications of these tea polyphenols.

Acknowledgments

We thank Dr. S. H. Wilson of the National Institutes of Health, U. S. A., for providing HIV-1 reverse transcriptase.

Literature Cited

1. Mitsuya, H.; Weinhold, K.J.; Furman, P.A.; St. Clair, M.H.; Nusinoff-Lehrman, S.; Gallo, R.C.; Bolognesi, D.; Barry, D.W.; Broder, S. *Proc. Natl. Acad. Sci. U.S.A.* **1985**, *82*, 7096.
2. Yarchoan, R.; Klecker, R.W.; Weinhold, K.J.; Markham, P.D.; Lyerly, H.K.; Durack, D.T.; Gelmann, E.; Nusinoff-Lehrman, S.; Blum, R.M.; Barry, D.W.; Shearer, G. M.; Fishcl, M.A.; Mitsuya, H.; Gallo, R.C.; Collins, J.M.; Bolognessi, D.P.; Myers, C.E.; Broder, S. *Lancet* **1986**, *i*, 575.
3. Mitsuya, H.; Broder, S. *Proc. Natl. Acad. Sci., U.S.A.* **1986**, *83*, 1911.
4. Richman, D.D.; Fischl, M.A.; Grieco, M.H.; Gottlieb, M.S.; Volberding, P.A.; Laskin, O.L.; Leedom, J.M., Groopman, J.E.; Mildvan, D.; Hirsch, M.S.; Jackson, G.G.; Durack, D.T.; Phil, D.; Nusinoff-Lehrman, S.; the AZT Collaborative Working Group. *New Engl. J. Med.* **1987**, *317*, 192.
5. Yarchoan, R.; Perno, C.F.; Thomas, R.V.; Klecker, R.W.; Allain, J.-P.; Wills, R.J.; McAtee, N.; Fischl, M.A.; Dubinsky, R.; McNeely, M.C.; Mitsuya, H.; Pluda, J.M.; Lawley, T.J.; Leuther, M.; Safai, B.; Collins, J.M.; Myers, C.E.; Broder, S. *Lancet* **1988**, *i*, 76.
6. Larder, B.A.; Darby, G.; Richman, D.D. *Science* **1989**, *243*, 1731.
7. Ono, K.; Nakane, H.; Fukushima, M.; Chermann, J.-C.; Barre-Sionoussi, F. *Biomed. Pharmacother.* **1990**, *44*, 13.

8. Ono, K.; Nakane, H.; Fukushima, M.; Chermann, J.-C.; Barre-Sionoussi, F. *Biochem. Biophys. Res. Commun.* **1989**, *160*, 982.
9. Ono, K. *Bull. Inst. Pasteur* **1987**, *85*, 3.
10. Oguni, I.; Nasu, K.; Kanaya, S.; Ota, Y.; Yamamoto, S.; Nomura, T. *Jpn. J. Nutr.* **1989**, *47*, 93.
11. Hara, Y; Matsuzaki, T.; Suzuki, T. *Nippon Nogeikagaku Kaishi* **1987**, *61*, 803.

RECEIVED May 17, 1993

Chapter 6

Mitogenic Activity of (−)-Epigallocatechin Gallate on B Cells and Its Structure—Function Relationship

Z.-Q. Hu, M. Toda, S. Okubo, and Tadakatsu Shimamura

Department of Microbiology and Immunology, Showa University School of Medicine, Tokyo 142, Japan

(-)-Epigallocatechin gallate (EGCg), the main component of tea catechins, enhanced *in vitro* plaque forming cell (PFC) response to sheep red blood cells (SRBC) and showed mitogenic activity for mouse splenic B cells but not for splenic T cells. Macrophages were not involved in the enhancement. Among the derivatives of catechin examined, only the derivatives with a galloyl group, (-)-epicatechin gallate (ECg), (-)-epigallocatechin gallate (EGCg), and theaflavin digallate (TF3), enhanced spontaneous proliferation of B cells. Among the structural analogues of catechin and galloyl group, gallic acid and tannic acid induced some enhancement, but rutin, pyrogallol and caffeine did not.

We have demonstrated that tea and tea catechins have antibacterial (*1,2*), antifungal (*3*), antiviral (*4*) and anti-hemolysin activities (*5,6*). Here, we report new evidence that (-)-epigallocatechin gallate (EGCg) (*7*) has mitogenic activity on B cells. The structure-function relationship of EGCg is also discussed.

Enhancement of Plaque Forming Cell Response by EGCg

In vitro plaque forming cell (PFC) response to sheep red blood cells (SRBC) in mice was used as described previously (*8,9*). Briefly, 5×10^6 normal spleen cells were incubated with 8×10^6 SRBC in RPMI-1640 medium supplemented with all the additives. The effect of EGCg on the PFC response was examined after 5 days incubation. EGCg (0.01-0.5 μg/ml) enhanced PFC response dose-dependently (0.05 μg, 2 times; 0.1 μg, 3 times; 0.5 μg, 3 times). Large doses of EGCg (50 μg/ml), however, suppressed PFC response.

We next determined the target cells of EGCg, since three kinds of cells—B cells, T cells and macrophages—are involved in PFC response.

Mitogenicity of EGCg on B Cells. T cell-depleted mouse spleen cells were used as B cells for mitogen response as described previously (*10*). Splenic B cells (or T

0097–6156/94/0547–0065$06.00/0

cells) (5×10^5 cells/well) were cultured in 96 well flat bottom plates in a total volume of 200 µl RPMI-1640 medium containing 10% fetal calf serum, 25 mM Hepes, 200 mM L-glutamine, 100 units/ml penicillin, 100 µg/ml streptomycin, and 5×10^{-5} M 2-ME. Cultures were incubated 48 hours with or without mitogens and pulsed for the final 4 hours with 0.5 µCi ^{3}H-TdR. The cells were then harvested and ^{3}H-TdR incorporation was determined by scintillation counting.

EGCg (0.5-10 µg/ml) enhanced both spontaneous and lipopolysaccharide (LPS, 0.4 µg/ml)-induced proliferation of splenic B cells. Large dose of EGCg (50 µg/ml) also caused suppression. The percent enhancement of both types of proliferation was almost equal for the same doses of added EGCg. This result confirmed that EGCg displayed mitogenic activity in splenic B cells and the activity did not depend on LPS.

Since macrophages are found among T cell-depleted spleen cells, they were further removed using a Sephadex G-10 column (*11*). After removal of macrophages, the remaining B cells retained the ability to respond to EGCg (5-50 µg/ml). Therefore, the enhancement of B cell proliferation by EGCg was not mediated by macrophages.

Effect of EGCg on Splenic T cells. Splenic T cells were further separated through a Nylon fiber column as described elsewhere (*12*). Small doses of EGCg (0.01-10 µg/ml) showed very weak enhancement of spontaneous proliferation of T cells. Mitogen response of T cells with 1 µg/ml concanavalin A was not affected by EGCg. These results revealed that the target cell of EGCg was the B cell, and not the T cell.

Mechanism of Suppression by Large Doses of EGCg

To investigate a possible mechanism of the suppression caused by large doses of EGCg, indomethacin was added to the culures to block the synthesis of prostaglandins. In contrast to our expectation, indomethacin (10^{-4}-10^{-5} M) enhanced the suppression induced by EGCg. This result indicates that the suppression by EGCg was not mediated by prostaglandin E.

Effects of Catechin Derivatives on B Cells.

EGCg is a derivative of catechin. To investigate a possible structure-function relationship, (+)-catechin (C), (-)-epicatechin (EC), (-)-epigallocatechin (EGC), (-)-epicatechin gallate (ECg), and EGCg (*7*), and theaflavin digallate (TF3) (*13*) were tested on spontaneous proliferation of B cells.

EGCg and TF3 were the most potent enhancers, ECg showed some enhancement, and the others had little effect. At large doses (50 µg/ml), most derivatives had suppressive activity. This result suggests that the galloyl group on EGCg, TF3, and ECg might be responsible for the enhancement.

Effects of Analogues on B cells. To confirm the above hypothesis, some analogues of catechins and galloyl group were further examined on the same system. Rutin, pyrogallol, and caffeine produced no enhancement at the levels tested (0.1-50 µg/ml). Gallic acid (25 µg/ml) and tannic acid (5 µg/ml) showed some enhancement. These results demonstrate that the enhancing function of EGCg is mainly due to its galloyl group.

Conclusions

- EGCg had mitogenic activity on B cells but not on T cells.
- Macrophages are unnecessary for the activity.
- The galloyl group on EGCg appears to be responsible for the enhancing activity.

Literature Cited

1. Toda, M.; Okubo, S.; Hiyoshi, R.; Shimamura, T. *Lett. in Appl. Microbiol.* **1989**, *8*, 123-125.
2. Toda, M.; Okubo, S.; Ikigai, H.; Suzuki, T.; Suzuki, Y.; Shimamura, T. *J. Appl. Bacteriol.* **1991**, *70*, 109-112.
3. Okubo, S.; Toda, M.; Hara, Y.; Shimamura, T. *Jpn. J. Bacteriol.* (in Japanese) **1991**, *46*, 509-514.
4. Nakayama, M.; Toda, M.; Okubo, S.; Shimamura, T. *Lett. in Appl. Microbiol.* **1990**, *11*, 38-40.
5. Okubo, S.; Ikigai, H.; Toda, M.; Shimamura, T. *Lett. in Appl. Microbiol.* **1989**, *9*, 65-66.
6. Ikigai, H.; Toda, M.; Okubo, S.; Hara, Y.; Shimamura, T. *Jpn. J. Bacteriol.* (in Japanese) **1990**, *45*, 913-919.
7. Matsuzaki, T.; Hara, Y. *Nippon Nogeikagaku Kaisi* (in Japanese) **1985**, *59*, 129-134.
8. Jerne, N. K.; Nordin, A. A. *Science* **1963**, *140*, 405.
9. Yoshida, T.; Shimamura, T.; Shigeta, S. *Int. J. Immunopharmac.* **1987**, *9*, 411-415.
10. Shimamura, T.; Hashimoto, K.; Sasaki, S. *Cellular Immunol.* **1982**, *68*, 104-113.
11. Ly, I. A.; Mishell, R. I. *J. Immunol. Methods* **1974**, *5*, 239-247.
12. Julius, M.; Simpson, E.; Herzenberg, L. A. *Eur. J. Immunol.* **1973**, *3*, 645-649.
13. Hara, Y.; Matsuzaki, T.; Suzuki, T. *Nippon Nogeikagaku Kaisi* (in Japanese) **1987**, *61*, 803-808.

RECEIVED May 17, 1993

Chapter 7

Suppression of the Formation of Advanced Glycosylation Products by Tea Extracts

N. Kinae, K. Shimoi, S. Masumori, M. Harusawa, and M. Furugori

Laboratory of Food Hygiene, School of Food and Nutritional Sciences, University of Shizuoka, Shizuoka 422, Japan

Three kinds of tea extracts were prepared from green tea (*Camellia sinensis*, unfermented), polei tea (*Camellia assamica*, fermented with yeast) and rooibos tea (*Aspalathus linearis*, irradiated with sunlight). Each tea extract was incubated with a mixture of D-glucose and human serum albumin under physiological conditions (pH 7.4, 37°C). All tea extracts showed suppression effects on the formation of glycated albumin including fluorescent advanced glycosylation endproducts (AGEs). The determination of ESR spectra of reaction mixture and tea extract suggested that the suppression activity is correlated to the radical scavenging potency of tea extracts. Several tea components such as catechins and flavonoids may play important roles in the disappearance of free radicals containing superoxide formed in the early stage of the Maillard reaction.

An amino-carbonyl reaction, the so-called Maillard reaction, is well known as one of the major phenomena observed in the process of cooking, manufacturing and storage of foodstuffs (*1*). Recent studies show that the reaction also occurs in our body tissues, especially in hemoglobin (*2*), lens crystallins (*3*) and collagens (*4*). The glycated Cu-Zn-superoxide dismutase was isolated from human erythrocytes and the content increased as a function of aging and diabetes development (*5*).

In the early stage of the reaction, Schiff's base and Amadori product are formed and then converted to deoxyglucosones, which are key substances in the intermediate stage. In the last stage, AGEs are formed *in vivo* and *in vitro* as yellow fluorescent condensates (*6*). The AGEs content in the body tissues of old people and diabetic patients is significantly higher than those of young and healthy individuals (*7,8*).

Tea extract is one of the most ancient and popular beverages in the world. Green tea is an attractive beverage because of its antioxidative, antimicrobial, antimutagenic and anticarcinogenic activities (*9–11*). In this paper, we report on the inhibitory effects of three kinds of tea extracts on the formation of AGEs and discuss the inhibitory constituents and the mechanism of inhibition.

0097–6156/94/0547–0068$06.00/0

Preparation of Tea Extracts

Green tea was supplied by Shizuoka Tea Experiment Station (Shizuoka, Japan). Polei tea (produced in China) and rooibos tea (produced in South Africa) were supplied by JACOS Co. (Tokyo, Japan) and Rooibos Tea Japan Co. (Osaka, Japan), respectively.

Ten grams of tea leaves were added to 300 ml boiling distilled water and kept at 100°C for an appropriate time (green tea: 0.1 min; polei tea: 15 min; rooibos tea: 15 min). Each tea infusion was filtered through gauze and the filtrate was lyophilized. The lyophilizate was referred to as tea extract.

Inhibitory Effects of Tea Extracts on AGEs Formation

Tea extract (10–50 mg) was added to 50 ml of phosphate buffered saline (PBS) containing 500 mg HSA and 5 g D-glucose then incubated at 37°C for 47 days. Aliquots of the reaction solution were taken out periodically and dialyzed against PBS. The fluorescence intensity of the lyophilizate (AGEs) was measured at the wavelength of 440 nm (excitation at 370 nm).

Time course of the relative fluorescence intensity of each reaction solution is shown in Figure 1. All tea extracts (10 mg) suppressed the formation of AGEs to 52–80% of the control. The order of decreasing inhibitory strength after 47 days was green tea, polei tea and rooibos tea.

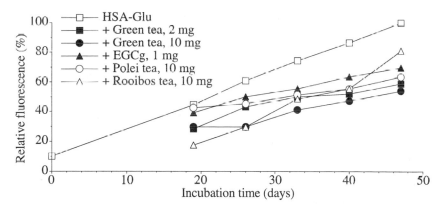

Figure 1. Inhibitory effect of tea extracts on the formation of fluorescent compounds.

Chemical Analysis of Tea Extract

In previous papers (*12,13*), we demonstrated that several catechins containing (-)-epigallocatechin gallate (EGCg) were isolated from the ethyl acetate-soluble fraction of green tea extract and possessed high inhibitory activity against AGE formation.

The other two kinds of tea extracts were submitted to HPLC analysis to determine catechin contents using the following conditions; column: TSK gel ODS-80 TM (ϕ4.6 x 150 mm), eluant: CH_3CN:50 mM KH_2PO_4 (1:9)→(4:6), flow rate: 0.96 ml/min, detection: UV (280 nm). The elution profile of the ethyl acetate-soluble fraction of polei tea extract is shown in Figure 2.

The catechin content of polei tea and rooibos tea was extremely low compared to green tea (Table I). Other chemical analysis was performed according to the Standard Method of Analysis for Hygienic Chemist (1990) in Japan.

Determination of Glycated Albumin by High Performance Affinity Chromatography

The content of glycated albumin, which contains mainly Amadori product, was determined by high performance affinity chromatography using TSK gel borate column (ϕ4.6 x 150 mm) (14). First, unreacted materials were removed with a solvent (0.25 M CH_3COONH_4/0.05 M $MgCl_2$, pH 8.5) and then glycated albumin was eluted with an eluant (0.2 M sorbitol/0.1 M Tris, pH 8.5). The elution profile and time course of glycated albumin content are shown in Figure 3. The formation of glycated albumin was inhibited in order of decreasing strength by rooibos tea, polei tea and green tea.

Table I. Chemical Analysis of Tea Extracts (%)

	Green tea	Polei tea	Rooibos tea
Total nitrogen	4.01	6.09	0.38
Crude protein	25.10	38.10	2.38
Reducing sugars[a]	16.30	11.40	17.30
Crude fat	1.06	1.10	0.79
Ash	6.59	14.50	6.88
Tannins[b]	30.20	1.14	
Gallic acid	0	0.08	
EC	4.19	0.28	trace
ECg	4.24	0.39	
EGC	8.79	0	trace
EGCg	13.20	0.39	

[a] Equivalent to D-glucose
[b] Equivalent to tannic acid

Determination of Radical Scavenging Activity of Tea Extract

1,1-Diphenyl-2-picrylhydrazil (DPPH) Method. DPPH contains a free radical in the molecule and combines with other free radicals to form a stable complex. The radical scavenging activity of each tea extract was evaluated by determination of the 50% inhibitory dose (ED_{50}) at $OD_{520\ nm}$ of DPPH-methanol solution according to the method of Fugita et al. (15). Compared with the ED_{50} of EGCg (2.58 μg), the relative activities of green tea, polei tea and rooibos tea were 0.24, 0.15, 0.16, respectively (Table II).

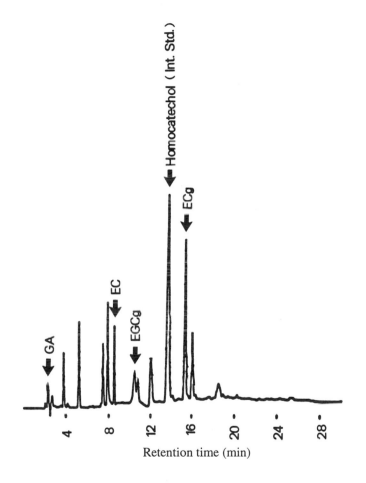

Figure 2. HPLC elution profile of the ethyl acetate fraction of polei tea extract.

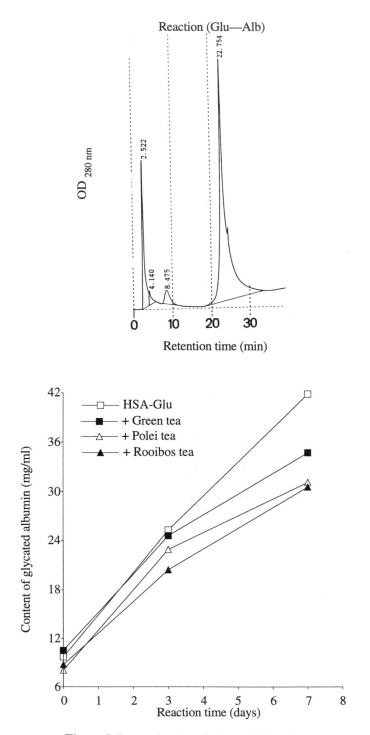

Figure 3. Determination of glycated albumin.

Table II. Radical Scavenging Activity of Tea Extracts by the DPPH Method

Procedure

DPPH (2.0×10^{-4} M) 4 ml

  +

Sample ($0–1.4 \times 10^{-3}$%) 1 ml

  | Mix and shake for 5 min.

  | Let stand for 30 min.

Mixture

  | Measure OD$_{520\ nm}$

ED_{50}

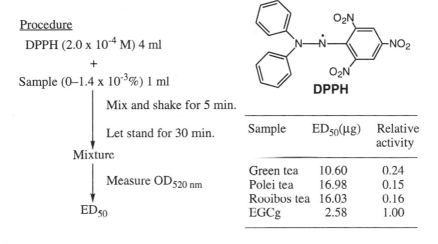

DPPH

Sample	$ED_{50}(\mu g)$	Relative activity
Green tea	10.60	0.24
Polei tea	16.98	0.15
Rooibos tea	16.03	0.16
EGCg	2.58	1.00

ESR Spin-trapping Method. Superoxide radical (O_2^-) supplied from a hypoxanthine-xanthine oxidase system (*16*) was trapped with 5,5-dimethyl-1-pyrroline-*N*-oxide (DMPO) to produce the spin adduct (DMPO-O_2^-). When each tea extract was added to the system, a decrease in ESR signal intensity was observed in the superoxide formation system. The inhibitory activity of green tea was the highest among the 3 kinds of tea extract (Figure 4).

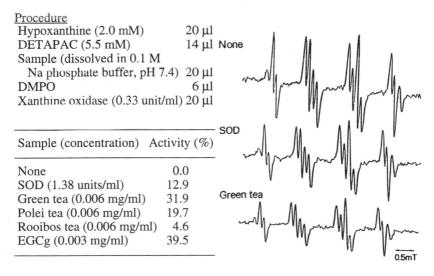

Procedure

Hypoxanthine (2.0 mM)	20 μl
DETAPAC (5.5 mM)	14 μl
Sample (dissolved in 0.1 M Na phosphate buffer, pH 7.4)	20 μl
DMPO	6 μl
Xanthine oxidase (0.33 unit/ml)	20 μl

Sample (concentration)	Activity (%)
None	0.0
SOD (1.38 units/ml)	12.9
Green tea (0.006 mg/ml)	31.9
Polei tea (0.006 mg/ml)	19.7
Rooibos tea (0.006 mg/ml)	4.6
EGCg (0.003 mg/ml)	39.5

Figure 4. Superoxide scavenging activity of tea extracts by ESR spin-trapping method.

Discussion

Recent reports have shown that glycated proteins are significantly accumulated in the several body tissues of aged individuals and diabetes patients (5). Amino-guanidine, a nucleophilic hydrazine compound, is noteworthy as an inhibitor of AGE formation *in vivo* and *in vitro* (17).

Aspirin (18) and Diclofenac (19) have also been reported as glycation blockers. Catechin-aluminum complex decreased the blood glucose level of diabetic mice (20). There is one report which showed a positive correlation between green tea production and average human life span (21).

In this study, we demonstrated that glycation of human serum albumin was inhibited by tea extract and polyphenols including catechins. In consideration of the generation of superoxides in the early stage of the reaction (22), these superoxides might be trapped into tea extract and therefore, AGE formation will be suppressed.

As the catechin content in polei tea and rooibos tea is very low, different effective components from catechins may play roles as radical scavengers.

Acknowledgements

We thank Dr. Y. Hara (Mitsui Norin Co., Shizuoka, Japan) for providing standard tea catechins and Dr. I. Oguni (Hamamatsu College, University of Shizuoka) for helpful discussion of this work. This research was supported by a grant from the Ministry of Education, Science and Culture of Japan.

Literature Cited

1. Hodge, J. E. *J. Agric. Food Chem.* **1953**, *1*, 928–943.
2. Bunn, H. F.; Gabbay, K. H.; Gallog, P. M. *Science* **1978**, *200*, 21–27.
3. Monnier, V. M.; Cerami, A. *Science* **1981**, *211*, 491–493.
4. Monnier, V. M. *Dev.Food Sci.* **1986**, *13*, 459–474.
5. Arai, K.; Maguchi, S.; Fujii, H.; Ishibashi, K.; Taniguchi, N. *J. Biol. Chem.* **1987**, *262*, 16969–16972.
6. Monnier, V. M.; Kohn, R. R.; Cerami, A. *Proc. Natl. Acad. Sci. USA* **1984**, *81*, 583–587.
7. Hayase, F.; Nagaraj, R.H.; Miyata, S.; Njoroge, F. G.; Monnier, V. M. *J. Biol. Chem.* **1989**, *264*, 3758–3764.
8. Monnier, V. M.; Sell, D. R.; Miyata, S.; Nagaraj, R. H. In *Maillard Reaction*; Finot, P. A.*et al.* Eds; *Adv. in Life Sci.*; Birkhauser Verlag: Basel, 1990; pp 393–414.
9. Matsuzaki, T.; Hara, Y. *Nippon Nogeikagaku Kaishi* (in Japanese) **1985**, *59*, 129–134.
10. Stich, H. F.; Rosin, M. P. *Adv. Exp. Med. Biol.* **1984**, *177*, 1–29.
11. Wang, Z. Y.; Khan, W. A.; Bicher, D. R.; Mukhtar, H. *Carcinogenesis* **1989**, *10*, 411–415.
12. Kinae, N.; Yamashita, M.; Esaki, S.; Kamiya, S. In *Maillard Reaction*; Finot, P. A. *et al.* Eds; *Adv. in Life Sci.*; Birkhauser Verlag: Basel, 1990; pp 221–226.
13. Kinae, N.; Masumori, S.; Nakada, J. Saito, T.; Furugori, M.; Esaki, S.; Kamiya, S.; Owada, K.; Masui, T. *Proc. Int. Symp. on Tea Science (Shizuoka, Japan)* **1991**, *14*, 649–661.
14. Mallia, A. K.; Hermanson, G. T.; Krohn, R. I.; Fujimoto, E. K.; Smith, P. K. *Anal. Letters* **1981**, *14*, 649–661.

15. Fujita, Y.; Uehara, I.; Morimoto, Y.; Nakajima, M.; Hatano, C.; Okuda, T. *Yakugaku Zassi* (in Japanese) **1988**, *108*, 129–135.
16. Miyagawa, H.; Yoshikawa, T.; Tanigawa, T.; Yoshida, N.; Sugino, S.; Kondo, M.; Nishikawa, H.; Kohno, M. *J. Clinic. Biochem. Nutr.* **1988**, *5*, 1–7.
17. Brownlee, M.; Vlassara, H.; Kooney, A.; Ulrich, P.; Cerami, A. *Science* **1986**, *232*, 1629–1632.
18. Ajiboye, R.; Harding, J. *J. Exp. Eye Res.* **1989**, *49*, 31–41.
19. van Boekel, M. A. M.; van den Bergh, P. J. P. C.; Hoenders, H. J. *Biochem. Biophys. Acta.* **1992**, *1120*, 201–204.
20. Asai, H.; Kunou, Y.; Ogawa, H.; Hara, Y.; Nakamura, K. *Kiso to Rinsho* (in Japanese) **1987**, *21*, 4601–4604.
21. Oguni, I. (personal communication)
22. Sakurai, T.; Tsuchiya, S. *FEBS Letters* **1988**, *236*, 406–410.

RECEIVED May 4, 1993

Chapter 8

Effects of Tea Polyphenols on Blood Rheology in Rats Fed a High-Fat Diet

F. Nanjo[1], Yukihiko Hara[1], and Y. Kikuchi[2]

[1]Food Research Laboratories, Mitsui Norin Company Ltd., Fujieda 426, Japan
[2]National Food Research Institute, Ministry of Agriculture, Forestry, and Fisheries, Tsukuba 305, Japan

Blood rheology plays a very important role in microcirculation. Making use of a new device that simulates capillaries, preliminary studies have been made on the influence of tea catechins in blood rheology. A high fat diet containing catechins was administered to rats for certain periods and the flow properties of red blood cells (RBCs) were examined. Tea catechins tended to prevent the transit time of RBCs-plasma suspension from increasing when the high fat diet was fed to rats for a long period. The transit time of RBCs-buffer suspension, however, was almost the same regardless of whether or not the animals received catechins. In addition, no significant differences in the hematological properties of the RBCs were observed. These results imply that the flow properties of RBCs are closely related to the disposition of plasma. It was found that tea catechins prevent plasma triglyceride levels from increasing by interrupting lipid absorption. It is likely that the plasma lipids influence the flow properties of RBCs.

Tea catechins, the main components of tea polyphenols, been demonstrated to have a variety of biological activities including hypocholesterolemic (1), antioxidant (2), hypotensive (3), and antitumor actions (4).

Blood flow properties are recognized as playing an important role in blood circulation. Particularly, red cell deformability is a critical determinant for capillary blood flow because the red cell must pass through small vessels of the micro-circulation which have a diameter less than that of the resting red cell. The deformability of the cell has been reported as reduced in hypercholesterolemia (5) and diabetes (6,7) which induce atherosclerosis.

Along with research which aims to elucidate the usefulness of tea catechins for maintaining health, preliminary studies have been made on the effects of catechins on the flow properties of red blood cells (RBCs) in rats fed a high fat diet.

0097—6156/94/0547—0076$06.00/0

Materials and Methods

Materials. Crude catechins having 90% purity were prepared from green tea leaves according to a method previously described (*8*). The crude catechins consist of (+)-catechin (1.4%), (-)-epigallocatechin (17.6%), (-)-epicatechin (5.8%), (-)-epicatechin gallate (12.5%), (-)-epigallocatechin gallate (53.9%), and others (9.8%).

Animals and diet. Male Wistar rats (5 weeks old) were fed a 30% lard diet containing 0 or 1% of the crude catechins for 4 or 10 weeks. The animals were housed individually in stainless-steel cages. Room temperature was controlled at 23 ± 2° C and lighting was on a 12 h light/dark cycle. Food and water were provided *ad libitum* throughout the experimental periods. The composition of the diets is given in Table I. Body weight and food consumption were recorded daily. The feces of each rat were collected at 2 day intervals and stored at -20° C until examination. The lipids excreted in the feces were determined by a method previously described (*1*).

Table I. Composition of Diets

	Control	1% Catechin
Corn starch	28.9 (%)	27.9 (%)
Casein	20.0	20.0
Lard	30.0	30.0
Sucrose	10.0	10.0
Cellulose	5.0	5.0
Mineral mix	4.0	4.0
Vitamin mix.	2.0	2.0
Choline chloride	0.1	0.1
Catechin	—	1.0

Analytical methods. Food was removed from rats the night before blood collection. Rats were anesthetized with ether, and blood was collected in heparinized syringes by means of heart puncture. Aliquots of the blood were used to measure plasma lipid concentration and hematological properties. The blood was centrifuged at 1800 x g for 10 min. The plasma and RBCs were separated, taking care that no contamination by white cells occurred. For the determination of the flow properties of RBCs, a suspension with a hematocrit of nearly 20% was prepared using the plasma or phosphate buffered saline (pH 7.4) containing 1% bovine serum albumin.

The transit time of RBCs-plasma or RBCs-buffer suspension was measured by an apparatus equipped with a newly designed filter instead of the Nuclepore filter generally used. The new filter consists of a silicon substrate and a glass plate (Figure 1). In the silicon substrate, a bank with microchannels (diameter 5 μm, length 14 μm, 2600 channels) in its upper surface encloses a square. The bank was formed using photolithography and orientation dependent etching, techniques used

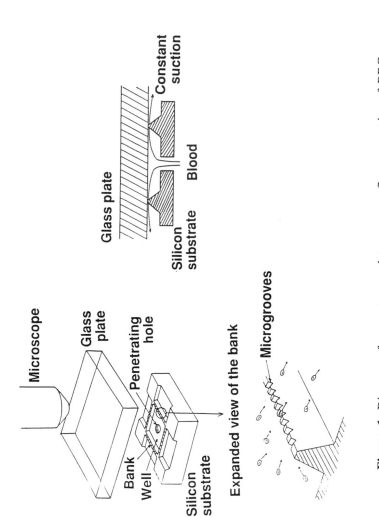
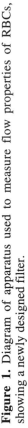

Figure 1. Diagram of apparatus used to measure flow properties of RBCs, showing a newly designed filter.

in the semiconductor manufacturing field. Simulated capillaries are formed by the microchannels when the substrate is mechanically pressed against an optically flat glass plate. Samples are introduced into the square enclosed by the banks through a hole penetrating the substrate. Flow is created by applying a negative pressure to the outside of the bank.

The atual measurement system and procedures used were the same as in the modified Nuclepore filtration method developed by Kikuchi *et al.* (9). The RBCs-plasma and RBCs-buffer suspensions were applied to the above system under a negative pressure of 20 cm of water. The time it took a 100 μl sample to pass through the simulated capillaries was determined and was taken as a indication of the flow properties of the RBCs.

Results and Discussion

Body weight gain and food intake of the rats fed with 30% lard diet containing 0 and 1% catechins for 10 weeks are shown in Figure 2. No significant differences between the two groups were observed.

Figure 3 shows the effects of tea catechins on the transit time of the RBCs-plasma and RBCs-buffer suspension in rats fed the high fat diet for 4 or 10 weeks. The transit time of the RBCs-plasma suspension increased in the 10 week groups compared to the 4 week groups. The addition of 1% catechins however, tended to prevent the transit time of the RBCs-plasma suspension from increasing in the rats fed the diet for 10 weeks. On the other hand, the transit time of the RBCs-buffer suspension was not influenced by the addition of tea catechins to the diet. These results imply that the flow properties of RBCs are affected by the interaction between the plasma and the RBCs, and tea catechins may work to counteract such an interaction.

In connection with this, we examined the hematological properties of the RBCs, total feces, fecal lipid excretion, and plasma lipid levels. There were no significant differences at 10 weeks between the control and the catechin group in hematocrit, RBC numbers, mean corpuscular volume, mean corpuscular hemo-globin, or mean corpuscular hemoglobin concentration (Table II).

Table II. Hematological properties in rats (10 weeks)

	Control	1% Catechin
Hematocrit (%)	43.6 ± 1.39	41.2 ± 1.52
RBCs ($10^4/mm^3$)	725.0 ± 18.6	687.0 ± 10.0
MCV (μ^3)	60.2 ± 1.83	61.7 ± 1.50
MCH (Pg)	19.7 ± 0.65	19.8 ± 0.48
MCH concentration (%)	32.7 ± 0.80	33.0 ± 0.20

MCV: mean corpuscular volume
MCH: mean corpuscular hemoglobin

The effects of tea catechins on the total amount of feces and fecal lipid excretion are indicated in Figure 4. Fecal lipid excretion significantly increased in

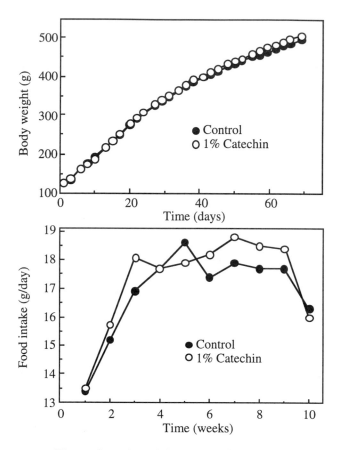

Figure 2. Body weight gain and food intake.

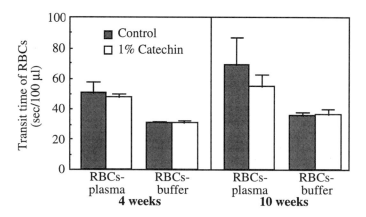

Figure 3. Effects of tea catechins on transit time of RBCs. Both diets contained 30% lard.

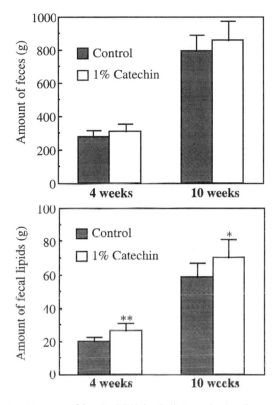

Figure 4. Effect of tea catechins in 30% lard diet on the total amount of feces and fecal lipids. * Significantly different from control, p<0.05. ** Significantly different from control, p<0.01.

the catechin group fed for both 4 and 10 weeks. These observations are thought to show that tea catechins inhibit lipid absorption. Similar results have also been obtained when rats were fed with a high cholesterol diet (*1*). Total lipid and triglyceride levels in the plasma were also examined (Figure 5). The total lipid content was hardly effected by the addition of catechins, whereas the triglyceride levels were significantly suppressed by catechins. Total cholesterol and phospholipid levels in the plasma were not altered significantly by the addition of catechins (data not shown). The decrease in triglyceride levels in the plasma appears to be closely related to the increase in fecal lipid excretion. Thus catechins seem to prevent the triglyceride levels from increasing by interrupting lipid absorption.

On the basis of the above results, we conclude that tea catechins may protect the flow properties of RBCs from deteriorating by restricting an undesirable interaction between the plasma and RBCs.

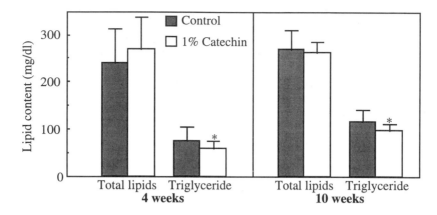

Figure 5. Effect of tea catechins in 30% lard diet on the total lipid and triglyceride levels in plasma. * Significantly different from control, $p<0.05$.

Literature Cited

1. Muramatsu, K.; Fukuyo, M.; Hara, Y. *J. Nutr. Sci. Vitaminol.* **1986**, *32,* 613–622.
2. Matsuzaki, T.; Hara, Y. *Nippon Nogeikagaku Kaishi* (in Japanese) **1985**, *59,* 129–134.
3. Hara, Y.; Tono-oka, F. *Nippon Eiyo Shokuryo Gakkaishi* (in Japanese) **1990**, *43,* 345–348.
4. Hara, Y.; Matsuzaki, S.; Nakamura, K. *Nippon Eiyo Shokuryo Gakkaishi* (in Japanese) **1989**, *42,* 39–45.
5. Kikuchi, Y.; Koyama, T. *Clin. Hemorheol.* **1983**, *3,* 375–382.
6. Kikuchi, Y.; Koyama, T.; Ohshima, N.; Oda, K. *Clin. Hemorheol.* **1988**, *8,* 171–181.
7. McMillan, D. E.; Utterback, N. G.; La Puma, J.; Barabara, S. *Diabetes* **1978**, *27,* 895–901.
8. Hattori, M.; Kusumoto, I. T.; Namba, T.; Ishigami, T.; Hara, Y. *Chem. Pharm. Bull.* **1990**, *38,* 717–720.
9. Kikuchi, Y.; Arai, T.; Koyama, T. *Med. & Biol. Eng. & Comput.* **1983**, *21,* 270–276.

RECEIVED May 4, 1993

Chapter 9

Inhibition of Saccharide Digestive Enzymes by Tea Polyphenols

M. Honda, F. Nanjo, and Yukihiko Hara

Food Research Laboratories, Mitsui Norin Company Ltd., Fujieda 426, Japan

The inhibition of salivary α-amylase, intestinal sucrase and maltase by polyphenolic components of tea and their specificity were investigated. Theaflavins isolated from black tea and galloyl catechins from green tea showed potent inhibitory effects. Free catechins and gallic acid showed little or no effect. Based on these results, *in vivo* experiments were conducted. Crude catechin powder, obtained by purifying the catechin fraction from green tea, was administered to rats, followed by the administration of starch or sucrose. Rats were sacrificed at time intervals and their intestinal enzyme activities and glucose concentrations in the blood were determined. The results showed that after catechin administration, the increase in intestinal α-amylase activity caused by starch was markedly suppressed and that glucose levels in the blood were also suppressed dose-dependently by tea catechins. Similar *in vivo* results were observed with the administration of catechins prior to sucrose administration.

Tea polyphenols — or more broadly, tannins — are known to have a strong affinity to proteins. In the case of tea polyphenols this characteristic is indicated by the astringency felt on our palates when we drink tea. The inhibition of enzymes by the polyphenolic fraction of tea, therefore, has long been assumed. We have isolated and purified individual polyphenolic components from tea and subjected them to a series of enzyme inhibition tests (1,2). The inhibition of α-amylase by tea polyphenols was also examined and fairly potent inhibition was confirmed (3). In this paper, the *in vitro* inhibitory potency of tea polyphenols on α-amylase as well as on sucrase or maltase and their *in vivo* effects on blood glucose levels in rats are presented.

In Vitro Inhibition of Saccharides by Tea Polyphenols

Inhibition of α–Amylase by Tea Polyphenols. α-Amylase from human saliva was purchased from Sigma Chemical Co. Representing tea polyphenols, individual

0097–6156/94/0547–0083$06.00/0

catechins from green tea (*4*) and individual theaflavins from black tea (*1*) were prepared. For comparison, gallic acid also tested. The inhibitory effects of samples were measured according to a method described elsewhere (*3*). α-Amylase solution was incubated with the inhibitor solution for 10 min at 37°C and the concentrations of samples that produced 50% inhibition of the enzyme were determined. The results are shown in Table I. Among the twelve samples, gallic acid, free catechins (EC and EGC) and their epimers did not have significant effects on the activity of α-amylase. All other samples were inhibitors and the inhibitory effects were in the descending order of TF3>TF2A>TF3B>TF1>Cg>GCg>EGCg.

Table I. Inhibition of α-Amylase by Tea Polyphenols and Gallic Acid

Sample	IC_{50} (μM)
(-)-Epicatechin (EC)	>1000
(-)-Catechin (C)	>1000
(-)-Epigallocatechin (EGC)	>1000
(-)-Gallocatechin (GC)	>1000
(-)-Epicatechin gallate (ECg)	130
(-)-Catechin gallate (Cg)	20
(-)-Epigallocatechin gallate (EGCg)	260
(-)-Gallocatechin gallate (GCg)	55
(-)-Theaflavin (TF1)	18
(-)-Theaflavin monogallate A (TF2A)	1.0
Theaflavin monogallate B (TF2B)	1.7
Theaflavin digallate (TF3)	0.6
Gallic acid	>1000

Inhibition of Sucrase and Maltase. To determine inhibition of sucrase and maltase, the small intestinal brush border was removed from rats, carefully homogenized in a buffer and centrifuged. The supernatant was a crude enzyme solution containing sucrase and maltase, and was preincubated with the polyphenol solution. The sucrase activity was measured according to the standard assay of Bernfeld (*5*). Maltase activity was assayed by adding the substrate *p*-nitrophenyl α-D-glucoside (NPG) to the enzyme-polyphenol solution. The inhibitory effects of tea polyphenols on the sucrase and maltase activity are shown in Table II. Among the samples, gallated polyphenols showed potent inhibition whereas non-gallated, free catechins and gallic acid showed little inhibitory activity. Their effects on sucrase were in the descending order of EGCg>TF3>ECg>TF2A>TF2B. The inhibition of maltase (determined as inhibition of NPG hydrolysis) by tea poly-phenols was in the order of TF3>TF2B>TF2A>ECg>EGCg.

The Effect of Tea Catechins on Starch Ingestion

Crude catechin powder fractionated from Japanese green tea was administered orally to rats followed by administration of starch and the influence of this catechin powder on intestinal α-amylase and the blood glucose level was studied. The composition of crude catechin powder (6) is shown in Table III. Crude catechin powder is abbreviated as catechin hereinafter. Male Wistar strain rats weighing 180–200 g (6 weeks of age) were fed a commercial diet for one week. Eighty rats were divided into 4 groups of 20 rats. One ml of a 40, 60 or 80 mg/ml catechin solution was administered orally to all the rats which were starved overnight. The control group was fed water instead of catechin. After 30 min, 4 ml of a 40% soluble starch solution was administered orally to the rats in each group. Following the administration of starch, five rats in each group were sacrificed at intervals of 0, 30, 60, and 120 min and the contents of the small intestine and the blood were collected.

Table II. Effect of Tea Polyphenols on Rat Small Intestinal Sucrase and Maltase

Sample	Concentration (mM)	Relative activity (%)	
		Sucrase	Maltase
None	0.5	100	100
Gallic acid	0.5	93	93
(+)-Catechin (+C)	0.5	86	99
(-)-Epicatechin (EC)	0.5	85	96
(-)-Epigallocatechin (EGC)	0.5	83	95
(-)-Epicatechin gallate (ECg)	0.5	38	59
	0.1	65	89
(-)-Epigallocatechin gallate (EGCg)	0.5	21	62
	0.1	51	91
Theaflavin (TF1)	0.1	95	93
Theaflavin monogallate A (TF2A)	0.1	81	52
Theaflavin monogallate B (TF2B)	0.1	87	39
Theaflavin digallate (TF3)	0.1	58	38

Table III. Composition of Crude Catechins

Catechin	Absolute (%)	Relative (%)
(+)-Gallocatechin (GC)	1.44	1.6
(-)-Epigallocatechin (EGC)	17.57	19.3
(-)-Epicatechin (EC)	5.81	6.4
(-)-Epigallocatechin gallate (EGCg)	53.90	59.1
(-)-Epicatechin gallate (ECg)	12.51	13.7
Total	91.23	100.0

α-Amylase Activity in the Intestine. The assay was carried out by the method of Willstatter and Schdel (7) with modification. The contents of the small intestine of starch-fed rats were filtered and the residue was removed to make an α-amylase solution. This sample solution was diluted with phosphate buffer and mixed with a fixed amount of soluble starch. The reaction mixture was incubated for 5 min at 37°C. One unit of α-amylase activity was defined as the amount of enzyme which liberated 1 μmol of maltose per min. The results are shown in Figure 1. As shown in the figure, α-amylase activity of the catechin-fed groups (40, 60 and 80 mg) scarcely increased in the 2 hrs after the starch-dosage, whereas the enzyme activity of the control group increased markedly peaking at a much higher point than the catechin groups.

Glucose Concentration in the Plasma. The starch-fed rats were sacrificed and their blood was collected at intervals as described above. The blood was centrifuged at 3,000 rpm for 15 min and the plasma was stored at -20°C. Concentration of glucose in the plasma was determined using the glucose oxidase kit. The results are shown in Figure 1. The figure shows an increase of concentration from the zero time. When 80 mg of catechin were given before the administration of starch, the increase of glucose concentrations in the plasma was significantly suppressed as compared with that of the control group. The suppression depended on the quantity of catechin administered. When 40 mg of catechin was administered, there was only a slight suppressive effect on plasma glucose level.

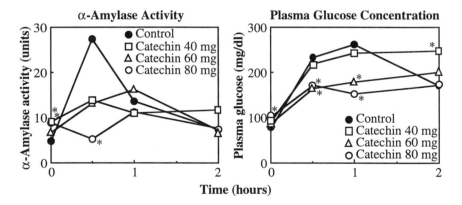

Figure 1. α-Amylase activity and plasma glucose levels in rats administered starch. *Statistically different from control ($p < 0.05$).

The Effect of Tea Catechins on Sucrose Ingestion

The influence of catechin on intestinal sucrase and on blood glucose levels in rats fed sucrose was also studied. The same strain of rats was used under the same conditions as stated above. Eighty rats were divided into 4 groups of 20 rats. One ml of a 5, 10 or 80 mg/ml catechin solution was administered orally to all the rats, which had been starved overnight. The control group was fed water instead of catechin. After 30 min, 4 ml of a 40% sucrose solution was administered orally to the rats in each group. Following the administration of sucrose, rats were sacrificed at same time intervals as stated previously and the membrane brush border of the small intestine and the blood was collected in the same manner as described above.

Sucrase Activity in the Intestine. The assay was conducted as described above. The results are shown in Figure 2. As is apparent from the figures, catechins suppressed the sucrase activity markedly. Even a 5 mg dose to a rat was sufficient to suppress the sucrase activity to a level lower than that of the control.

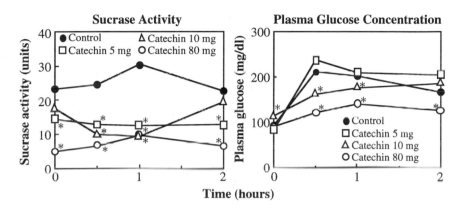

Figure 2. Sucrase activity and plasma glucose levels in rats administered sucrose. *Statistically different from control, $p < 0.05$.

Effects of Tea Catechins on Body Weight Gains and on Food Intake

Tea polyphenols have inhibitory effects on α-amylase or sucrase. Moreover, in previous feeding tests, it was observed that the amount of feces increased noticeably in rats fed catechin (data not shown). Accordingly, it may be assumed that these enzyme inhibitory effects might eventually lead to malnutrition. Because of this possibility, long term feeding tests were conducted in order to observe the influence of catechin on body weight gain and on food intake. As the results in Figure 3 show, as much as 1 or 2% addition of tea catechins to a normal MF diet (Oriental Co.) affected neither food intake nor body weight gain.

Concluding Remarks

The fact that plant polyphenols have an affinity to combine with protein led us to investigate the inhibitory effect of tea polyphenols on α-amylase. Among the tea polyphenols tested, theaflavins were much stronger inhibitors than catechins. In the case of catechins, a galloyl (3,4,5-trihydroxybenzoyl) group at the 3-OH showed potent inhibition and their inhibitory activity was enhanced 10 times by epimerization of these gallated compounds (ECg→Cg and EGCg→GCg). The inhibitory potency of theaflavins was enhanced as the number of galloyl moieties increased (TF3> TF2>TF1). In any case, it appears that a 3,4,5-trihydroxybenzoyl moiety at the 3-OH position is essential for the inhibition of α-amylase.

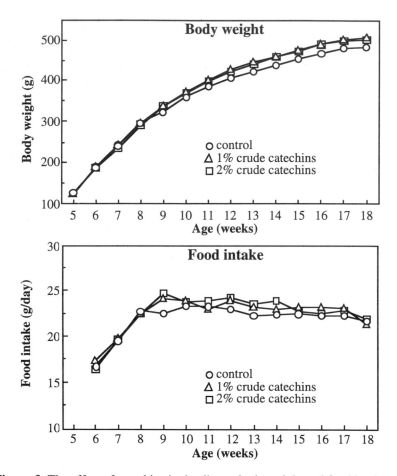

Figure 3. The effect of catechins in the diet on body weight and food intake.

The above findings led us to further examine the possibility that tea poly-phenols may suppress α-amylase activity in the intestine, resulting in the reduction of the glucose level in the blood. Since theaflavins are minor elements and hard to obtain, the catechin fraction from green tea was used for animal experiments.

Oral administration of tea catechin prior to starch administration suppressed α-amylase activity otherwise elevated by starch. In addition, the increase in glucose concentration after starch administration was also arrested by the prior administration of catechin. Likewise, tea polyphenols suppressed sucrase and maltase activity *in vitro* as well as *in vivo*. Here again the galloyl moiety of tea polyphenols seems to determine the degree of inhibitory potency, and the effects are dose dependent.

From these results we conclude that tea polyphenols are effective in suppressing the activity of α-amylase and sucrase in the intestine when an excess amount of starch or sucrose is fed, thereby moderately suppressing the increase of the blood glucose level. Furthermore, it was confirmed that catechin supplementation in a normal diet over an extended period of time will not suppress the food intake nor body weight gain. In another series of experiments, various prophylactic functions were noted with long term feeding of catechins (Hara, Y., "Prophylactic Functions of Tea Polyphenols," in this proceeding). These experimental facts seem to invite wider utilization of tea polyphenols for human health.

Literature Cited

1. Hara, Y.; Matsuzaki, T. *Nippon Nogeikagaku Kaishi* (in Japanese) **1987**, *61*, 803–808.
2. Hattori, M.; Kusumoto; I. T., Namba, T.; Ishigami, T.; Hara, Y. *Chem. Pharm. Bull.* **1990**, *38*, 717–720.
3. Hara, Y.; Honda, M. *Agric. Biol. Chem.* **1990**, *54*, 1939–1945.
4. Hara, Y.; Matsuzaki, S.; Nakamura, K. *Nippon Eiyo Shokuryo Gakkaishi* (in Japanese) **1989**, *42*, 39–45.
5. Bernfeld, P. In *Methods of Enzymology*; Colwick, S. B.; Kaplan, N. O., Eds.; Academic Press: New York, 1955; pp 149–150.
6. Hara, Y.; Tono-oka, F. *Nippon Eiyo Syokuryo Gakkaishi* (in Japanese) **1990**, *43*, 345–348.
7. Willstatter, R.; Schdel, G. *Berichte* **1918**, *51*, 780–781.

RECEIVED May 4, 1993

Chapter 10

Interactions of Green Tea Catechin with Polyamides

H. Li[1], C. Fisher[1], R. W. Keown[1], and C. P. Malone[2]

Departments of [1]Food Science and [2]Textiles, Design and Consumer Economics, University of Delaware, Newark, DE 19716

The interaction of extracts from green tea with polyamides have been measured by their ability to accept the acidic dye, FD&C Red 40, and the basic dye, Basic Green 3, as a stain. The acidic dye interacts with the protein amine during the staining process. If this amine is not free to react, the protein will not be stained by the acidic dye. The green tea extracts behave like a dye resist agent which reduces the acidic dyeability and enhances the basic dyeability. The basic dye interacts with the protein carboxyl groups during the staining process. Measurement of the level of enhancement of basic dye staining relates directly to the amount of interaction between the green tea phenols and polyamides. Dyeing the green tea extract-polyamide complex with Basic Green 3 is a simple and sensitive analytical tool for measuring the quantity of the active ingredients which react with the polyamide. This application of textile dye technology to tannin/protein interactions may provide an insight into the mechanism of biological activity.

Tannins or polyphenols interact with proteins at low pH through hydrogen bonding between the protein amide carbonyls and the phenolic hydroxyl (1). The fact that tannins bind to synthetic polyamides (nylons), which contain -CONH- as their only reactive group, was put forth as evidence for the hydrogen bonding mechanism (2). The aromatic portions of tannins also hydrophobically interact with nonpolar amino acids. This type of interaction is pH independent (3). At high pH and under oxidative conditions, covalent interactions result between the amino acid side chains and the quinonoid tannins (4). All of these interactions have been reviewed by A.E. Hagerman (5,6).

Many of the polyphenols are considered anti-nutrients because they reduce the amount of proteins that are digestible by animals. The phenols from tea have recently been found to have many healthful biological functions such as being anti-carcinogenic, anti-microbial, and possibly effective against AIDS (see chapters by C-T. Ho, Y. Hara and H. Nakane).

0097–6156/94/0547–0090$06.00/0

The adsorption of tea catechins or phenols onto proteins cannot be measured by normal methods such as spectrophotometry or HPLC, since proteins absorb at the same wavelengths as the catechins. Protein precipitation on guard and analytical HPLC columns have necessitated the development of a high performance thin layer chromatography (HPTLC) method using densitometry to measure the protein/tea catechin interaction.

Polyamides can be dyed with acidic dyes in a reaction with the basic amine group or with basic dyes by reaction with the carboxylic acid in the polyamide. If the polyamide is treated with a family of materials known as stain resists, the amine groups become inaccessible so that the treated polyamide will not be dyed with acidic dyes.

FD&C Red 40 (Figure 1a) is an acidic or anionic dye which consists of a sodium salt of a diaryl azo chromophore and a sulfonic acid auxochrome. In acidic solutions, the sulfonic acid groups react quantitatively with the accessible amine groups in polyamides (7).

$$\text{dye-SO}_3\text{H} + \text{H}_2\text{N-polyamide} \rightarrow \text{dye-SO}_3^- {}^+\text{H}_3\text{N-polyamide}$$

In a polyamide such as nylon, which contains one amine group at the end of each polymer chain, the molecular weight of the polymer can be determined by the intensity of the color (8,9). In more complex molecules such as proteins, the FD&C Red 40 reacts with the accessible amine groups; the intensity of the color is determined by the available amine groups of the protein.

A basic or cationic dye such as Basic Green 3 (Figure 1b) consists of the sodium salt of a triaryl methane chromophore with tertiary amine auxochromes. In basic solutions, the tertiary amine groups react quantitatively with the acid groups in the polyamide.

$$\text{dye-NR}_2 + \text{HOOC-polyamide} \rightarrow \text{dye-NR}_2\text{H}^+ \, {}^-\text{OOC-polyamide}$$

In the case of 6,6 nylon which contains no acid groups, the nylon is not dyed with basic dyes. In more complex molecules such as a protein, the carboxylic acid groups in the protein are dyed with basic dyes.

Stain resists (Figure 1c) have been developed in the textile industry to prevent staining by the acidic synthetic and natural food dyes. These materials are extensively used on nylon in carpeting to achieve food stain resistance. The stain resists are acidic, which when applied to a polyamide, forms an ion double layer repulsion preventing the anionic dye from reacting with the polyamide, in addition to forming a "crosslinked membrane" at or near the surface of the polyamide. The existence of the crosslinked membrane is demonstrated by diffusion of non-ionic molecules into the stain resist treated polyamide (10). For example, if a polyamide is treated with a dye resist containing phenolic and sulfonic acid groups, the resist blocks the accessible amine groups and prevents the subsequent reaction with the acidic dye (11). A polyamide treated with a commercial dye resist retards or reduces its reaction with acidic dyes. Acidic groups in the dye resist, however, provide the dye sites for the reaction with basic dyes, so a resist-treated polyamide will stain more intensely with basic dye than an untreated polyamide. It is expected that other acidic materials similar to the acidic dyes would also be less reactive with the polyamide/resist complex.

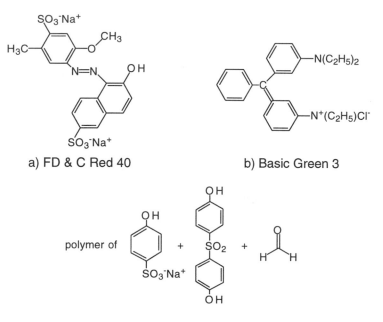

a) FD & C Red 40   b) Basic Green 3

c) Mesitol NBS

Figure 1. Structures of a) the acidic dye FD&C Red 40, b) the basic dye Basic Green 3, and c) the textile resist Mesitol NBS, a complex reaction mixture.

Green tea contains relatively large amounts of the phenolic compounds epicatechin (EC), epigallocatechin (EGC), epicatechin gallate (ECG) and epigallocatechin gallate (EGCG). As phenolic materials are known to react with proteins, it is believed that the phenolic/protein interaction and resist/polyamide interaction are similar and potentially provide a technique to study this interaction. Thus, utilizing the stain resist technology, a simple and sensitive assay was developed that can determine the extent of interaction of phenols with polyamides. The polyamides can be simple synthetic polymers such as nylon, or complex natural polymers like proteins.

Experimental

Materials. All fabric materials — nylon, wool and silk (Test Fabrics, Inc.) — were washed in 0.1% aqueous Alkanol A-CN (DuPont Co.) solution for 30 minutes, rinsed well with distilled water and air dried in order to remove grease and dirt from the fibers.

Soy protein was prepared by the method of Thanh *et al.* (*12*). Tris-HCl buffer solution (200 ml, 0.64 mM, pH 7.8) was added to soy protein concentrate (50 g, Staley Co.), mixed and allowed to settle for 30 minutes. The precipitate was collected and air dried. A 10% aqueous stock solution gave a colloidal suspension of the collected precipitate.

Green tea extract was prepared using a modified procedure obtained from C-T. Ho, Rutgers University. Green tea leaves (50 g, Chinese) were extracted twice for 30 minutes each at room temperature with 80:20 acetone:water (2 x 500 ml) and filtered. The filtered extracts were combined and evaporated under vacuum at 50°C to the raw extract in water and then decolorized with chloroform.

The four major catechins in green tea — epicatechin (EC), epigallocatechin (EGC), epicatechin gallate (ECG) and epigallocatechin gallate (EGCG) — were isolated by extracting the decolorized raw green tea extract in water with ethyl acetate. All the ethyl acetate extracts were combined and concentrated under vacuum at 50°C. The concentrated ethyl acetate solution was applied in a band on a Prep TLC plate (Analtech Silica, 70 μm, 20 cm x 20 cm). After elution with 1:1 ethyl acetate:chloroform, the four separated compounds were detected under UV lamp (284 nm) and identified by R_f comparison with standards (supplied by C-T. Ho, Rutgers University). The bands were scraped off and the catechins extracted from the silica gel with ethyl acetate and freeze dried. Stock solutions of EC, EGC, ECG and EGCG (100 ppm) were prepared with distilled water.

Treatment. Nylon (0.3 g), wool (0.3 g), silk (0.3 g), and soy protein (2 ml) were added to 40 ml of the commercial stain resist Mesitol NBS (Mobay Co.; 100 ppm in distilled water at pH 3.0) and mixed well. The textile fibers were removed manually and rinsed with distilled water. The soy protein was vacuum filtered and rinsed with distilled water. All samples were divided into two equal portions and stained with FD&C Red 40 and Basic Green 3 respectively. For the green tea extract treatment, the procedure and materials are the same as in the above stain resist treatment except that decolorized raw tea extract (about 2%) was used instead of Mesitol NBS. For treatment with the pure catechins, the procedure and materials are the same as in the above stain resist treatment except that 100 ppm levels of the four purified compounds (EC, ECG, EGC & EGCG) were used instead of Mesitol NBS.

Isotherm Study. A series of stock ECG solutions (0, 10, 19, 38, 75, 150, 300, and 600 ppm) were prepared in citric acid buffer (pH 2.8).

For the 6,6 nylon study, each 5 μl ECG solution was spotted onto an Analtech HPTLC-GHLF (Silica w/UV 254; 10 x 20 cm scored) plate using a capillary tube. Then 6,6 nylon was reacted for one hour, removed and rinsed. Each nylon sample was divided in two parts and separately stained by FD&C Red 40 or Basic Green 3. The solutions containing unreacted ECG were then also spotted (5 μl) onto the HPTLC plate. The separated spots were scanned on a densitometer. The difference between ECG concentration before and after nylon treatment gives the amount of ECG that reacted with the nylon.

For the soy protein study, a 2% soy protein solution was prepared by making a colloidal suspension from the air dried soy protein prepared above (2 g) with 100 ml of citric acid buffer (pH 2.8). This colloidal suspension (0.2 ml) was added to the series of ECG solutions for one hour and 5 μl of the solutions spotted immediately onto the HPTLC plate. After elution and densitometric scanning, the amount of ECG reacted with the protein was calculated. Each protein solution was vacuum filtered, washed well with distilled water, divided into two parts and stained by FD&C Red 40 or Basic Green 3.

Dyeing. To half of the stain resist treated samples, 40 ml of the FD&C Red 40 stock solution (0.025 g FD&C Red 40 in 1 liter citric acid buffer, pH 2.8) was added and stirred for 20 minutes. The samples were either removed manually for fibers or vacuum filtered for proteins, rinsed with distilled water until no color was removed and air dried.

To the other half of the stain resist treated samples, 40 ml of the Basic Green 3 stock solution (0.1 g Basic Green 3 in 1 liter distilled water; pH adjusted to 1.5 with HCl) was added and stirred for 20 minutes. The samples were rinsed with distilled water until clear and air dried. The basic dyeing was performed under acidic conditions to assure that the polyamides were in the acid form. The dyeings were not done under conditions of electrical neutrality.

Measurement. All air dried samples were ground to a fine powder and spread evenly onto double coated Scotch tape and the reflective color intensities [Hunter color system (13); L, a, b values] were measured on a Minolta CR-200 Chroma-Meter. For HPTLC/Densitometry, 5 μl of each solution was spotted onto an Analtech HPTLC-GHLF (Silica w/UV 254; 10 x 20 cm scored) plate. The plate was eluted with ethyl acetate:chloroform (4:1), dried and scanned at 254 nm on a Shimadzu CS 9000-U densitometer. A standard curve was obtained from the areas under the densitometer scans of the solutions before treatment. The difference between concentration before and after treatment is the amount adsorbed.

Results and Discussion

Nylon, wool, silk and soy protein dye to a deep red color with FD&C Red 40. Nylon and wool, however, remain virtually undyed when treated with a commercial stain resist, Mesitol NBS. The silk and soy protein have more accessible amine groups and when treated with a stain resist dye to a light red color. These results (Table I) demonstrate the blockage of amine groups with the stain resist agent. The extract of green tea, which contains phenolic compounds, also behaves like a dye resist and blocks the amine groups in the polyamides. This results in the tea extract-treated polyamide dying to a lighter shade than the untreated polyamide. Comparison of Mesitol NBS and green tea extract shows the green tea extract appearing less active than the commercial resist, however, a quantitative estimate cannot be made because both the Mesitol NBS and the extract are mixtures of phenolic compounds.

The increase in the basic dyeability of nylon, wool, silk and soy protein resulting from the stain resist treatment is shown in Table II. The stain resist agent has provided "acidic" sites to react with the basic dye. Under the acidic conditions of the basic dying, the non-treated polyamides are only slightly stained. These non-conventional conditions were chosen to maximize the difference between treated and untreated polyamides.

The increase in the basic dyeability of nylon, wool, silk and soy protein treated with a green tea extract is also shown in Table II. The green tea extract-polyamide complex provides the additional sites for the basic dye reaction. This extract sterically blocks the accessible amine groups in the polyamide. The basic dye is free to react with the available acid groups in the green tea-polyamide complex. The increased dyeability may in part be due to the carboxylic groups becoming more accessible to the basic dye when the amine groups in the proteins are blocked.

Table I. Reduction of Acidic Dyeability

FD&C Red 40 dye — a values

Treatment	Nylon		Wool		Silk		Protein	
	C^a	T^b	C	T	C	T	C	T
Mesitol	28	1	20	2	41	23	58	48
Green tea extract	25	4	25	11	42	38	54	52

[a] Untreated control, [b] Treated sample

Table II. Enhancement of Basic Dyeability

Basic Green 3 dye — b values

Treatment	Nylon		Wool		Silk		Protein	
	C^a	T^b	C	T	C	T	C	T
Mesitol	2	12	12	22	28	30	19	26
Green tea extract	2	14	5	25	7	27	2	28

[a] Untreated control, [b] Treated sample

Soy protein has a very large number of accessible amine groups. The tea catechins block only a small proportion of the total amine groups and thus give a low level of resistance (Table III). Note, in particular, that the gallates — ECG and EGCG — are more effective at equal concentrations as a resist than the simple catechins — EC and EGC. In addition, protein treated with the gallates pick up basic dye more effectively than protein treated with the simple catechins. This may be due to the fact that the gallates are larger and contain more phenolic groups.

Table III. Relative Resist Properties of Green Tea Catechins

	Untreated control	EC	EGC	ECG	EGCG
FD&C Red 40 — a value	57	58	52	52	42
Basic Green 3 — b value	6	11	11	16	13

The amount of catechin adsorbed onto the polyamide was determined in order to quantify these dyeability effects. The adsorption isotherm of the pure gallate ECG adsorbed onto the polyamides nylon and soy protein as a function of ECG concentration is shown in Figure 2. These adsorption isotherms are of the Freundlich type (7). In this type of adsorption, when the surface is only partially occupied, the amount adsorbed is directly proportional to the concentration of the adsorbate. As seen in Tables I and II, the adsorbate forms a complex with the polyamide which reacts with the basic dye.

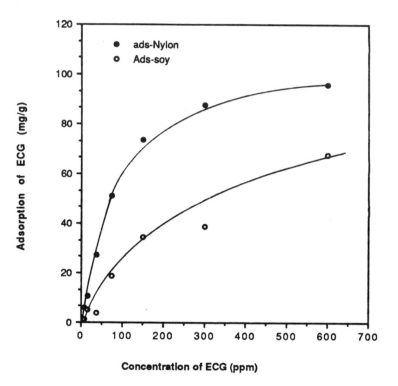

Figure 2. The adsorption of ECG by nylon and soy protein.

At low concentrations of ECG, the color intensity is directly proportional to the concentration of ECG (Figure 3). A similar relationship is shown in Figure 4 in the curve depicting the reduced acidic dyeability. At low concentrations of ECG a linear relationship is found between the amount of ECG adsorbed and the color intensity of the basic dye. This is shown in Figure 5 which was calculated from the data found in Figures 3 and 4 using the adsorption isotherm (Figure 2).

The curves showing the effects of adsorption of ECG on the staining of 6,6 nylon and soy protein are different (Figures 5 and 6). Nylon contains only terminal amine groups. Soy protein contains numerous amine groups in different domains.

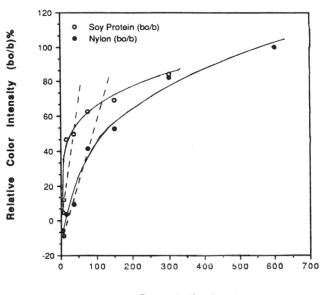

Figure 3. The effect of ECG on the staining of nylon and soy protein by Basic Green 3. The relative color intensity is b_0/b, where b_0 is the value of the most intensely dyed sample and b is the value of ECG treated sample.

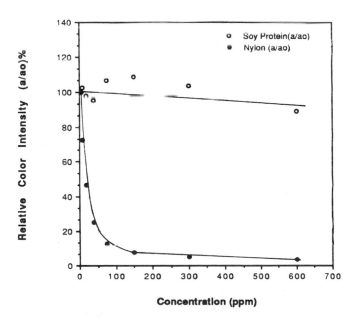

Figure 4. The effect of ECG on the staining of nylon and soy protein by FD&C Red 40. The relative color intensity is a_0/a, where a_0 is the value of the most intensely dyed sample and a is the value of ECG treated sample.

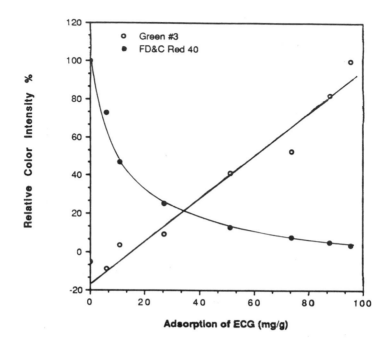

Figure 5. The effect of adsorption of ECG on the staining of nylon by Basic Green 3 and FD&C Red 40.

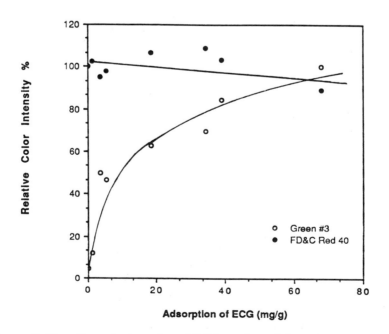

Figure 6. The effect of adsorption of ECG on the staining of soy protein by Basic Green 3 and FD&C Red 40.

The nylon and protein have similar reactivities, but differ in their responses: nylon shows only the response to the terminal domain, whereas protein shows a summation of the many different responses to the various domains in soy protein. The soy protein may be undergoing a conformational change with the adsorption of ECG, as indicated in Figure 6, where the soy protein dyes more intensely than the control between the 100–300 ppm ECG treatments.

Examination of the basic dyed nylon fibers treated with ECG by optical microscopy shows the fibers to be uniformly stained throughout the interior of the fibers at low concentrations. At high concentrations of ECG treatment, an intensely dyed irregular coating was seen on the surface of the nylon. All of the acidic dyed nylon fibers were uniformly dyed. The basic dyed soy protein treated with ECG also showed this intensely dyed surface coating.

Conclusions

The tea catechins behave like textile resists; after the polyamide has been treated with tea catechins, the polyamides no longer react with the acidic dyes. The tea catechin/polyamide complex is now dyeable with basic dyes at low pH. The color intensity of the basic dyed catechin/polyamide complex is a very sensitive method for quantitatively measuring the interaction of individual catechins with polyamides. The gallates EGCG and ECG are more effective dye resists than the simple catechins EGC and EC.

Acknowledgements

The authors wish to thank Cynthia W Ward for her advice on basic dyeing and Chi-Tang Ho for his generous gift of the pure tea catechins.

Literature Cited

1. Hagerman, A. E.; Butler, L. G. *J. Agric. Food Chem.* **1980**, *28*, 944.
2. Gustavson, K. H. *J. Polym. Sci.* **1954**, *12*, 317.
3. Oh, H. I.; Hoff, J. E.; Armstrong, G. S.; Haff, L. A. *J. Agric. Food Chem.* **1980**, *28*, 394.
4. Pierpoint, W. S. *Biochem. J.* **1969**, *112*, 619.
5. Hagerman, A. E. In *Chemistry and Significance of Condensed Tannins*; Plenum Publishing Corp., 1989; pp 323–333.
6. Hagerman, A. E.; Klucher, K. M. In *Plant Flavonoids in Biology and Medicine: Biochemical, Pharmacological, and Structure-Activity Relationships*; 1986; pp 67–76.
7. Zollinger, H. *Color Chemistry: Syntheses, Properties and Applications of Organic Dyes & Pigments*; VCH Verlagsgebellshaft mbH: Weinheim, Germany, 1987; pp 215–237.
8. Oehrl, L.L. *Interactions of the Food, Drug and Cosmetic (FD&C) Dyes with Polymer Systems*; M.S. Thesis, University of Delaware, 1989.
9. Oehrl, L. L.; Malone, C. P.; Keown, R. K. In *Food and Packaging Interactions II*; Risch S. J.; Hotchkiss, J. H., Eds.; ACS Symposium Series No. 473, American Chemical Society: Washington, DC, 1991; pp. 37–52.
10. Weigmann, H-D. H. *Mechanism of Stain Blocker Function in Nylon Carpet Yarns*; AATCC International Conference; October 6, 1992.

11. Harris, P. W.; Hangey, D. A.*Textile Chem. & Colorist* **1989**, *21(11)*, 25.
12. Thanh, V. H.; Shibasaki, K. *J. Agric. Food Chem.* **1976**, *24*, 1117.
13. Hunt, R. W. G. *Measuring Color*; Ellis Horwood Ltd.: Chichester, England, 1987; p 100.

RECEIVED June 7, 1993

Chapter 11

Inhibition of Influenza Virus Infection by Tea Polyphenols

Tadakatsu Shimamura

Department of Microbiology and Immunology, Showa University School of Medicine, Tokyo 142, Japan

Influenza is a disease with a high mortality rate throughout the world. Despite efforts to develop effective vaccines and therapeutic agents against influenza virus infection, it is still virtually uncontrolled. The use of vaccines against influenza is presently limited because of the frequent conversion of viral antigens. Symptomatic therapy is the only treatment possible for influenza virus infection except in a few countries where the antiviral compound amantadine is in current use. Amantadine, however, is effective as prophylaxis and therapy only against influenza A virus. Furthermore, it has adverse side effects, and resistant virus mutants arise. Therefore, new vaccines and antiviral strategies are being explored. Tea, one of the most popular beverages in the world, is well known to possess many medicinal properties — numerous laboratories have reported many therapeutic effects of tea In this chapter, the effects of tea and its constituents against influenza virus infection are discussed.

We have previously reported that tea possesses antibacterial activity against various bacteria that cause diarrheal (*1–3*), respiratory (*4,5*) and skin (*6*) diseases. In ancient China tea was first used as an antidote. There were, however, no scientific reports of antitoxin activity by tea until our laboratory found that tea inhibited the activity of bacterial exotoxins (*7*). We have further demonstrated that polyphenols found in tea are among the components responsible for these activities (*4–6,8–11*).

More recently, we found that tea markedly inhibited the infectivity of both influenza A and B viruses in cultured cells by blocking adsorption of virus to the cells (*12*). This led us to study the inhibitory activity of polyphenols against influenza viruses.

In Vitro Studies

Inhibition of Plaque Formation by Tea Polyphenols. (-)-Epigallocatechin gallate (EGCg) and theaflavin digallate (TF3) were purified from green tea (*13*) and

0097–6156/94/0547–0101$06.00/0

black tea (*14*), respectively. The plaque assay was performed using Madin-Darby canine kidney (MDCK) cells as the influenza virus target.

We first studied the capacity of EGCg and TF3 to inhibit infection of influenza A/Yamagata/120/86(HINI) or influenza B/USSR/100/83 viruses in MDCK cells. The virus was mixed with EGCg or TF3 for either 5 min or 60 min before attempting adsorption to the cells. Contact for a short time of EGCg and TF3 with either virus effectively inhibited infectivity of the virus *in vitro*. Even concentrations of polyphenols as low as about 1 μM inhibited the plaque formation by 100%. A control experiment using amantadine mixed with influenza A virus for 10 min before addition to MDCK cells showed that the concentration of amantadine required to inhibit plaque formation was approximately 100 times that of polyphenols.

We next determined if polyphenols are effective when added during adsorption of virus to cells. Influenza A virus was adsorbed to cells at 4°C for 30 min, then polyphenol was added to virus-adsorbed cells for 15 min, and the cells were cultured. Although the effective concentration of polyphenols was higher than when added before adsorption, inhibition of plaque formation was observed. When polyphenol was added 30 min after adsorption of the virus to cells at 37°C, however, plaque forming activity of the virus was not inhibited at any concentration. These findings suggest that polyphenol prevents the virus from entering into cells, but not virus multiplication in the cells. These results also suggest that polyphenols may bind to surface proteins of influenza virus.

Observation by Electron Microscopy

We used electron microscopy to study the possibility that polyphenols may bind to surface proteins of influenza virus. When the capacities of polyphenols and anti-influenza A virus antibody to bind to influenza A virus was compared, both EGCg and TF3 agglutinated virus particles after a short time, but anti-influenza A virus antibody agglutinated the viruses only weakly. We further examined the capacity of virus pretreated with polyphenol or antibody to bind to MDCK cells by scanning electron microscopy. Untreated virus bound to MDCK cells, whereas virus pretreated with either EGCg or antibody could not. These observations by electron microscopy suggest that polyphenols bind to hemagglutinin of influenza virus and prevent the virus from adsorbing to cells.

Inhibition of Hemagglutination by Tea Polyphenols

Influenza virus can bind to chicken red blood cells by their hemagglutinin and agglutinate the cells. Examination of the capacity of polyphenols to inhibit hemagglutination by the virus revealed that, like anti-influenza A virus antibody, EGCg and TF3 inhibited the hemagglutination dose-dependently.

In Vivo **Studies**

Inhibition of Influenza Virus Infection in Mice.   The capacity of tea extracts containing polyphenols to inhibit influenza virus infection was examined in C3H/He mice. Two groups of 10 mice were intranasally administered 4×10^5 influenza A viruses or a mixture of virus and black tea extract at the beverage concentration. The body weight of mice administered virus alone decreased gradually from four days after infection but increased in mice administered the mixture of virus and black tea extract. All mice administered virus alone died within 10 days after infection while all other mice survived. Anti-influenza virus antibodies in surviving mice were examined by the hemagglutination inhibition test 3 weeks after virus administration. Nine of ten mice administered the mixture of virus and black tea extract did not develop specific antibody, even though in this mouse model specific antibody appears in mice administered only 100 virus particles. These findings suggest that tea extracts inhibit the infectivity of influenza virus in nasal epithelial cells.

Inhibition of Natural Influenza Virus Infection in Swine

It is well known that swine are as susceptible to influenza virus as human. Field trials to test the inhibitory activity of polyphenols against natural influenza virus infection in swine were performed in the northern Japanese prefecture of Miyagi from October 1990 to March 1991 using a polyphenol mixture prepared from Japanese green tea (*15*). Two thousand one-month-old shoats were exposed to 0.1% polyphenols sprayed from a sprinkler for 20 seconds every 30 minutes for six months. Specific anti-influenza virus antibodies in the sera of 40 shoats were examined once a month using swine influenza A virus isolated at that time. In the case of the control group, specific antibodies that were naturally transferred from sows decreased gradually for two months after birth. Then antibodies increased in the third month, meaning that natural infection had occurred. On the other hand, specific antibodies of shoats exposed to polyphenols continued to decrease—by the end of the third month, no antibody was detected in any polyphenol-treated shoat. Unfortunately, the sprinkler system broke down at the end of the third month. After that, specific antibodies appeared at higher levels. These results suggest that polyphenols can prevent natural infection in swine.

Use of Tea Polyphenols as a New Strategy Against Influenza

Tea polyphenols may be useful as a strategy against influenza virus infection, for example, in a gargle. We compared black tea to four kinds of gargles now available in Japan in their capacities to inhibit the infectivity of influenza virus in MDCK cells. Unlike tea, most gargles could not inhibit the infectivity of the virus.

All pupils in one of the primary schools in the Shizuoka district of Japan used Japanese green tea as a gargle several times per day during one winter season. It was reported that tea as a gargle was effective at preventing influenza infection and that no class in the school was closed down by influenza outbreak.

The Japanese Buddhist priest Eisai wrote a book entitled <u>Tea Drinking For Health</u> about eight hundred years ago. In the first paragraph of the book, he described tea as "a wonder drug for health." Echoing Eisai, I would say that tea is a wonder drug against influenza.

Literature Cited

1. Toda, M.; Okubo, S.; Hiyoshi, R.; Shimamura, T. *Letters in Appl. Microbiol.* **1989**, *8*, 123–125.
2. Toda, M.; Okubo, S.; Ohnishi, R.; Shimamura, T. *Jpn. J. Bacteriol.* **1989**, *44*, 669–672.

3. Toda, M.; Okubo, S.; Ikigai, H.; Suzuki, T.; Suzuki, Y.; Shimamura, T. *J. Appl. Bacteriol.* **1991**, *70*, 109–112.
4. Horiuchi, Y.; Toda, M.; Okubo, S.; Hara, Y.; Shimamura, T. *J. J. A. Inf. D.* **1992**, *66*, 599–605.
5. Chosa, H.; Toda, M.; Okubo, S.; Hara, Y.; Shimamura, T. *J. J. A. Inf. D.* **1992**, *66*, 606–611.
6. Okubo, S.; Toda, M.; Hara, Y.; Shimamura T. *Jpn. J. Bacteriol.* **1991**, *46*, 509–514.
7. Okubo, S.; Ikigai, H.; Toda, M.; Shimamura, T. *Letters in Appl. Microbiol.* **1989**, *9*, 65–66.
8. Toda, M.; Okubo, S.; Ikigai, H.; Shimamura, T. *Jpn. J. Bacteriol.* **1990**, *45*, 561–566.
9. Ikigai, H.; Toda, M.; Okubo, S.; Hara, Y.; Shimamura, T. *Jpn. J. Bacteriol.* **1990**, *45*, 913–919.
10. Toda, M.; Okubo, S.; Hara, Y.; Shimamura, T. *Jpn. J. Bacteriol.* **1991**, *46*, 839–845.
11. Toda, M.; Okubo, S.; Ikigai, H.; Suzuki, T.; Suzuki, Y.; Hara, Y.; Shimamura, T. *Microbiol. Immunol.* **1992**, *36*, 999–1001.
12. Nakayama, M.; Toda, M.; Okubo, S.; Shimamura, T. *Letters in Appl. Microbiol.* **1990**, *11*, 38–40.
13. Matsuzaki, T.; Hara, Y. *Nippon Nogeikagaku Kaishi* **1985**, *59*, 129–134.
14. Hara, Y.; Matsuzaki, T.; Suzuki, T. *Nippon Nogeikagaku Kaishi* **1987**, *61*, 803–808.
15. Hara, Y.; Watanabe, M. *Nippon Shokuhin Kogyo Gakkaishi* **1989**, *36*, 951–955.

RECEIVED June 7, 1993

Chapter 12

Inhibitory Effect of Rooibos Tea (*Aspalathus linearis*) on the Induction of Chromosome Aberrations In Vivo and In Vitro

K. Shimoi[1], Y. Hokabe[1], Y. F. Sasaki[2], H. Yamada[2], K. Kator[3], and
N. Kinae[1]

[1]Laboratory of Food Hygiene, School of Food and Nutritional Sciences,
University of Shizuoka, Shizuoka 422, Japan
[2]Biological Laboratory, School of Science, Kwansei Gakuin University,
Nishinomiya, Hyogo 662, Japan
[3]Zoological Institute, Faculty of Science, University of Tokyo, Tokyo 113,
Japan

The treatment of Chinese hamster ovary cells with Rooibos tea
extract (RT, lyophilizate of Rooibos tea infusion, 100–1000 µg/ml)
simultaneously with or subsequent to mitomycin C or benzo[*a*]py-
rene (BP) significantly suppressed chromosome aberrations induced
in the presence or absence of a metabolic activation system. Furthermore, gastric intubation of RT (0.05–0.1%) to ICR male mice at 1
ml per mouse per day for 28 days before intraperitoneal mitomycin
C or BP treatment reduced micronucleus induction in peripheral
blood reticulocytes. When mice received oral administration of RT
(0.05%) in the drinking water *ad libitum* for 28 days, the frequency
of micronuclei induced by mitomycin C (i.p.) or gamma rays was
also decreased. As the dose of tea extract used in the experiments
was almost the same as that of typical human daily consumption, RT
intake might be a useful tool for chemoprevention of human cancer.

Catechins in green tea leaves have been known to possess antimutagenic and
anticarcinogenic potency (*1–6*). In 1992, Wang *et al.* (*7*) reported that green tea
extract as the sole source of drinking water inhibited UV-induced skin tumorigenesis. Antimutagenic and anticarcinogenic effects of Rooibos tea, an herb tea
from South Africa, however, have not been reported so far. In this study, we
investigated the suppressing effect of a crude aqueous extract of Rooibos tea, at a
concentration equivalent to typical human daily intake, on the induction of chromosome aberrations *in vivo* (ICR male mice) and *in vitro* [Chinese hamster ovary
(CHO) cells] in order to evaluate its chemopreventive activity.

Rooibos Tea (*Aspalathus linearis*)

Rooibos tea is one of the major crops cultivated in the western districts of the Cape
Province of South Africa and is consumed as an herb beverage in Africa and

0097–6156/94/0547–0105$06.00/0

Europe. This tea contains several polyphenols and flavonoids like quercetin, but no caffeine or theaflavins (8–9). As shown in Figure 1, we demonstrated from HPLC analysis that Rooibos tea contains little catechins which were characteristic to green tea.

Preparation of Tea Extract

Rooibos tea leaves (50 g) were boiled in 1.5 l distilled water for 15 min. The infusion was cooled, filtered with a glass fiber filter and then freeze dried. About 7 g of solids was obtained. This solid extract, called RT, was dissolved in distilled water for CHO cells or tap water for mice just before use.

Suppressing Effect of RT on Chromosome Aberration in Cultured CHO Cells

CHO cells originally obtained from American Type Cell Collection were used. Cells were grown in Ham's F12 medium (Nissui Pharmaceuticals Inc., Tokyo) supplemented with 10% fetal bovine serum (HyClone Laboratories Inc., USA) and several antibiotics. In the simultaneous treatment study, the cultured CHO cells were exposed to 100 μM benzo[a]pyrene (BP) for 3 h or to 1 μM mitomycin C for 1 h with or without S9 mix and RT (50–1000 μg/ml). Subsequent treatment with RT was performed for 20 h in the absence of S9 mix and for 3 h in the presence of S9 mix after washing the cells treated with mutagens. Chromosome preparations were made according to the air drying method and then were stained with Giemsa. One hundred metaphases per culture were analyzed for chromosome aberrations. The numbers of aberrant cells were statistically analyzed using the χ^2 test.

As shown in Figures 2 and 3, simultaneous or subsequent treatment of RT with or without S9 mix suppressed the induction of chromosome aberrations by BP or mitomycin C. The suppressing effects of RT were stronger in simultaneous treatment than in subsequent treatment.

Suppressing Effect on Micronucleus Induction in Mice

Male ICR mice (8 weeks old at the beginning of the study, Japan SLC Inc., Hamamatsu, Japan) were used. They were housed in an air conditioned room, where ambient lightning was controlled to provide 12 h light/dark cycles, and were given CE-2 commercial food pellets (Crea Japan Co., Tokyo). Tap water or RT dissolved in tap water was administered *ad libitum* or by gastric intubation. Five mice were assigned to each experimental group.

The effects of preadministration of RT on the induction of micronuclei in mouse peripheral blood reticulocytes by BP (125 mg/kg, i.p.), mitomycin C (2 mg/kg, i.p.) and γ-rays (1.5 Gy) were examined according to the acridine orange-coated slide method of Hayashi *et al.* (10). One thousand peripheral reticulocytes per mouse were scored and the number of micronucleated peripheral reticulocytes (MNRETs) was statistically analyzed using Student's *t*-test.

Gastric intubation of RT (0.05–0.1%) at 1 ml/mouse/day for 28 days reduced the frequency of MNRETs induced by BP and mitomycin C, but no effect was observed on γ-ray-induced MNRETs (Table I). On the other hand, oral administration of RT (0.05%) *ad libitum* for 28 days decreased the frequency of mitomycin C and γ-ray-induced MNRETs (Table II). As shown in these Tables, γ-ray treatment exhibited different results between gastric intubation and oral

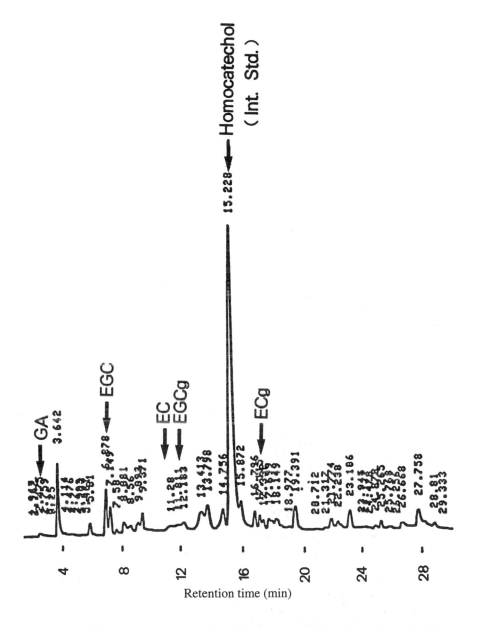

Figure 1. HPLC elution pattern of AcOEt fraction of RT. GA: gallic acid, EC: (-)-epicatechin, ECg: (-)-epicatechin gallate, EGC: (-)-epigallocatechin, EGCg: (-)-epigallocatechin gallate. Conditions: column, TSK gel ODS-80TM (ø4.6 x 150 mm); eluant, CH₃CN-50 mM KH₂PO₄ (1:9→4:6), 0.96 ml/min; detection, UV 280 nm.

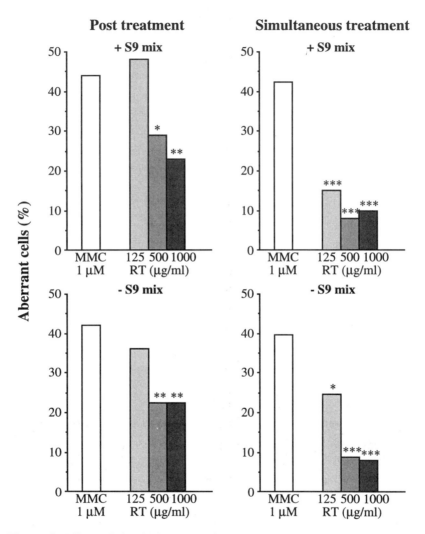

Figure 2. Effects of simultaneous or subsequent treatment of RT on mitomycin C-induced chromosome aberrations in CHO cells.

  * Significantly different from control (p<0.05, χ^2 test).
 ** Significantly different from control (p<0.01, χ^2 test).
*** Significantly different from control (p<0.001, χ^2 test).

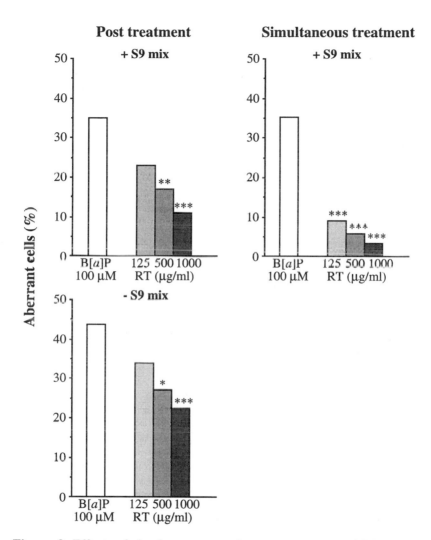

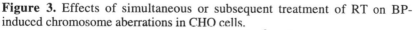

Figure 3. Effects of simultaneous or subsequent treatment of RT on BP-induced chromosome aberrations in CHO cells.

   * Significantly different from control (p<0.05, χ^2 test).
  ** Significantly different from control (p<0.01, χ^2 test).
*** Significantly different from control (p<0.001, χ^2 test).

administration. This may be due to a difference in total intake of RT in each experiment.

Table I. Effects of Intragastric RT on the Induction of Micronuclei in Mouse Peripheral Blood Reticulocytes

Mutagen	Dose	Tea extract (%)	MNRETs (%)	
			0 hr	42 or 48 hrs
BP[a]	125 mg/Kg (i.p.)	0	0.16 ± 0.11	1.30 ± 0.25
		0.1	0.18 ± 0.14	0.66 ± 0.18**
Mitomycin C[a]	2 mg/Kg (i.p.)	0	0.24 ± 0.15	5.22 ± 1.27
		0.1	0.20 ± 0.10	3.46 ± 0.54*
γ-rays[b]	1.5 Gy (whole body)	0	0.04 ± 0.09	3.22 ± 0.92
		0.1	0.12 ± 0.11	2.96 ± 1.20

Five mice were used per group, 1 ml tea extract was administered by gastric intubation each day for 28 days and the data are expressed as the mean $\pm$ S.D.
   [a] 48 hrs, [b] 42 hrs
  * Significantly different from control ($p < 0.05$, Student's t- test).
 ** Significantly different from control ($p < 0.01$, Student's t- test).

Table II. Effects of Oral RT on the Induction of Micronuclei in Mouse Peripheral Blood Reticulocytes

Mutagen	Dose	Tea extract (%)	MNRETS (%)	
			0 hr	42 or 48 hrs
Mitomycin C[a]	2 mg/Kg (i.p.)	0	0.16 ± 0.09	8.47 ± 1.32
		0.05	0.10 ± 0.07	5.73 ± 0.77*
γ-rays[b]	1.5 Gy (whole body)	0	0.24 ± 0.13	6.44 ± 0.75
		0.05	0.14 ± 0.09	3.92 ± 1.92**

Five mice were used per group, the tea extract was provided *ad libitum* for 28 days and the data are expressed as the mean $\pm$ S.D.
   [a] 48 hrs, [b] 42 hrs
  * Significantly different from control ($p < 0.05$, Student's t- test).
 ** Significantly different from control ($p < 0.01$, Student's t- test).

Changes in Daily Food Intake, Water Consumption and Body Weight

During preadministration of RT, daily food intake, daily water consumption and body weight were determined every week (Figure 4). Administration of 0.05% RT as drinking water had a tendency to slightly increase the three parameters.

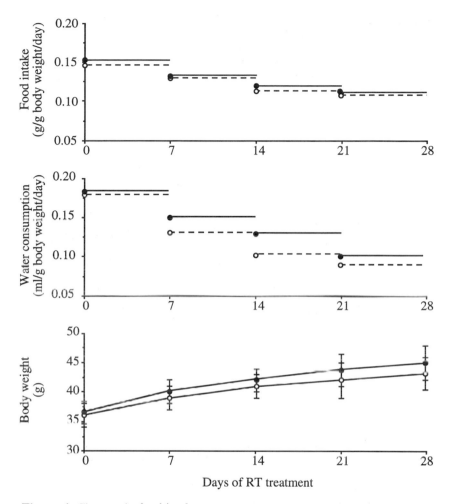

Figure 4. Changes in food intake, water consumption and body weight. RT in tap water (0.05%) was given to mice for 28 days *ad libitum*. Five mice were used for each group. (○: control, ●: RT)

Antioxidative Activity of RT in Mouse Blood Plasma

It has been reported that free radicals are generated *in vivo* by exposure to ionizing radiation (*11*) and that lipid peroxide levels are changed (*12*).

In order to get insight into the assumed mechanism of the inhibition of γ-ray-induced chromosome aberrations by RT, antioxidative activity in mouse blood plasma was examined. Before or 3 h after a single exposure to γ-rays (4.5 Gy), the contents of thiobarbituric acid reactive substances (TBARS) and non-protein SH levels in the plasma of ICR male mice treated with or without RT (0.05%) for 28 days were determined by the modified methods of Yagi (*13*) and

Elleman (*14*), respectively. Since only three mice were used for each group, statistical analysis was not performed.

Figure 5 shows following results: (1) TBARS levels in the plasma of non-irradiated mice decreased with preadministration of RT for 28 days. (2) Non-protein SH content decreased after γ-ray irradiation with or without RT intake, but the non-protein SH content in the RT-treated group was almost the same level as the non-irradiated control group.

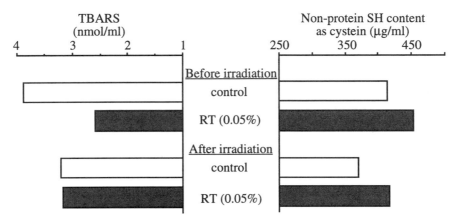

Figure 5. Effect of RT on the antioxidative levels in mouse plasma before and after γ-ray irradiation. Three mice were used for each group.

Discussion

In this chapter, it was suggested that daily intake of RT might be a useful tool to diminish the attack of environmental mutagens and/or carcinogens including active oxygen species. The effective components in RT have not been identified so far, but, since RT contains few catechins, other components must play an important role in the suppression of induced chromosome aberrations. The modes of action of RT against BP, mitomycin C and γ-rays are different. The reduction of the frequency of micronuclei induced by γ-rays *in vivo* seems to be due to the antioxidative activity of RT. More study is needed to elucidate the mechanisms of RT, its effects on proliferative potency of bone marrow cells and antioxidative levels in bone marrow cells.

Acknowledgment

The authors are grateful to the Rooibos Tea Japan Co. (Osaka, Japan) for providing Rooibos tea leaves.

Literature Cited

1. Kada, T.; Kaneko, K.; Matsuzaki, T.; Hara, Y. *Mutat. Res.* **1985**, *150*, 127–132.
2. Shimoi, K.; Nakamura, Y.; Tomita, I.; Hara, Y.; Kada, T. *Mutat. Res.* **1986**, *173*, 239–244.

3. Ito, Y.; Ohnishi, S.; Fujiie, K. *Mutat. Res.* **1989**, *222*, 253–261.
4. Imanishi, H.; Sasaki, Y. F.; Ohta, T.; Watanabe, M.; Kato, T.; Shirasu, Y. *Mutat. Res.* **1991**, *259*, 79–87.
5. Yoshizawa, S.; Horiuchi, T.; Fujiki, H.; Yoshida, T.; Okuda, T.; Sugimura, T. *Phytother. Res.* **1987**, *1*, 44–47.
6. Wang, Z. Y.; Khan, W. A.; Bickers, D. R.; Mukhtar, H. *Carcinogenesis* **1989**, *10*, 411–415.
7. Wang, Z. Y.; Huang, M. T.; Ferraro, T.; Wong, C. Q.; Lou, Y. R.; Reuhl, K.; Iatropoulos, M.; Yang, C. S.; Conney, A. H. *Cancer Res.* **1992**, *52*, 1162–1170.
8. Blommaert, K. L. J.; Steenkamp, J. *Agroplantae* **1978**, *10*, 49.
9. Joubert, E. M. Sc.Thesis, University of Stellenbosch, Rep. South Africa, 1984.
10. Hayashi, M.; Morita, T.; Kodama, Y.; Sofuni, T.; Ishidate, M.Jr. *Mutation Res.* **1990**, *245*, 245–249.
11. Lai, E. K.; Crossley, C.; Sridhar, R.; Misra, H. P.; Janzen, E. G.; McCay, P. B. *Arch. Biochem. Biophys.* **1986**, *244*, 156–160.
12. Yamaoka, K.; Edamatsu, R.; Mori, A. *Free Radicals Biol. Med.* **1991**, *11*, 299–306.
13. Yagi, K. *Biochem.Med.* **1976**, *15*, 212–216.
14. Elleman, G. L. *Arch. Biochem. Biophys.* **1959**, *82*, 70–77.

RECEIVED June 22, 1993

ANTIOXIDANTS

Chapter 13

Prevention of Cancer by Agents That Suppress Production of Oxidants

W. Troll, J. S. Lim, and K. Frenkel

Department of Environmental Medicine, New York University Medical Center, New York, NY 10016

Structurally different cancer preventive agents suppress superoxide anion radical ($O_2^{-}\bullet$) and hydrogen peroxide (H_2O_2) production by phorbol ester tumor promoter-activated human neutrophils. This inhibition of H_2O_2 generation can serve as a facile system for identifying and measuring the activity of cancer preventive agents. Those agents inhibit inflammation and oxidative DNA damage as well as tumor promotion. H_2O_2 causes formation of strand breaks in DNA and oxidation of DNA bases. Cancer preventive agents include retinoids, garlic oil, protease inhibitors (PIs) isolated from natural sources (*e.g.*, soybeans, potatoes and tomatoes), (-)•epigallo-catechin gallate (EGCG) (isolated from green tea), vitamin B_3 (nicotinic acid), and *trans*-tamoxifen (TAM). Nicotinic acid and TAM are the new additions to this growing group of inhibitors. Nicotinic acid may act through formation of the cofactor [*e.g.*, nicotine-adenine-dinucleotide (NAD^+)] of many dehydrogenases. Phorbol ester-mediated H_2O_2 induction in neutrophils — which is not in-hibitable by estradiol — is suppressed by the antiestrogen TAM, which points to additional properties of TAM that contribute to the prevention of breast cancer but are not estrogenic in nature.

The epidemiological data on cancer identified vegetarian populations as having significantly lower incidences of breast, colon and prostate cancer than do non-vegetarian popuations. The cause of this lower occurrence may be related to their low-fat diet. The formation of oxidized DNA bases elicited by the consumption of a diet with a high meat/fat content has been shown in human volunteers (*1*). Dietary fat has been identified as a tumor promoter and acts in a manner similar to the phorbol ester tumor promoter 12-*O*-tetradecanoylphorbol-13-acetate (TPA). The promoting action of fat may be mediated by fatty acid metabolites such as arachidonic acid, which mimics the action of TPA by inducing H_2O_2 formation (*2*). The contribution of fat to the increased rate of breast cancer in the world population was shown by the epidemiological studies of Carroll (*3*). Armstrong and Doll (*4*)

0097–6156/94/0547–0116$06.00/0

reported the lowest rate of breast cancer to be in the population in Thailand, where the food staples are rice and soybeans, and the highest rate in the Netherlands, where meat and dairy products contribute to a high fat content in the diet.

Another factor, which is perhaps more important than fat as a promoting agent, is the abundance of chemopreventive agents in vegetarian diets. These compounds that can prevent cancer were recognized and described by Wattenberg (*5*). The first agents to be identified as suppressors of tumor promotion were PIs, which occur in all seeds presumably to prevent their consumption by insects. PIs have been shown to interfere with insect digestive enzymes (*6*). While PIs are widely distributed in vegetable products, they are not available in pure form in sufficient quantites to use in human studies.

Dietary Protease Inhibitors Suppress Cancer

The 'western' dict, which consists of a high proportion of meat and a relatively low proportion of vegetables, appears to be responsible for the higher occurrence rate of breast, colon and prostate cancers. People who eat diets rich in rice, maize and beans have lower incidences of these cancers. Since seeds contain high concentrations of PIs, they may limit the occurrence of these cancers in humans (*3,4,7–9*).

The initial clue for this possibility was discovered when the synthetic PIs trypsin or chymotrypsin were applied to mouse skin where they interfered with TPA-induced tumor promotion in the two-stage carcinogenesis model (*10*). Further observations demonstrated that tumor development was suppressed by feeding PIs to animals. ε-Amino caproic acid, an inhibitor of plasminogen activator, blocks dimethylhydrazine-induced colon cancer in mice when added to the drinking water (*11*). Diets containing soybean PIs reduce the size of tumors in mouse skin treated with 4-nitroquinoline-N-oxide and TPA (*12*), breast tumors in Sprague-Dawley rats subjected to ionizing radiation (*13*), spontaneous liver cancer in C_3H mice (*14*) and colon cancer in mice (*15*). The possible mechanisms of this anticarcinogenesis include interference with oxyradical formation by neutrophils, suppression of oncogene expression and modulation of adenosine diphosphate ribosyltransferase, which are described in the next section.

Interference with Formation of Reactive Oxygen Species (ROS)

Tumor promoters, including phorbol esters, teleocidin and aplysiatoxin, induce an oxidative burst in polymorphonuclear leukocytes (PMNs), which results in the formation of $O_2^-\bullet$, H_2O_2, hydroxyl radicals and singlet oxygen (*16–21*). Phorbol derivatives that are inactive as tumor promoters (*i.e.*, phorbol, phorbol diacetate and 4-O-methyl-TPA) fail to elicit production of $O_2^-\bullet$ and H_2O_2 by PMNs (*22,23*). PIs have been shown to suppress the formation of $O_2^-\bullet$ and H_2O_2 in human neutrophils treated with tumor promoters. In an extensive study by Frenkel *et al.* (*23*), the PIs that inhibited chymotrypsin were identified as the primary agents responsible for this effect.

Nicotinamide, benzamide and 3-aminobenzamide preferentially inhibit chymotrypsin and suppress the production of $O_2^-\bullet$ by TPA-treated human neutrophils, as shown by measuring SOD-inhibitable cytochrome reduction (Table I) (*24,25*). Oxygen radicals, H_2O_2 and organic peroxides have been identified as agents that contribute to tumor promotion, perhaps by forming oxidized DNA bases [*e.g.*, 5-hydroxymethyl uracil (HMU), thymine glycol (TG) or 8-hydroxylguanine

(8-OHG)] (26,27). Moreover, all chemopreventive agents tested, including PIs, garlic oil, retinoids, sarcophytol A, EGCG, vitamin B_3, nicotinamide and TAM, interfere with oxidative processes by suppressing the generation of $O_2^{-\bullet}$, H_2O_2 and other active oxygen species induced by promoting carcinogens in various cell systems (27).

Table I. Inhibition of Carcinogenesis, ROS and Oxidative DNA Base Damage by Chemopreventive Agents

Agent	Carcinogenesis	ROS	HMU, TG or 8-OHG	Reference
Sarcophytol A	+	+	+	21,28–30
EGCG	+	+	+	27,31–33
Tamoxifen	+	+	+	27,34,35
Protease inhibitors	+	+	+	15,23,24,28
Potato inhibitors 1 & 2				
Chicken ovoinhibitor				
Bowman-Birk inhibitor				
Garlic and onion oils	+	+	nd	36–38
Nicotinic acid	+	+	nd	39
Nicotinamide	+	+	nd	39
Benzamide	+	+	nd	39
3-Aminobenzamide	+	+	nd	39

nd: Not determined.

Tumor promoter-induced oxidative modification of DNA bases is a consequence of H_2O_2 production due to TPA-mediated stimulation of neutrophils, which may lead to the activation of oncogenes (22,27,40,41). The contribution of oncogenes to promotion was demonstrated by Balmain and his colleagues (42), who showed in mouse skin that transfection by some oncogenes mimicked promotion. Like oxidant formation, oncogene expression also can be suppressed by chemopreventive agents. For example, ras oncogene transformation of NIH-3T3 cells is inhibited by PIs, retinoids, sarcophytol A and TAM. The decrease in oncogene-induced transformation serves as a useful method for measuring and identifying these anticarcinogenic agents, and may present another important mechanism of chemoprevention that is related to the formation of oxidized DNA bases. Fos and jun oncogenes have been shown to be induced by TPA and by oxidants (27,43).

Tamoxifen — A Novel Chemopreventive Agent

Clinicians have successfully used TAM as therapy for estrogen-dependent human breast cancer and in the prevention of its recurrence. This prevention was thought to be due to TAM's interference with estrogen-mediated tumor promotion, but TAM has wider anticarcinogenic properties, placing it in the class of chemopreventive agents that suppress tumor promotion in two-stage carcinogenesis by interfering with protein kinase C activity.

TAM at a low dose (5 µM) totally inhibits H_2O_2 formation by TPA-treated human neutrophils. Interestingly, β-estradiol (10 µM) also slightly inhibits the oxidative burst of neutrophils. TAM and β-estradiol additively inhibit H_2O_2 formation with pretreatment of the phagocytes (*34*).

TPA-treated neutrophils (*44*) and HeLa cells (*45*) contain increased levels of HMU. Dietary fat, which is a risk factor for breast cancer, also induces the formation of HMU in the DNA of human white blood cells (*1*). TAM prevents the TPA-induced cellular formation of HMU, both in cell culture and mouse skin (*46* and unpublished data). Rather than acting as an antiestrogen, TAM may suppress the dietary fat-induced HMU in the same manner as it inhibits HMU formation induced by TPA-activated neutrophils.

Conclusions

The formation of oxidized DNA bases by active oxygen species is qualitatively similar to damage caused by ionizing radiation, which can act as a cancer initiator as well as a promoter. The formation of oxidized DNA bases may lead to mutations, a characteristic of initiation of cancer by oxidants and other carcinogens. Therefore, preventing the formation of these oxidized DNA bases may block the process of tumorigenesis.

Chemopreventive substances are a large group of structurally varied agents that interfere with one or more steps of carcinogenesis. They may suppress the activation of carcinogens or the promotion and progression of cancer, some of them by suppressing the formation of H_2O_2 and $O_2^{-\bullet}$. TAM inhibits neutrophil infiltration, as well as formation of H_2O_2 and oxidized DNA bases in TPA-treated mouse skin (*27*). It also suppresses formation of H_2O_2 by TPA activated human neutrophils when used at a concentration that is comparable to that present in the plasma of patients who received TAM (*34*). Hence TAM may be capable of preventing initiation as well as promotion of breast cancer, a possibility that can be explored by further epidemiological studies of TAM's role in cancer prevention.

Literature Cited

1. Djuric, Z.; Heilbrun, L. K.; Reading, B. A.; Boomer, A.; Valeriote, F. A.; Martino, S. *J. Natl. Cancer Inst.* **1991**, *83*, 766–769.
2. Badwey, J. A.; Curnutte, J. T.; Robinson, J. M.; Berde, C. B.; Karnovsky, M. J.; Karnovsky, M. L. *J. Biol. Chem.* **1984**, *259*(12), 7870–7877.
3. Carroll, K. *Cancer Res.* **1975**, *35*, 3374–3383.
4. Armstrong, B.; Doll, R. *Int. J. Cancer* **1975**, *15*, 617–631.
5. Wattenberg, L. W. In *Carcinogenesis*; Slaga, T. J., Ed.; Raven Press: New York, 1980; Vol. 5, pp. 85–98.
6. Birk, Y. *Int. J. Peptide Protein Res.* **1985**, *25*, 113–131.
7. Wynder, E.; Mabuchi, K.; Whitmore, W. *Cancer* **1971**, *28*, 344–360.
8. Phillips, R. L. *Cancer Res.* **1975**, *35*, 3513–3522.
9. Correa, P. *Cancer Res.* **1981**, *41*, 3685–3690.
10. Troll, W.; Klassen, A.; Janoff, A. *Science* **1970**, *169*, 1211–1213.
11. Corasanti, J. G.; Hobika, G.H.; Markus, G. *Science* **1982**, 216, 1020–1021.
12. Troll, W.; Belman, S.; Wiesner, R.; Shellabarger, C. J. In *Biological Function of Proteinases*; Holzer, H.; Tschesche, H., Eds.; Springer-Verlag: Berlin, 1979; pp 165–170.

13. Troll, W.; Wiesner, R.; Shellabarger, C. J.; Holtzman, S.; Stone, J. P. *Carcinogenesis* **1980**, *1*, 469–472.
14. Becker, F. F. *Carcinogenesis* **1981**, *2*, 1213–1214.
15. Weed, H. G.; McGandy, R. B.; Kennedy, A. R. *Carcinogenesis* **1985**, *6*, 1239–1241.
16. Hozumi, M.; Ogawa, M.; Sugimura, T.; Takeuchi, T.; Umezawa, H. *Cancer Res.* **1972**, *32*, 1725–1729.
17. Badwey, J. A.; Karnovsky, M. L. *Ann. Rev. Biochem.* **1980**, *49*, 695–726.
18. Fantone, J. C.; Ward, P. A. *Am. J. Pathol.* **1982**, *107*, 397–418.
19. Formisano, J.; Troll, W.; Sugimura, T. *Ann. N. Y. Acad. Sci.* **1983**, *407*, 429–431.
20. Kinzel, V.; Fürstenberger, G.; Loehrke, H.; Marks, F. *Carcinogenesis* **1986**, *7*, 779–782.
21. Narisawa, T.; Takahashi, M.; Niwa, M.; Fukaura, Y.; Fujiki, H. *Cancer Res.* **1989**, *49*, 3287–3289.
22. Frenkel, K.; Chrzan, K. *Carcinogenesis* **1987**, *8*, 455–460.
23. Frenkel, K.; Chrzan, K.; Ryan, C.; Wiesner, R.; Troll, W. *Carcinogenesis* **1987**, *8*, 1207–1212.
24. Troll, W.; Wiesner, R.; Frenkel, K. *Adv. Cancer Res.* **1987**, *49*, 265–283.
25. Troll, W.; Garte, S.; Frenkel, K. In *Antimutagenesis and Anticarcinogenesis Mechanisms II*; Kuroda, Y.; Shankel, D. M.; Waters, M. D., Eds.; Plenum: New York, 1990; pp 225–232.
26. Slaga, T. J.; Klein-Szanto, A. J. P.; Triplett, L. L.; Yotti, L. P.; Trosko, J. E. *Science* **1981**, *213*, 1023–1025.
27. Frenkel, K. *Pharmac. Ther.* **1992**, *53*, 127–166.
28. Frenkel, K.; Zhong, Z.; Rashid, K.; Fujiki, H. In *Anticarcinogenesis and Radiation Protection, 2: Strategies in Protection from Radiation and Cancer*; Nygaard, O. F.; Ed.; Plenum: New York, 1991; pp 357–366.
29. Frenkel, K. In *Protease Inhibitors as Cancer Chemopreventige Agents*; Troll, W.; Kennedy, A. R.; Eds.; Plenum: New York, 1993; pp 227–249.
30. Wei, H.; Frenkel, K. *Cancer Res.* **1992**, *52*, 2298–2303.
31. Yoshizawa, S.; Horiuchi, T.; Fujiki, H.; Yoshida, T.; Okuda, T.; Sugimura, T. *Phytotherapy Res.* **1987**, *1*, 44–47.
32. Bhimani, R.; Frenkel, K. *Proc. Am. Assoc. Cancer Res.* **1991**, *31*, 126.
33. Zhong, Z.; Tius, M.; Troll, W.; Fujiki, H.; Frenkel, K. *Proc. Am. Assoc. Cancer Res.* **1991**, *32*, 127.
34. Lim, J. S.; Frenkel, K.; Troll, W. *Cancer Res.* **1992**, *52*, 4969–4972.
35. Buckley, M. M.-T.; Goa, K. L. *Drugs* **1989**, *37*, 451–490.
36. Belman, S. *Carcinogenesis* **1983**, *4*, 1063–1065.
37. Sparnins, V. L.,; Mott, A. W.; Barany, G.; Wattenberg, L. W. *Nutr. Cancer* **1986**, *8*, 211–215.
38. Belman, S.; Solomon, J.; Segal, A.; Block, E.; Barany, G. *J. Biochem. Toxicol.* **1989**, *4*, 151–160.
39. Troll, W. In *Protease Inhibitors as Cancer Chemopreventige Agents*; Troll, W.; Kennedy, A. R., Eds.; Plenum: New York, 1993; 177–189.
40. Garte, S. J.; Currie, D. D.; Troll, W. *Cancer Res.* **1987**, *47*, 3159–3162.
41. Cox, L. R.; Motz, J.; Troll, W.; Garte, S. J. *J. Cancer Res. Clin. Oncol.* **1991**, *117*, 102–108.
42. Quintanilla, M.; Brown, K.; Ramsden, M.; Balmain, A. *Nature* **1986**, *322*, 78–80.

43. Amstad, P. A.; Krupitza, G.; Cerutti, P. A. *Cancer Res.* **1992**, *52*, 3952–3960.
44. Bhimani, R.; Zhong, Z.; Stern, A.; Frenkel, K. *Proc. Am. Assoc. Cancer Res.* **1992**, *33*, 161.
45. Frenkel, K.; Chrzan, K. In *Anticarcinogenesis and Radiation Protection*; Cerutti, P. A.; Nygaard, O. F.; Simic, M. G., Eds.; Plenum: New York, 1987; pp 97–102.
46. Wei, H.; Frenkel, K. *Proc. Am. Assoc. Cancer Res.* **1992**, *33*, 179.

RECEIVED July 6, 1993

Chapter 14

Cancer Chemoprevention by Antioxidants

Masao Hirose, Katsumi Imaida, Seiko Tamano, and Nobuyuki Ito

First Department of Pathology, Nagoya City University Medical School, 1 Kawasumi, Mizuho-cho, Mizuho-ku, Nagoya 467, Japan

Modulation effects of antioxidants on chemical carcinogenesis were investigated using three experimental approaches. First, naturally occurring antioxidants such as N-tritriacontane-16,18-dione (TTAD), tannic acid, phytic acid, green tea catechins (GTC), diallyl sulfide (DAS) and diallyl disulfide (DDS) were examined in two rat multi-organ carcinogenesis models in which both enhancing and inhibitory effects of chemicals could be investigated at the whole body level. In one model, TTAD and phytic acid inhibited hepatocarcinogenesis, tannic acid inhibited large intestinal carcinogenesis and phytic acid enhanced bladder carcinogenesis when they were administered after carcinogen treatment. In the second multi-organ model, GTC clearly inhibited small intestinal carcinogenesis when it was given either during or after carcinogen treatment. DDS potently inhibited large intestine and kidney carcinogenesis, but DAS enhanced hepatocarcinogenesis when applied subsequent to carcinogen treatment. Another approach used the analysis of chemical carcinogen enhancement of liver preneoplastic GST-P positive foci in partially hepatectomized rats. GTC, phenethyl isothiocyanate, α-tocopherol and β-carotene were found to strongly inhibit the Glu-P-1-enhancement of GST-P positive foci, but not dimethylnitrosamine enhancement. In an third approach, female SD rats fed 1% GTC in the diet for 35 weeks, starting one week after a single intragastric administration of DMBA, did not have significantly different incidences or average numbers of mammary carcinomas, but the survival rate was clearly higher in the GTC group (94%) as compared with controls (33%).

It is generally accepted that environmental factors are appreciable causes of human cancer. This is partly reflected in the increasing number of man-made or naturally occurring chemicals which have been shown to have carcinogenic potential. Since cancer mortality is rising with prolongation of the life span, it is important to detect

0097–6156/94/0547–0122$06.00/0

human carcinogens and to eliminate them, wherever possible, from our environment. This approach, however, is clearly limited by the number of experimental facilities, and the involved political and social expense. Another important possible way to decrease human cancer mortality is through cancer chemoprevention. Although much attention has recently been paid to this area, ideal chemopreventors have not yet been developed. This is largely because chemicals inhibit or enhance carcinogenesis depending on the organ site, timing of administration (before carcinogen exposure, with carcinogen or after carcinogen exposure), and species used (*1–4*). Therefore, the development or detection of chemopreventive compounds in the case of natural agents, requires analysis of potential at the whole body level. The present investigations of possible chemopreventive antioxidants were conducted using different three experimental systems with this requirement in mind.

Materials and Methods

Effects of Antioxidants in Rat Multi-Organ Carcinogenesis Models. In the DED model, groups of 6-week-old F344 male rats (Charles River Japan, Inc.) were given, at intervals of 3-4 days, two i.p. injections of 1000 mg/kg body weight (bw) 2,2'-dihydroxy-di-*n*-propylnitrosamine (DHPN), two i.g. administrations of 1500 mg/kg bw *N*-ethyl-*N*-hydroxyethylnitrosamine (EHEN) and three s.c. injections of 75 mg/kg bw 3,2' dimethyl 4 aminobiphenyl (DMAB) over the initial 3 week period. The animals were housed 5 or 6 to a plastic cage with wood chips for bedding in an air-conditioned room at $24 \pm 2°C$ with a 12 hr light-dark cycle. In the first experiment, 4 groups of 15 to 17 animals each were fed Oriental MF powdered basal diet containing 0.2% *N*-tritriacontane-16,18-dione (TTAD, Eisai Co. Ltd., Tsukuba, Japan), 1% tannic acid (Wako Pure Chemical Industries Ltd., Osaka, Japan), 2% phytic acid (Tokyo Kasei Kogyo Co. Ltd., Tokyo, Japan) or basal diet alone starting one day before the EHEN treatment until one week after the last DMAB treatment. The animals were then placed on basal diet until sacrifice at week 36. In the second experiment, starting one week after the last DMAB treatment, 4 groups of 15 rats were treated with basal diet containing 0.2% TTAD, 1% tannic acid, 2% phytic acid or basal diet alone for 32 weeks. A further 4 groups of 10 animals each were treated with antioxidants or basal diet alone without carcinogen treatment. Animals were killed at week 36.

In the DMBDD model, animals were given combined treatment with a single i.p. administration of 100 mg/kg bw diethylnitrosamine (DEN), 4 i.p. administrations of 20 mg/kg bw methylnitrosourea (MNU), 4 s.c. doses of 40 mg/kg bw 1,2-dimethylhydrazine (DMH), 0.05% *N*-butyl-*N*-(hydroxybutyl)nitrosamine (BBN) in the drinking water for 2 weeks and 0.1% DHPN in drinking water for 2 weeks during the initial 4 week period for initiation. In the first experiment, 3 groups of animals each were continuously administered 1% green tea catechins (GTC, Mitsui Norin Co. Ltd., Japan) in powdered basal diet starting one day before until the completion of carcinogen exposure, or starting 3 days after the carcinogen exposure until week 36. Further groups of animals were given the DMBDD treatment alone, or 1% GTC alone throughout the experiment without carcinogen exposure. In the second experiment, 3 groups of 20 animals each were administered i.g. 200 mg/kg bw diallyl sulfide (DAS), 50 mg/kg bw diallyl disulfide (DDS) or corn oil alone 3 times a week after DMBDD treatment. DAS and DDS were obtained from Tokyo Kasei Kogyo Co. Ltd., Tokyo. Further groups of animals were treated with DAS, DDS or basal diet alone without DMBDD treatment. All animals were killed for autopsy at week 28.

In each of these multi-organ carcinogenesis models, all major organs were removed, fixed in neutral buffered formalin solution, or cold acetone for liver, and routinely processed for H&E staining or immunohistochemical staining with anti-glutathione S-transferase placental form (GST-P) antibody.

Effects of Antioxidants in a Rat Medium-term Bioassay for the Detection of Hepatocarcinogens. Six-week-old F344 male rats were given a single i.p. injection of 100 mg/kg bw DEN and starting 2 weeks later received diet containing 300 ppm 2-amino-6-methyldipyrido[1,2-a:3′,2′-d]imidazole (Glu-P-1) or 20 ppm dimethylnitrosamine (DMN) in the drinking water, alone, or in combination with 1% GTC, 0.1% 2-phenylethyl isothiocyanate (PEITC), 1.5% α-tocopherol or 0.1% β-carotene for 6 weeks. Glu-P-1 was kindly supplied by Drs. Takashi Sugimura and Minako Nagao of the National Cancer Center Research Institute, Tokyo, Japan. PEITC was obtained from Tokyo Kasei Kogyo Co. Ltd., Tokyo, α-tocopherol from Eisai Co. Ltd., Tokyo, and β-carotene from Riken Vitamins Co. Ltd., Tokyo.

Animals underwent partial hepatectomy at week 3 and the experiment was terminated at week 8. Livers were removed, fixed in cold acetone, and paraffin-embedded sections were stained immunohistochemically with anti-GST-P antibody.

Effects of an Antioxidant in a DMBA-induced Rat Mammary Carcinogenesis Model. Seven-week-old female SD rats (Charles River Japan Inc., Kanagawa, Japan) were given a single i.g. administration of 7,12-dimethylbenz[a]anthracene (DMBA) at a dose of 50 mg/kg bw. Starting one week after DMBA administration, groups of 15 or 16 animals each were fed powdered basal diet containing 1% GTC, or basal diet alone for 35 weeks. Further groups of 10 animals each were fed GTC or basal diet alone without the carcinogen treatment. Presence of palpable mammary tumors was recorded once every 2 weeks. Animals were killed at week 36, when mammary tumors, Zymbal's glands, liver, kidneys and spleen were removed and routinely processed for examination of sections stained with H&E. Student's t test, cumulative χ^2 test and Fisher's exact test were used for statistical analysis of the data.

Results

Effects of Antioxidants in Rat Multi-organ Carcinogenesis Models.

DED Model. The final body weights of animals treated with carcinogens simultaneously with antioxidants were not significantly different from those receiving DED treatment alone. Those treated with DED followed by TTAD, however, were significantly lower.

The incidences of histopathologically assessed preneoplastic and neoplastic lesions in the thyroid, esophagus, large intestine, liver, pancreas, lung, kidney and urinary bladder of rats treated with antioxidants during DED treatment were not significantly different from those with DED alone.

Incidences of preneoplastic and neoplastic lesions in major organs of rats treated with antioxidants after DED treatment are shown in Table I. In the group given phytic acid, the incidence (50%, p<0.01) and the number per rat (0.75 ± 0.92, p<0.01) of urinary bladder papillomas were significantly higher than in the controls (which had no urinary bladder papillomas), though data for carcinomas did not show any difference. The incidences of liver hyperplastic nodules (42%) and hepatocellular carcinomas (8%, p<0.01) were lower than in the DED alone group (64 and 43%, respectively).

In the group treated with TTAD, development of pancreatic eosinophilic foci (20%, p<0.01) and hepatocellular carcinomas (7%, p<0.05) was significantly lower than in the controls (71 and 43%, respectively).

In the tannic acid group, slight inhibition of colon tumor development was observed. In particular the number of adenomas per rat (0.17 ± 0.37, p<0.05) was significantly reduced as compared with DED alone values (0.79 ± 0.86) (5).

Table I. Histopathological Findings in Major Organs of DED-treated Rats

Organs and lesions	Chemicals (%)			
	Phytic acid n = 12[a]	TTAD n = 15	Tannic acid n = 12	Basal diet n = 14
Large intestine				
Adenoma	5 (42)	7 (47)	2 (17)	7 (50)
Carcinoma	4 (33)	2 (13)	1 (8)	3 (21)
Pancreas				
Eosinophilic focus	4 (33)	3 (20)[b]	9 (75)	10 (71)
Liver				
Hyperplastic nodule	5 (42)[d]	10 (67)	5 (42)	9 (64)
Hepatocellular carcinoma	1 (8)[d]	1 (7)[c]	3 (25)	6 (43)
Lung				
Adenoma	9 (75)	6 (40)	7 (58)	7 (54)
Carinoma	3 (25)	3 (20)	2 (17)	1 (7)
Kidney				
Atypical tubule	12 (100)	15 (100)	12 (100)	14 (100)
Adenoma	7 (58)	5 (33)	7 (58)	6 (43)
Urinary bladder				
Papilloma	6 (50)[b]	3 (20)	3 (25)	0 (0)
Carcinoma	2 (17)	1 (7)	2 (17)	1 (7)
Thyroid gland	3 (25)	3 (20)	2 (17)	3 (21)
Carcinoma	2 (17)	2 (13)	2 (17)	3 (21)

[a] Number of animals (percent of animals)
[b] p<0.01, [c] p<0.05 versus basal diet group by the Fisher's exact test
[d] p<0.01 versus basal diet by the cumulative χ^2 test

DMBDD Model. Final average body weights of animals treated with GTC simultaneously with or after carcinogen exposure tended to be higher than those receiving DMBDD treatment alone.

Significant decreases in the incidences of small intestinal tumors (adenomas and adenocarcinomas) were evident in the group treated with GTC during DMBDD exposure (13%, p<0.02) as compared with DMBDD alone (57%). Multiplicities were also lower in groups treated with GTC both during (0.13 ± 0.35, p<0.01) and after (0.31 ± 0.48, p<0.05) DMBDD exposure than in the DMBDD alone case (1.07 ± 1.21) (Table II). No significant difference in the incidences of neoplastic and/or preneoplastic lesions in the esophagus, forestomach, colon, liver, kidney, urinary bladder, lung and thyroid gland were observed.

The average final body weights of animals given DAS or DDS were reduced about 10–12% as compared with DMBDD alone. On the other hand their relative liver weights were significantly increased.

Histopathologically, the incidences of adenomas in the large intestine were decreased in groups treated with DAS (5.3%) or DDS (5.3%) as compared with that given DMBDD treatment alone (30%). The incidence of carcinomas was also lower in rats treated with DDS (15.8%) than in animals given DMBDD alone (40%). DDS significantly decreased the incidences of neoplastic lesions in the large intestine as assessed using the cumulative chi-square test, but no equivalent modifying effects on colonic carcinogenesis were observed in the DAS treated group (Figure 1). DDS treated animals also showed significant reduction in the incidences of kidney lesions (atypical tubules, 21%; adenomas, 0%; carcinomas, 0%) as compared with DMBDD alone (35, 20 and 5%, respectively) (Figure 1).

On the other hand, the number per cm^2 and size (mm^2) of liver GST-P positive foci in rats treated with DAS (15.3 ± 5.6, p<0.001 and 2.1 ± 1.8, p<0.01, respectively) were significantly higher than in the DMBDD alone group (6.9 ± 2.4 and 0.6 ± 0.3, respectively) (Figure 2). The incidence of preneoplastic and neoplastic lesions in other organs was not affected by treatment with either DAS or DDS (16).

Table II. Incidences and Numbers of Tumors in the Small Intestine of DMBDD-treated Rats

Treatment	Adenomas Incidence (%)	Adenomas Number per rat[a]	Carcinomas Incidence (%)	Carcinomas Number per rat	Total Incidence (%)	Total Number per rat
DMBDD with GTC	1 (7)*	0.07 ± 0.26*	1 (7)*	0.07 ± 0.26*	2 (13)**	0.13 ± 0.35***
DMBDD then GTC	1 (8)*	0.08 ± 0.28*	3 (23)	0.23 ± 0.44	4 (31)	0.31 ± 0.48*
DMBDD alone	6 (43)	0.57 ± 0.76	5 (36)	0.50 ± 0.76	8 (57)	1.07 ± 1.21

[a]Mean ± SD; DMBDD with GTC, 15 animals; DMBDD then GTC, 13; alone, 14.
***p<0.01, **p<0.02, *p<0.05 versus respective initiation alone group value.

Effects of Antioxidants in a Rat Medium-term Bioassay for the Detection of Hepatocarcinogens. The average final body weights of animals treated with

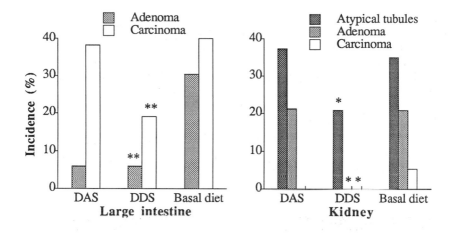

Figure 1. Incidences of preneoplastic and neoplastic lesions in the large intestine and kidney of rats treated with DAS or DDS after DMBDD treatment.
**Significantly different from basal diet group values (p<0.01, by cumulative χ^2 test).
*Significantly different from basal diet group values (p<0.05, by cumulative χ^2 test).

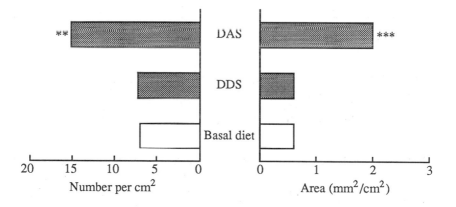

Figure 2. Quantitative data for GST-P positive liver foci in rats treated with DAS or DDS after DMBDD treatment.
***Significantly different from basal diet group values at p<0.001
**Significantly different from basal diet group values at p<0.01.

Glu-P-1 with or without antioxidants did not significantly differ. Liver weights, however, tended to be lower in the antioxidant-treated groups.

The average number per cm^2 and size (mm^2) of liver GST-P positive foci was significantly decreased in animals treated with Glu-P-1 combined with GTC (25.5 ± 6.0, $p<0.001$; 4.4 ± 1.8, $p<0.001$), PEITC (28.2 ± 7.0, $p<0.001$; 4.0 ± 1.2, $p<0.001$), α-tocopherol (30.5 ± 14.4, $p<0.001$; 10.0 ± 2.17, not significant) or β-carotene (34.4 ± 10.5, $p<0.001$; 6.2 ± 2.8, $p<0.001$) as compared with Glu-P-1 alone (47.5 ± 8.9; 11.1 ± 4.7) (Figure 3).

On the other hand, the average number of foci was slightly increased in animals treated with DMN simultaneously with PEITC (21.2 ± 5.5, $p<0.05$) or β-carotene (20.9 ± 5.1, $p<0.05$) as compared with the DMN alone group (16.5 ± 3.9), though the size of foci did not significantly differ.

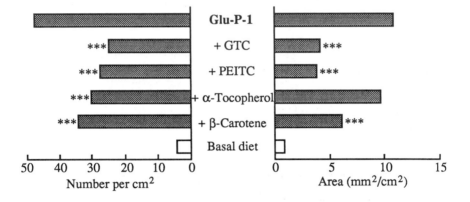

Figure 3. Quantitative data for GST-P positive liver foci in rats treated with Glu-P-1 with or without simultaneous treatment with antioxidants, or basal diet alone without Glu-P-1 treatment in a medium term liver bioassay.
***Significantly different from Glu-P-1 group values at $p<0.001$.

Effects of Antioxidants in the DMBA-induced Rat Mammary Carcinogenesis Model. The average body weights of animals treated with GTC were not significantly different from those receiving DMBA alone at any point during the experimental period. The incidence of palpable mammary tumors in animals treated with DMBA followed by GTC was 31% at week 10, 69% at week 22 and finally 94%, and in those given DMBA alone, 47% at week 10, 87% at week 22 and finally 93%. There were no statistically significant differences between these incidences, but a pronounced difference in survival rate was observed. In the DMBA alone group, the first animal died of mammary tumors at week 22 and the percentage of animals surviving decreased to 53.3% at week 28 and finally to only 33.3%. In contrast, only one animal treated with DMBA followed by GTC died of an unknown cause at week 32 and the rest of the animals survived until the end of the experiment. In addition, the average tumor diameter of animals treated with GTC was significantly smaller (1.1 ± 0.5 cm, $p<0.05$) than those with DMBA alone (1.8 ± 1.5 cm) at week 18, at a point when all animals were alive.

Histopathological analysis showed that the final incidences and the average number of tumors per rat were not significantly different between the groups treated with DMBA followed by GTC (87%, 4.0 ± 2.4 per rat) and DMBA alone (93%, 4.0 ± 2.2 per rat).

Discussion

It is now generally accepted that the carcinogenic process can be divided into initiation, promotion and progression stages. Although experimental chemoprevention has long occupied the attention of several investigators, most trials have been limited to a single organ or one stage of multi-step carcinogenesis, despite the finding that different modifying effects on chemical carcinogenesis depend on the organ and time of administration. For example, the phenolic antioxidant BHA enhances second stage forestomach (7,8) and urinary bladder (8,9) carcinogenesis, but inhibits liver (10,11) and mammary gland (12) carcinogenesis. Although pheno-barbital strongly enhances second stage rat hepatocarcinogenesis (13,14) it can inhibit the initiating activities of some carcinogens in the same organ (15). Therefore, for assessment of potential chemopreventors, it is necessary that their effects be analyzed at a whole body level and at all stages of carcinogenesis.

Antioxidants are a particularly interesting class of chemicals since they possess a variety of biological activities including the induction of drug metabolizing enzymes, inhibition of prostaglandin synthesis, inhibition of carcinogen-induced mutagenesis and scavenging of free radicals (16). Such activities could all play a part in the inhibition of chemical carcinogenesis at either the initiation or promotion stages.

Many different antioxidants have been shown to alter carcinogenesis. Although the effects of the strong antioxidant TTAD on chemical carcinogenesis have not previously been examined, sodium phytate has been shown to inhibit azoxymethane- or DMH-induced colon carcinogenesis of rats or mice when given either during or after carcinogen exposure (17,18). In addition to skin carcinogenesis (19), tannic acid has been shown to inhibit benzo[a]pyrene (B[a]P)-induced mouse lung and forestomach carcinogenesis at the initiation stage (20). (-)-Epigallocatechin gallate, a component of GTC, was found effective in inhibiting second stage N-ethyl-N'-nitro-N-nitrosoguanidine-induced mouse duodenal carcinogenesis (21), and a green tea infusion inhibited DEN- or 4-(methylnitrosamino)-1-(3-pyridyl)-1-butanone (NNK)-induced mouse lung and forestomach carcinogenesis when applied during either the initiating or the promoting stage (22). GTC also possesses anti-initiation and anti-promotion activity in mouse skin carcinogenesis (23–25). DAS is a potent inhibitor of B[a]P-induced mouse forestomach and lung carcinogenesis (26), DMH-induced mouse colon carcinogenesis (17) and N-nitrosomethylbenzylamine-induced rat esophageal carcinogenesis (28) when given to animals during the initiating stage. The effects of DAS and DDS on promotion have not yet been ascertained.

In the present DED multi-organ model, TTAD and phytic acid inhibited hepatocarcinogenesis, tannic acid inhibited colon carcinogenesis, and phytic acid enhanced urinary bladder carcinogenesis when the chemicals were applied after the multiple initiation treatments. The reason why phytic acid did not, as expected from the sodium phytate results, inhibit colon carcinogenesis is not known, but differences in the experimental system, including the colon carcinogen (DMAB vs

AOM) or sodium ion (phytic acid vs sodium phytate) were presumably contributing factors.

In the DMBDD multi-organ carcinogenesis model, GTC clearly inhibited small intestinal carcinogenesis either during or after DMBDD treatment without appreciable enhancing effects on any other organs. DDS selectively inhibited kidney and large intestinal carcinogenesis, and although DAS enhanced hepato-carcinogenesis when administered after DMBDD treatment, if DDS possesses anti-initiating activity regardless of the carcinogen, it might be a good chemopreventor candidate.

In the rat medium-term liver bioassay, GTC, PEITC, α-tocopherol and β-carotene all inhibited Glu-P-1-enhanced development of preneoplastic GST-P positive foci. Of these antioxidants, GTC appeared to exert the most potent influence. Since Glu-P-1 is a carcinogenic heterocyclic amine present in cooked foods which targets the rat liver, colon, small intestine and Zymbal's glands in males (29), the present results are clearly interesting in that they indicate a beneficial effect of naturally occurring antioxidants. The fact that no equivalent inhibition of DMN-enhancement of GST-P positive foci development suggests important differences in the action of these two carcinogens. Further work in ascertaining mechanistic aspects is clearly warranted.

The effects of GTC on DMBA-induced 2-stage mammary carcinogenesis model were also investigated since the multi-organ carcinogenesis models did not cover the mammary gland. Although the eventual incidences and multiplicities of mammary tumors did not appear to be reduced, GTC strongly enhanced survival, suggesting that it retarded the growth of DMBA-initiated rat mammary tumors.

Among the antioxidants assessed in these experiments, GTC appears particularly interesting. Inhibition of small intestinal carcinogenesis, lowered carcinogenic potential of Glu-P-1 in the liver, and increased survival of DMBA-treated female rats were all found without any appreciable enhancing effect in other organs when administered during either the initiation or promotion stages. Although no inhibition of colon carcinogenesis was observed in the present experiment, a different source of GTC at doses 0.1 and 0.01% in the diet was documented to inhibit the second stage of AOM-induced two-stage rat colon carcinogenesis (30). Whether the discrepancy between findings might be due to differences in the GTC composition warrants attention. One cup of green tea contains 0.1–0.15 g GTC. Therefore the 1% dose level applied in the present experiments was equivalent to 30- 50 times the normal human ingestion level. Since 0.1% in the diet was still effective in inhibiting colon and small intestinal carcinogenesis (data not shown), however, this daily intake of GTC, along with other antioxidants, might indeed contribute to chemoprevention of some human cancers. In support of this conclusion, there are epidemiological findings demonstrating a correlation between increased green tea ingestion and decreased risk of gastric and colorectal neoplasms (31–33).

While it is much easier to establish carcinogenicity than to find good chemopreventors, our present examination of modifying effects of chemicals at the whole body level does indicate that the latter may be possible with the right approach. Multi-organ models (34–37) in combination with medium term liver bioassays (34,38) and 2-stage carcinogenesis models are clearly useful tools for this endeavor.

Literature Cited

1. Wattenberg, L. W. *J. Natl. Cancer Inst.* **1978**, *60*, 11–18.
2. Wattenberg, L. W.; Lam, L. K. T. *Inhibition of Chemical Carcinogenesis by Phenols, Coumarins, Aromatic Isothiocyanates, Flavones, and Indoles*; Plenum Press: New York, 1981; pp 1–22.
3. Ito, N.; Hirose, M. *Jpn. J. Cancer Res.* **1987**, *78*, 1011–1026.
4. Ito, N.; Hirose, M. *Adv. Cancer Res.* **1989**, *53*, 247–302.
5. Hirose, M.; Ozaki, K.; Takaba, K.; Fukushima, S.; Shirai, T.; Ito, N. *Carcinogenesis* **1991**, *12*, 1917–1921.
6. Takahashi, S.; Hakoi, K.; Yada, H.; Hirose, M.; Ito, N.; Fukushima, S. *Carcinogenesis* **1992**, *13*, 1513.
7. Hirose, M.; Kagawa, M.; Ogawa, K.; Yamamoto, A.; Ito, N. *Carcinogenesis* **1989**, *10*, 2223–2226.
8. Imaida, K.; Fukushima, S.; Shirai, T.; Masui, T.; Ogiso, T.; Ito, N. *Gann* **1984**, *75*, 769–775.
9. Imaida, K.; Fukushima, S.; Shirai, T. *Carcinogenesis* **1983**, *4*, 1895–899.
10. Tsuda, H.; Sakata, T.; Masui, T.; Imaida, K.; Ito, N. *Carcinogenesis* **1984**, *5*, 525–531.
11. Thamavit, W.; Tatematsu, M.; Ogiso, T.; Mera, Y.; Tsuda, H.; Ito, N. *Cancer Lett.* **1985**, *27*, 295–303.
12. Hirose, M.; Masuda, A.; Inoue, T.; Fukushima, S.; Ito, N. *Carcinogenesis* **1986**, *7*, 1155–1159.
13. Shirai, T.; Hosoda, K.; Hirose, K.; Hirose, M.; Ito, N. *Cancer Lett.* **1985**, *28*, 127–133.
14. Ito, N.; Tsuda, H.; Hasegawa, R.; Imaida, K. *Toxicol. Pathol.* **1982**, *10*, 37–49.
15. Homburger, F. *The Physiopathology of Cancer*; Kargar: Basel, 1974; pp 110–154.
16. Kahl, R. *Toxicology* **1984**, *33*, 185–228.
17. Shamsuddin, A. M.; Elsayed, A. M.; Ullah, A. *Carcinogenesis* **1988**, *9*, 577–580.
18. Shamsuddin, A. M.; Ullah, A.; Chakravarthy, A. K. *Carcinogenesis* **1989**, *10*, 1461–1463.
19. Mukhtar, H.; Das, M.; Khan, W. A.; Wang, Z. Y.; Bik, D. P.; Bickers, D. R. *Cancer Res.* **1988**, *48*, 2361–2365.
20. Athar, M.; Khan, W. A.; Mukhtar, H. *Cancer Res.* **1989**, *49*, 5784–5788.
21. Fujita, Y.; Yamane, T.; Tanaka, M.; Kuwata, K.; Okuzumi, J.; Takahashi, T.; Fujiki, H.; Okuda, T. *Jpn. J. Cancer Res.* **1989**, *80*, 503–505.
22. Wang, Z. Y.; Hong, J.-Y.; Huang, M.-T.; Reuhl, K. R.; Conney, A. H.; Yang, C. S. *Cancer Res.* **1992**, *52*, 1943–1947.
23. Yoshizawa, S.; Horiuchi, T.; Fujiki, H.; Yoshida, T.; Okuda, T.; Sugimura, T. *Phytother. Res.* **1987**, *1*, 44–47.
24. Khan, W. A.; Wang, Z. Y.; Athar, M.; Bickers, D. R.; Mukhtar, H. *Cancer Lett.* **1988**, *42*, 7–12.
25. Huang, M.-T.; Ho, C.-T.; Wang, Z. Y.; Ferraro, T.; Finnegan-Olive, T.; Lou, Y.-R.; Mitchell, J. M.; Laskin, J. D.; Newmark, H.; Yang, C. S.; Conney, A. H. *Carcinogenesis* **1992**, *13*, 947–954.
26. Sparnins, V. L.; Barany, G.; Wattenberg, L. W. *Carcinogenesis* **1988**, *9*, 131–134.

27. Wargovich, M. J.; Woods, C.; Eng, V. W. S.; Stephens, L. C.; Gray, K. *Cancer Res.* **1988**, *48*, 6872–6875.
28. Wargovich, M. J. *Carcinogenesis* **1987**, *8*, 487–489.
29. Takayama, S.; Masuda, M.; Mogami, M.; Ohgaki, H.; Sato, S.; Sugimura, T. *Gann* **1984**, *75*, 207–213.
30. Yamane, T.; Hagiwara, N.; Tateishi, M.; Akachi, S.; Kim, M.; Okuzumi, J.; Kitao, Y.; Inagake, M.; Kuwata, K.; Takahashi, T. *Jpn. J. Cancer Res.* **1991**, *82*, 1336–1339.
31. Oguni, I.; Nasu, K.; Kanaya, S.; Ota, Y.; Yamamoto, S.; Nomura, T. *Jpn. J. Nutr.* **1989**, *47*, 93–102.
32. Kono, S.; Ikeda, M.; Tokudome, S.; Kuratsune, M. *Jpn. J. Cancer Res.* **1988**, *79*, 1067–1074.
33. Kato, I.; Tominaga, S.; Matsuura, A.; Yoshii, Y.; Shirai, M.; Kobayashi, S. *Jpn. J. Cancer Res.* **1990**, *81*, 1101–1108.
34. Ito, N.; Tsuda, H.; Hasegawa, R.; Tatematsu, M.; Imaida, K.; Asamoto, M. In *Biologically-based Methods for Cancer Risk Assessment*; Travis, C.C., Ed.; Plenum Publishing Co.: New York, 1989, 209–230.
35. Ito, N.; Imaida, K.; Tsuda, H.; Shibata, M. A.; Aoki, T.; Camargo, J. L. V.; Fukushima, S. *Jpn. J. Cancer Res.* **1988**, *79*, 413–417.
36. Shibata, M. A.; Fukushima, S.; Takahashi, S.; Hasegawa, R.; Ito, N. *Carcinogenesis* **1990**, *11*, 1027–1031.
37. Fukushima, S.; Hagiwara, A.; Hirose, M.; Yamaguchi, S.; Tiwawech, D.; Ito N. *Jpn. J. Cancer Res.* **1991**, *82*, 642–649.
38. Ito, N.; Tsuda, H.; Tatematsu, M.; Inoue, T.; Tagawa, Y.; Aoki, T.; Uwagawa, S.; Kagawa, M.; Ogiso, T.; Masui, T.; Imaida, K.; Fukushima, S.; Asamoto, M. *Carcinogenesis* **1988**, *9*, 387–394.

RECEIVED May 4, 1993

Chapter 15

Chemistry and Antioxidative Effects of Phenolic Compounds from Licorice, Tea, and Composite and Labiate Herbs

Takuo Okuda, Takashi Yoshida, and Tsutomu Hatano

Faculty of Pharmaceutical Sciences, Okayama University, Tsushima, Okayama 700, Japan

Licorice phenolics and several other types of plant polyphenolics which are comparatively small molecules, often called by names such as tea tannin, caffeetannin or labiataetannin, have been isolated from various plants used for food and medicine. The phenolics isolated from licorice of several different origins, green tea, and Composite and Labiate herbs have shown antioxidant activity in several assay systems — inhibition of leukotriene biosynthesis in human polymorphonuclear neutrophils, xanthine oxidase, and mono amine oxidase, and radical-scavenging activity.

A large number of plants used for health care in the world are rich in phenolics, particularly in polyphenolic compounds. This is also true for some beverages such as tea. Although there often have been erroneous images of the effects of polyphenols because of delay in their isolation and structural elucidation, recent investigations show significant effects on human health (*1*).

One of the activities most widely found in plant polyphenols of various types is antioxidant effects based on radical scavenging activity (*2,3*). This activity may underlie various effects of polyphenols in plant tissues, and also their medicinal effects such as those related to the inhibition of lipid peroxidation (*4*) and tumor promotion (*5*).

The plant polyphenols comprise numerous types of polyhydroxylated compounds, including flavonoids, lignans, coumarins, stilbenes, cinnamic acid derivatives, etc., besides various metabolites of gallic acid and flavans, which are the tannin compounds. Among them, some gallic acid metabolites with antioxidant activity (*6*) and antitumor and antiviral activities (*7*) were presented by the authors at the 1991 ACS symposium "Phenolic Compounds in Food and Their Effects on Health." The present paper describes the antioxidant effects of some other types of polyphenolic compounds that are rather small molecules found in plants used for food and beverage, and also for health care. There are considerable differences between the chemical structures these polyphenols and the other types of polyphenols found in "tannin-containing plants" mainly used for medicines.

0097–6156/94/0547–0133$06.00/0

Phenolic Constituents in Licorice

Licorice, the root or rhizome of *Glycyrrhiza* plants, has long been used as a medicine in human history. The main chemical constituent in licorice is known to be glycyrrhizin (*1*), and this triterpenoid glucuronide has been used to treat stomach ulcer and other conditions. Glycyrrhizin is often regarded as if it is the only active component in licorice, but recent research has shown significant pharmacological effects of other licorice constituents (*8,9*). Notable differences in the phenolic constituents of licorice specimens from different origins have also been shown (*10*). These findings may be correlated with the fact that some licorice specimens brought from certain area of China have been preferably used in the traditional medicine in Japan, in spite of abundant presence of glycyrrhizin in almost all licorice species (*11*).

Differences in Phenolic Structures Due to Difference of Licorice Species. The structures of the main phenolic compounds in various species of licorice, such as *Glycyrrhiza uralensis*, *G. glabra* and *G. inflata* (*10*) are shown in Figure 1. They contain many phenolics which are flavonoids, and some 3-arylcoumarins which are biogenetically related to flavonoids. One of these investigations has shown that the licorice species used as traditional medicine in Japan (*G. uralensis*) contains lico-pyranocoumarin (**2**), glycycoumarin (**3**) and/or licocoumarone (**4**), which are not found in *G. glabra* and *G. inflata*. The other *Glycyrrhiza* species contain several other phenolics in place of **2**, **3** and **4**. They are glabridin (**5**), and glabrene (**6**) in *G. glabra*, and licochalcones A (**7**) and B (**8**) in *G.inflata* (*10*) (Table I).

Antioxidant Effects of Licorice Phenolics. The authors and their collaborators have investigated the effects of licorice phenolics on leukotriene biosynthesis in human polymorphonuclear neutrophils, radical scavenging activity, and inhibition of xanthine oxidase (XOD) (10) and monoamine oxidase (MAO) (*10*).

Leukotriene Biosynthesis in Human Polymorphonuclear Neutrophils. As shown in Figure 2, potent inhibition of 5-lipoxygenase-mediated arachidonic acid metabolism induced by A23187 by licochalcones A (**7**) and B (**8**) at concentrations of 10^{-6} to 10^{-3} M was observed in human polymorphonuclear neutrophils, while inhibition of the products from cyclooxygenase was not so strong. Liquiritigenin (**9**) and isoliquiritigenin (**11**) also inhibited the formation of 5,12-diHETE, although their effects at concentration between 10^{-6} to 10^{-5} M were low (*11*). These inhibitory activities are attributable to the radical scavenging activity of the phenolics, since licochalcone B and several others exhibited the radical-scavenging activity stronger than that of the other phenolics from several licorice species (*12*), as described in the following paragraph.

Radical Scavenging Activity on DPPH Radical. The radical-scavenging activity of licorice phenolics on DPPH (1,1-diphenyl-2-picrylhydrazyl) radical are shown in Table II. Among the activities, those by licochalcone B and several others were markedly stronger than other licorice phenolics (*12*).

Inhibitory Effects on Xanthine Oxidase and Monoamine Oxidase. The inhibitory effects of licorice phenolics on the activity of xanthine oxidase (XOD), which catalyzes the formation of uric acid from xanthine, are shown in Table III

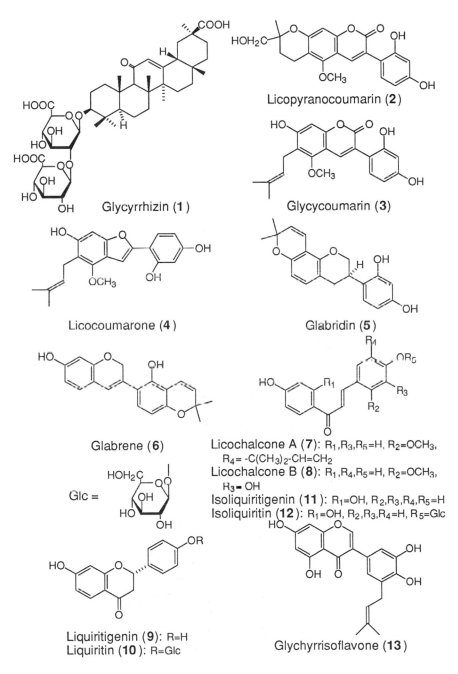

Figure 1. Structures of the main phenolic compounds in licorice.

Table I. Phenolic Content in Licorice from Various Sources

Sample	Extract (%)	10 (%)	12 (%)	9 (%)	11 (%)	2 (%)	3 (%)	4 (%)	5 (%)	6 (%)	7 (%)	8 (%)	Type
A1 Glycyrrhiza uralensis (root)	24.9	1.54	0.23	0.06	0.01	0.06	0.16	0.09	—	—	—	—	A
A2 Glycyrrhiza uralensis (root)	33.7	2.40	0.34	0.03	—	0.05	0.06	0.03	—	—	—	—	A
A3 Glycyrrhiza uralensis (root)	33.6	2.74	0.40	0.07	0.01	Trace	0.02	—	—	—	—	—	A
A4 Glycyrrhiza uralensis (rhizome)	27.5	0.79	0.08	—	—	—	0.06	—	—	—	—	—	A
A5 Glycyrrhiza uralensis (root)	13.2	0.47	0.06	0.07	0.01	—	0.04	0.02	—	—	—	—	A
A6 Glycyrrhiza uralensis (rhizome)	22.7	0.54	0.08	0.08	0.02	—	0.04	0.04	—	—	—	—	A
B1 Glycyrrhiza glabra (root)	22.6	0.23	0.14	—	0.01	—	—	—	0.15	0.01	—	—	B
B2 Glycyrrhiza glabra (rhizome)	16.2	0.01	0.04	—	0.01	—	—	—	0.57	0.06	—	—	B
C Glycyrrhiza inflata (root)	24.2	0.03	0.10	—	0.04	—	—	—	—	—	1.68	0.35	C

—: not detected

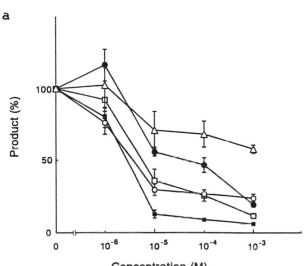

Licochalcone A

Licochalcone B

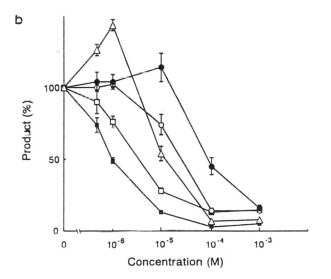

Figure 2. Effects of licochalcones A and B on A23187-induced [1-^{14}C]arachidonic acid metabolism in human polymorphonuclear neutrophils. Values are the means ± standard errors for three experiments.

Table II. Scavenging Effects of Licorice Phenolics on 1,1-Diphenyl-2-picrylhydrazyl Radical

Compound	EC_{50} (x 10^{-5} M)
Licochalcone B (8)	2.2
Glycyrrhisoflavone (13)	3.8
Isolicoflavonol	4.0
Glycycoumarin (3)	4.1
Licochalcone A (7)	12
Glycyrrhisoflavanone	36
Isoliquiritigenin (11)	96
Liquiritigenin (9)	>100
Liquiritin (10)	>100
Isoliquiritin (12)	>100
4',7-Dihydroxyflavone	>100

Table III. Inhibitory Effects of Licorice Phenolics on Xanthine Oxidase

Compound	IC_{50} (x 10^{-5} M)
Licocoumarone (4)	1.3
Licochalcone B (8)	3.0
Glycyrrhisoflavone (13)	5.3
Licochalcone A (7)	5.6
Licopyranocoumarin (2)	>10
Glycycoumarin (3)	>10
Glisoflavone	>10
Glycyrrhisoflavanone	>10
Kaempferol-3-O-methyl ether	>10
Licoarylcoumarin	>10
Allopurinol	0.016

(*8*). Among the tested compounds, licocoumarone (**4**), glycyrrhisolfavone (**13**), and licochalcones A (**7**) and B (**8**) showed stronger inhibitory activities against XOD than other licorice phenolics. It is noticeable that licocoumarone (**4**), which was the strongest inhibitor of XOD, is also the most potent inhibitor of monoamine oxidase (MAO), among the tested compounds as shown in Table IV. Licochalcones A (**7**) and B (**8**), which are fairly strong inhibitors of XOD, however, are weak or negligible inhibitors of MAO (*9*). The kinetic analysis using Lineweaver-Burk plots revealed that the inhibition of licocoumarone and the other licorice polyphenolics on both the XOD and MAO activities is noncompetitive (*9*).

Antioxidant Effects of Green Tea Polyphenols

Green tea is a nonfermented tea used as the main beverage in Japan and China. It is prepared by steaming or baking fresh tea (*Camellia sinensis*) leaves, to prevent the fermentation which causes chemical transformation of constituents, mainly condensation of (-)-epigallocatechin gallate (EGCG), the major tea polyphenol, and accompanying polyphenols, (-)-epicatechin gallate (ECG), (-)-epigallocatechin (EGC), (-)-epicatechin (EC), among others (Figure 3). The antitumor promoting activity of EGCG, exhibited in several experimental systems (*5,13,14*) is accompanied by inhibition of protein kinase C, implying participation of a suppressing effect on active oxygen in the inhibition of tumor promotion (*14*). EGCG and ECG possess noticeable inhibitory effects on the lipid peroxidation induced in rat liver mitochondria by ADP and ascorbic acid, and that induced in rat liver microsomes by ADP and NADPH (*15*). Antioxidant activities of these polyphenols on autoxidation of methyl linoleate were also observed, although the potencies of the inhibition by these compounds, calculated by molar basis, were somewhat lower than those by several other types of polyphenols isolated from some medicinal plants (*16*). An inhibitory effect by EGCG was also observed on oxidative damage of mouse ocular lens induced by the xanthine-xanthine oxidase system in an oxidative damage model, although it was somewhat lower than that of several hydrolyzable tannins (geraniin and pentagalloylglucose) (*17*) (Table V).

Suppression of Superoxide Anion Radical by Plant Polyphenols. The antioxidative activity of the plant polyphenols, including EGCG, have been found to have positive correlation with the potencies of each polyphenol to form a stable polyphenol radical after scavenging active oxygen species, lipid peroxy radical, etc. ESR measurement in the presence of DMPO showed that the generation of superoxide anion radical in the hypoxanthine-XOD system was remarkably suppressed by the plant polyphenols, the potency of which is positively correlated with antioxidant activity. When the concentration of each polyphenol was high enough to scavenge the superoxide radical, the signals assignable to a DMPO adduct of the C-centered radical from the polyphenol were recognized in the ESR spectrum (*3*).

Antioxidant Effects of Polyphenols from Composite and Labiate Plants

A large group of polyphenolic compounds containing caffeic acid, which are sometimes called caffeetannins and labiataetannins (Figure 3), are found in various plants used for food, beverage, herbs and spices.

Table IV. Inhibitory Effects of Licorice Phenolics on Monoamine Oxidase

Compound	IC_{50} (x 10^{-5} M)
Licocoumarone (**4**)	6.0
Licofuranone	8.7
Glycyrrhisoflavone (**13**)	9.5
Genistein	9.5
Licopyranocoumarin (**2**)	14
Glicoricone	14
Glycycoumarin (**3**)	>20
Licochalcone A (**7**)	>20
Licochalcone B (**8**)	>20
Isoliquiritigenin (**10**)	>20
(-)-Medicarpin	>20
Glycyrrhizin (**1**)	16
Harmane hydrochloride	5.7
Quinine sulfate	6.4

Table V. Effects of Tannins on Lipid Peroxidation Level in Intact Lens

Polyphenol (10^{-4} M)	MDA (nmol/g wet wt.)	Elimination of lipid peroxide (%)
Control (untreated lens)	1.31 ± 0.12	—
Without polyphenol[a]	3.10 ± 0.14	—
Catechin	3.02 ± 0.33	4
Gallic acid	3.01 ± 0.19	5
Chlorogenic acid (**18**)	2.50 ± 0.20	34
Ellagic acid	2.43 ± 0.40	37
Epigallocatechin gallate (**14**)	2.24 ± 0.19	48
Geraniin	1.85 ± 0.26	69
Penta-O-galloyl-β-D-glucose	1.50 ± 0.16	89

Each value represents means (± S.D.) of five samples.
[a] Post-incubated without polyphenol after treatment with the xanthine-xanthine oxidase system.

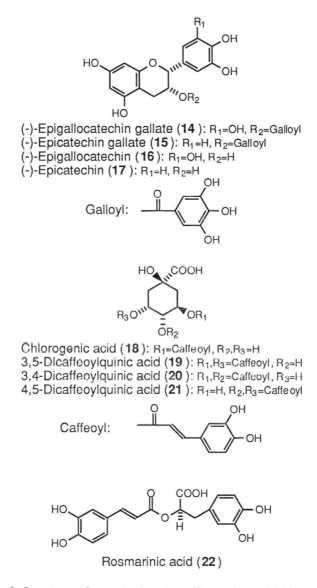

(-)-Epigallocatechin gallate (**14**): R₁=OH, R₂=Galloyl
(-)-Epicatechin gallate (**15**): R₁=H, R₂=Galloyl
(-)-Epigallocatechin (**16**): R₁=OH, R₂=H
(-)-Epicatechin (**17**): R₁=H, R₂=H

Galloyl:

Chlorogenic acid (**18**): R₁=Caffeoyl, R₂,R₃=H
3,5-Dicaffeoylquinic acid (**19**): R₁,R₃=Caffeoyl, R₂=H
3,4-Dicaffeoylquinic acid (**20**): R₁,R₂=Caffeoyl, R₃=H
4,5-Dicaffeoylquinic acid (**21**): R₁=H, R₂,R₃=Caffeoyl

Caffeoyl:

Rosmarinic acid (**22**)

Figure 3. Structures of tea polyphenols, caffeetannins and labiataetannins.

Caffeetannins. The main constituent of caffeetannins contained in coffee is chlorogenic acid (3-caffeoylquinic acid) (**18**), which has low binding affinity for protein (*18*). The main polyphenols in several *Artemisia* species of plants, which are popular as medicines and foods in Japan and China, are dicaffeoylquinic acids [3,5-dicaffeoylquinic acid (**19**), 3,4-dicaffeoylquinic acid (**20**) and 4,5-dicaffeoylquinic acid (**21**)] which have stronger binding activities, and are accompanied by a small amount of chlorogenic acid in the plant (*19*).

Labiataetannins and Analogs. The name labiataetannin was originally given to rosmarinic acid (**22**) which was isolated from rosemary, a plant used as medicine and a spice mainly in Europe (*20*). It is a condensate of caffeic acid with 3-(3,4-dihydroxyphenyl)lactic acid, and was found to have antioxidant activity (*21,22*). Several compounds related to rosmarinic acid (Figure 3), which also may be called labiataetannin, have been isolated recently from Labiate medicinal plants (*23,24*).

Antioxidant Activities of Caffeetannins and Labiataetannins. Noticeable inhibition by 3,5-dicaffeoylquinic acid on lipoxygenase-dependent peroxidation of linoleic acid was observed, while the effects of chlorogenic acid and caffeic acid were lower (*25*). 3,5-, 4,5- and 3,4-Dicaffeoylquinic acid, caffeoylmalic acid, caffeoyltartaric acid, rosmarinic acid and caffeic acid also inhibited the formation of leukotriene B_4 (LTB_4) to varying extent. Rosmarinic acid potently inhibited the 5-lipoxygenase products 5-HETE and LTB_4. The formation of prostaglandin E_2, however, was enhanced by several caffeoyl esters (*26*), as found in several other caffeoyl esters. These observations show that caffeetannins and labiataetannins, like several polyphenols of other types (*27*), can specifically inhibit the activity of lipoxygenase in arachidonic acid metabolism.

Acknowledgments

The authors thank Prof. Y. Fujita and Prof. A. Mori (Okayama University), the late Prof. S. Arichi and Dr. Y. Kimura (Kinki University), the late Prof. S. Iwata (Meijyo University), Prof. T. Noro (University of Shizuoka), Dr. H. Fujiki (National Cancer Center Research Institute), and their coworkers for their collaboration.

Literature Cited

1. Okuda, T.; Yoshida, T.; Hatano, T. In *Economic and Medicinal Plant Research*; Wagner, H.; Farnsworth, M. R., Eds.; Academic Press: London, 1991; Vol. 5, pp 129–165.
2. Okuda, T.; Yoshida, T.; Hatano, T.; Fujita, Y. In *Free Radical and Sino-Japanese Medicine*; Okuda, T.; Yoshikawa, T., Eds.; Kokusai-ishoshuppan: Tokyo, 1990; pp 42–70.
3. Hatano, T.; Edamatsu, R.; Hiramatsu, M.; Mori, A.; Fujita, Y.; Yasuhara, T.; Yoshida, T.; Okuda, T. *Chem. Pharm. Bull.* **1989**, *37*, 2016.
4. Kimura, Y.; Okuda, H.; Okuda, T.; Hatano, T.; Agata, I.; Arichi, S. *Planta Medica* **1984**, 473.
5. Fujita, Y.; Yamane, T.; Tanaka, M.; Takahashi, T.; Fujiki, H.; Okuda, T. Jpn. J. *Cancer Res.* **1989**, *80*, 503.

6. Okuda, T.; Yoshida, T.; Hatano, T. In *Phenolic Compounds in Foods and Their Effects on Health II: Antioxidants & Cancer Prevention*; Huang, M-T.; Lee, C-Y.; and Ho, C-T., ACS Symposium Series No. 507; American Chemical Society: Washington D.C., 1992; pp 87–97.
7. Okuda, T.; Yoshida, T.; Hatano, T. In *Phenolic Compounds in Foods and Their Effects on Health II: Antioxidants & Cancer Prevention*; ACS Symposium Series No. 507; American Chemical Society: Washington, D.C., 1992; pp 160–183.
8. Hatano, T.; Yasuhara, T.; Fukuda, T.; Noro, T.; Okuda, T. *Chem. Pharm. Bull.* **1989**, *37*, 3005.
9. Hatano, T.; Fukuda, T.; Miyase, T.; Noro, T.; Okuda, T. *Chem. Pharm. Bull.* **1991**, *39*, 1238.
10. Hatano, T.; Fukuda, T.; Liu, Y. -Z.; Noro, T.; Okuda, T. *Yakugaku Zasshi* **1991**, *111*, 311.
11. Kimura, Y.; Okuda, H.; Okuda, T.; Arichi, S. *Phytotherapy Res.* **1988**, *2*, 140.
12. Hatano, T.; Kagawa, H.; Yasuhara, T.; Okuda, T. *Chem. Pharm. Bull.* **1988**, *36*, 2090.
13. Yoshizawa, S.; Horiuchi, T.; Suganuma, M.; Nishiwaki, S.; Yatsunami, J.; Okabe, S.; Okuda, T.; Muto, Y.; Frenkel, K.; Troll, W.; Fujiki, H. In *Phenolic Compounds in Foods and Their Effects on Health II: Antioxidants & Cancer Prevention*; Huang, M-T.; Lee, C-Y.; and Ho, C-T., ACS Symposium Series No. 507; American Chemical Society: Washington, D.C., 1992; pp 316–325.
14. Yoshizawa, S.; Horiuchi, T.; Fujiki, H.; Yoshida, T.; Okuda, T.; Sugimura, T. *Phytotherapy Res.* **1987**, *1*, 44.
15. Okuda, T.; Kimura, Y.; Yoshida, T.; Hatano, T.; Okuda, H.; Arichi, S. *Chem. Pharm. Bull.* **1983**, *31*, 1625.
16. Fujita, Y.; Komagoe, K.; Niwa, Y.; Uehara, I.; Hara, R.; Mori, H.; Okuda, T.; Yoshida, T. *Yakugaku Zasshi* **1988**, *108*, 528.
17. Iwata, S.; Fukaya, Y.; Nakazawa, K.; Okuda, T. J. *Ocular Pharmacol.* **1987**, *3*, 227.
18. Okuda, T.; Mori, K.; Hatano, T. *Chem. Pharm. Bull.* **1985**, *33*, 1424.
19. Okuda, T.; Hatano, T.; Agata, I.; Nishibe, S.; Kimura, K. *Yakugaku Zasshi* **1986**, *106*, 894.
20. Hegnauer, R. *Chemotaxonomie der Pflanzen*; Bd. 4, Birkhäuser Verlag: Basel, 1966; pp 327-328.
21. Gracza, L.; Koch, H.; Loffler, E. *Arch. Pharm. (Weinheim)* **1985**, *318*, 1090.
22. Fujita, Y.; Uehara, I.; Morimoto, Y.; Nakajima, M.; Hatano, T.; Okuda, T. *Yakugaku Zasshi* **1988**, *108*, 129.
23. Okuda, T.; Hatano, T.; Agata, I.; Nishibe, S. *Yakugakub Zasshi* **1986**, *106*, 1108.
24. Agata, I.; Hatano, T.; Nishibe, S.; Okuda, T. *Phytochemistry* **1989**, *28*, 2447.
25. Fujita, Y.; Uehara, I.; Morimoto, Y.; Nakashima, M.; Hatano, T.; Okuda, T. *Yakugaku Zasshi* **1988**, *108*, 129.
26. Kimura, Y.; Okuda, H.; Okuda, T.; Hatano, T.; Arichi, S. *J. Nat. Prod.* **1987**, *50*, 392.
27. Kimura, Y.; Okuda, H.; Okuda, T.; Arichi, S. *Planta Medica* **1986**, 337.

RECEIVED October 4, 1993

Chapter 16

Chemistry of Antioxidants from Labiatae Herbs

Nobuji Nakatani

Department of Food and Nutrition, Osaka City University, Sumiyoshi, Osaka 558, Japan

Antioxidants have been widely used to delay or prevent oxidation of fats and oils in a variety of foods. In the process of lipid oxidation, many radical species, such as lipid alkoxyl radical, are generated and cause not only food deterioration but DNA, cell and tissue damage. Great interest in this process has prompted us to search for anti-oxidants from natural sources and to elucidate their chemistry for further research concerning chemoprevention of inflammation, tumors and aging. In our search for effective antioxidants we focused on edible plants, especially herbs and spices.

Several studies on the antioxidants of spices had appeared before Chipault *et al.* examined systematically more than 70 spices for antioxidative activity (*1*). They reported that rosemary and sage were remarkably effective spices. As shown in Table I, we also found that spices belonging to the Family Labiatae exhibited the highest activity in a wide range of polarity. In addition, Myristicaceae and Myrtaceae showed activity in each fraction. Zingiberaceae showed the strongest activity in the slightly polar fraction. Cumin and fennel, Umbelliferae family members, contain effective substances in their polar fractions.

Less Polar Antioxidants from Labiatae

We chose to search rosemary (*Rosmarinus officinalis* L.) first for antioxidative compounds (*2,3*). Active compounds were isolated by a combination of chromatographies from the weakly acidic fraction of *n*-hexane extract. The HPLC chromatogram is shown in Figure 1. Purification of peak 3 afforded carnosol (**1**, Figure 2). The most effective compound isolated from Peak 1 was named rosmanol (**2**), the structure of which was determined by IR, ^{1}H- and ^{13}C-NMR analyses. H-H COSY and COLOC spectra (Figure 3) precisely confirmed the H-H, C-H correlation. NOE and long range coupling was measured. Two isomers of rosmanol were isolated and determined to be epirosmanol (**3**) and isorosmanol (**4**). Structure correlation between **2** and **3** was confirmed by chemical transformation as shown in Figure 4.

0097–6156/94/0547–0144$06.00/0

Table I. Antioxidant Activities of Spices

Family	Spice	Plant part	CH_2Cl_2 extract	Methanol extract	
				EtOAc sol.	H_2O sol.
Labiatae	Basil	Leaves	+++	+++	+++
	Marjoram	Leaves	+++	+++	+++
	Oregano	Leaves	+++	+++	+++
	Perilla	Leaves	+++	+++	+++
	Rosemary	Leaves	+++	+++	+++
	Sage	Leaves	+++	+++	+++
	Thyme	Leaves	+++	+++	+++
Zingiberaceae	Cardamom	Fruits	−	+++	−
	Dried ginger	Rhizomes	+++	+++	−
	Fresh ginger	Rhizomes	++	+++	+++
	Turmeric	Rhizomes	−	+++	−
Myristicaceae	Mace	Aril	+++	+++	+++
	Nutmeg	Seeds	+	+++	+++
Lauraceae	Bay leaf	Leaves	+	+++	−
	Cinnamon	Bark	−	+	+++
Myrtaceae	Allspice	Fruits	+++	+++	+++
	Cloves	Floral parts	++	+++	+++
Umbelliferae	Caraway	Seeds	−	−	−
	Coriander	Seeds	−	−	−
	Cumin	Seeds	−	++	+++
	Fennel	Seeds	−	++	+++
Magnoliaceae	Star–anise	Fruits	−	−	++

+++: remarkably effective
 ++: moderately effective
  +: slightly effective
  −: ineffective

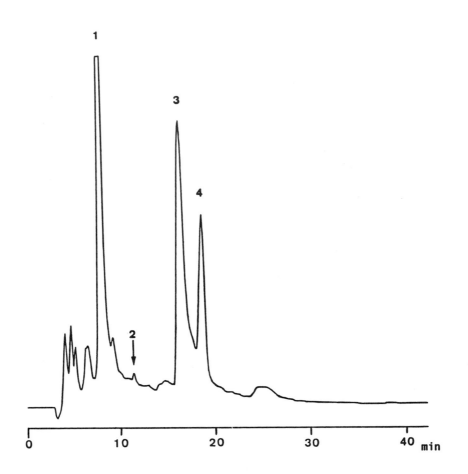

Figure 1. HPLC chromatogram of the weakly acidic fraction of *n*-hexane extract from rosemary (*Rosmarinus officinalis* L.)
Column: Develosil-ODS (φ4.6 x 250 mm); Solvent: MeOH: H_2O (80:20); Flow rate: 1 ml/min; monitored at 285 nm.
   Peak 1. isorosmanol+rosmanol; Peak 2. epirosmanol; Peak 3. carnosol; Peak 4. rosmadial.

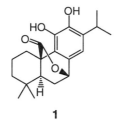
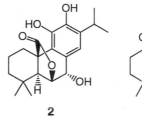
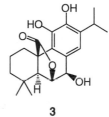

1

2

3

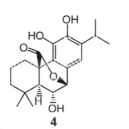
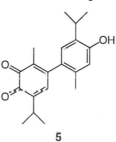

4

5

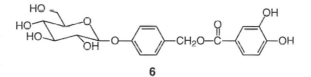

6

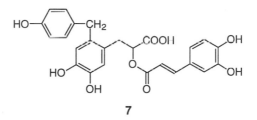

7

Figure 2. Antioxidants isolated from Labiatae.

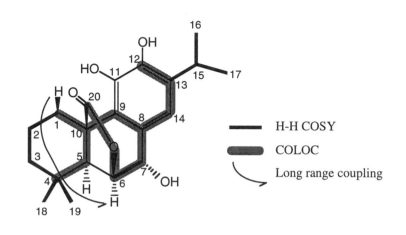

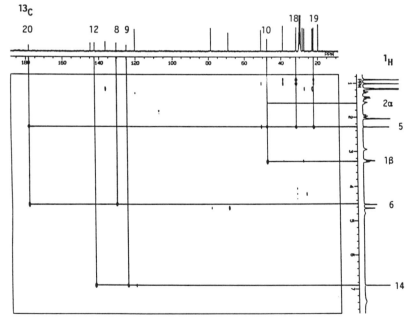

Figure 3. COLOC spectrum of rosmanol.

The four diterpene antioxidants (**1–4**) were also isolated from sage (*Salvia officinalis* L.). Antioxidant efficacy was measured by the ferric thiocyanate method to reveal much higher activity of all compounds than α-tocopherol, BHA and BHT (Figure 5). Several flavonoids were isolated from sage in addition to the diterpenoids. One of them, a new glucuronoside (**F-5**, Figure 6) was determined based on spectroscopic methods.

Four methylated flavonoids (**F-1–F-4**) were isolated from the antioxidant active fractions from thyme (*Thymus vulgaris* L.) (Figure 6) (*4*). **F-4** was also isolated from rosemary and sage. Because the solubility of flavonoids differs in various media, different antioxidant activities have been reported. Five new biphenyls with significant activity were isolated (*5–7*) and compound **5** was synthesized by Miura *et al.*

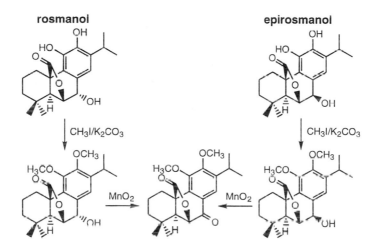

Figure 4. Structure correlation between rosmanol and epirosmanol.

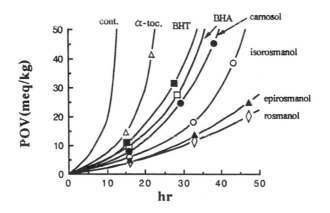

Figure 5. Antioxidant activity of diterpenoids isolated from rosemary and sage (AOM; α-tocopherol, BHA, BHT, 0.02%; diterpenoids, 0.01%).

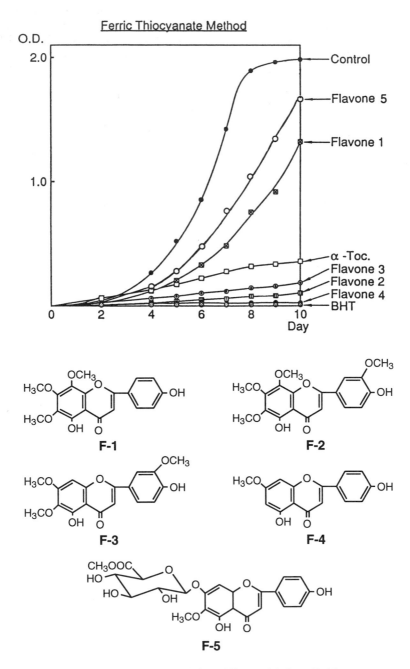

Figure 6. Antioxidant activity of flavonoids from Labiatae.

Polar Antioxidants from Labiatae

In practical use, water soluble antioxidants are strongly required. Oregano (*Origanum vulgare* L.) showed high antioxidative properties in the polar fraction. Purification with polyamide chromatography afforded five phenol carboxylic acid derivatives (*8,9*). A new glucoside (**6**), which exhibited remarkable activity, was determined by chemical degradation (Figure 7) and synthesis. Rosmarinic acid and a related new compound (**7**) were obtained accompanied by protocatequic acid and caffeic acid (*9*).

Marjoram (*Origanum majorana* L.), a herb close to oregano in botanical classification, contains phenol carboxylic acids in the MeOH extract. As shown in Figure 8, rosmarinic acid and acylated arbutin are major active products in polar fraction.

Conclusion

Herbs and spices are one of the most important targets to search for natural antioxidants from the point of view of safety. It is expected that natural antioxidants which are investigated for basic and applied experiments will lead to chemoprevention of inflammation, cancer and aging.

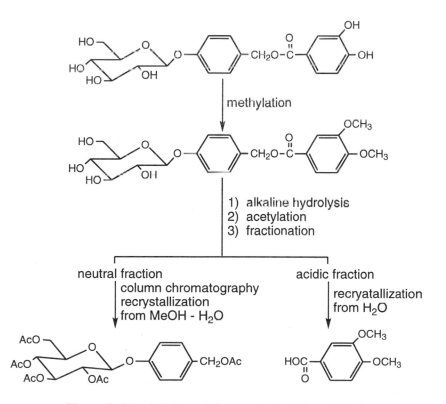

Figure 7. Chemical degradation products of glucoside (**6**).

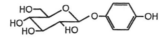

A : arbutin

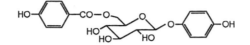

B : 6-o-p-hydroxybenzoylarbutin

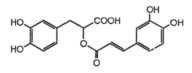

C : rosmarinic acid

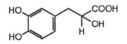

D : 2-hydroxy-3-(3,4,-dihydroxyphenyl)-
propionic acid

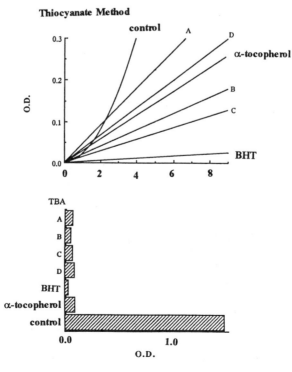

Figure 8. Antioxidant activity of the phenolic constituents of marjoram.

Literature Cited

1. Chipault, J. R.; Hawkins, J. M.; Lundberg, W. O. *Food Res.* **1952**, *17*, 46.
2. Inatani, R.; Nakatani, N.; Fuwa, H.; Seto, H. *Agric. Biol. Chem.* **1982**, *46*, 1661.
3. Nakatani, N.; Inatani, R. *Agric. Biol. Chem.* **1984**, *48*, 2081.
4. Miura, K.; Nakatani, N. *Agric. Biol. Chem.* **1989**, *53*, 3043.
5. Miura, K.; Nakatani, N.; *Chem. Express* **1989**, *2*, 237.
6. Miura, K.; Inagaki, T.; Nakatani, N. *Chem. Pharm. Bull.* **1989**, *37*, 1816.
7. Nakatani, N.; Miura, K.; Inagaki, T. *Agric. Biol. Chem.* **1989**, *53*, 1375.
8. Nakatani, N.; Kikuzaki, H. *Agric. Biol. Chem.* **1987**, *51*, 2727.
9. Kikuzaki, H.; Nakatani, N. *Agric. Biol. Chem.* **1989**, *53*, 519.

RECEIVED October 4, 1993

Chapter 17

Flavonoid with Strong Antioxidative Activity Isolated from Young Green Barley Leaves

Takayuki Shibamoto[1], Yoshihide Hagiwara[1], Hideaki Hagiwara[2], and Toshihiko Osawa[3]

[1]Department of Environmental Toxicology, University of California, Davis, CA 95616
[2]Hagiwara Institute of Health, 1173 Maruyama, Asazuma-cho, Kasai 679—01, Japan
[3]Department of Food Science and Technology, Faculty of Agriculture, Nagoya University, Chikusa, Nagoya 464—01, Japan

Ethanol extract from young green barley leaves has been known to possess potent pharmacological activies, including antioxidative activity, anti-inflammatory effects, and antiallergic acitivity. Antioxidative activity testing on column chromatographic fractions from 80% ethanol extract of green barley leaves using the thiobarbituric acid method showed that a 60% methanol eluate fraction inhibited lipid peroxidation by over 90%, which was almost equivalent to the activity of α-tocopherol. A flavonoid, 2"-O-glycosyl isovitexin (2-O-GIV) was isolated and identified as the major component of a 60% methanol eluate fraction. α-Tocopherol inhibited formation of malonaldehyde (MA) from arachidonic acid oxidized with Fenton's reagent by approximately 95%, whereas 2-O-GIV inhibited MA formation by over 99%. In addition, 2-O-GIV inhibited the formation of 4-hydroxynonenal approximately 80% in contrast to α-tocopherol, which did not inhibit 4-hydroxynonenal formation at all.

The role of polyunsaturated fatty acids in biological systems has been extensively studied to understand the damage to cells caused by oxidation. Lipid peroxidation initiated by free radical reactions is believed to damage cells by the loss of polyunsaturated fatty acids. In addition, the formation of lipid peroxides and their secondary products such as reactive carbonyl compounds may damage cellular constituents, including enzymes. Figure 1 shows the summary of the role of lipid peroxidation in biological systems.

Administration of lipid peroxides to rats causes several types of biological damage, such as oxidative hemolysis of red blood cells and a decrease in the glutathione peroxidase activity in liver (1). Rats fed oxidized oils exhibited necrosis due to the toxic effect of the secondary oxidation products, such as low molecular weight carbonyl compounds (2).

The intake of oxidized lipids can cause alterations in the fluidity of plasma membranes (3). Alteration of membrane lipids caused by lipid peroxidation may result in loss of membrane function and damage to membrane bound enzymes (4).

0097—6156/94/0547—0154$06.00/0

Consumption of oxidized fat by experimental animals was accompanied by weight loss, alteration in the size of organs, and changes in triglyceride levels (5). Lipid peroxidation *in vivo* may be initiated by the intake of oxidized fats and subsequently cause damage.

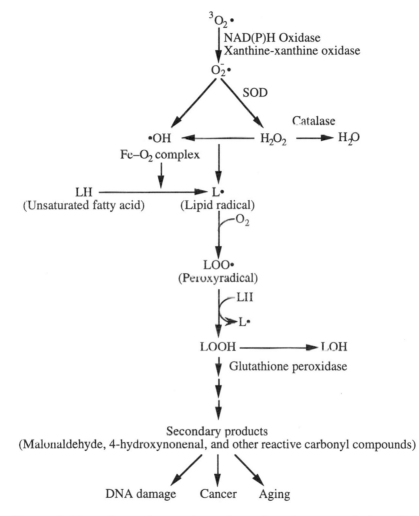

Figure 1. Formation pathway of reactive carbonyl compounds from lipid peroxidation.

Proteins and amino acids exposed to lipid peroxidation undergo cross-linking and polymerization resulting in an increase in molecular weight (6,7). The most sensitive amino acids are methionine, lysine, tryptophan, and histidine (7).

Lipid peroxidation *in vivo* would disrupt of the normal precise arrangement of proteins and enzymes in membrane systems and subsequent loss or impairment of their biological activities.

Among the toxic secondary products of lipid peroxidation, malonaldehyde (MA) is one of the most studied products because of its extreme reactivity. DNA reacts with MA to form fluorescent products (8). The fluorescent products are attributed to cross-linking of amino groups of the bases of DNA with MA which form conjugated amino-iminopropenes. The formation of fluorescent products has been reported to correlate with the loss of DNA template activity (8). In addition, the fluorescent compounds have the same spectral characteristics as the lipofuscin pigment that accumulates in animal tissues as a function of age (9).

Various antioxidants have been most commonly used to prevent lipid peroxidation. Recently, some antioxidants occurring naturally in plants have begun to receive much attention because people and animals regularly consume them.

In the present study, young green barley was chosen because there are many reports on biological activities associated with its extract, including anti-inflammatory effects (10), antiulcer activity (11), and effects on hypercholesterolemia (12).

Materials and Methods

Cultivation of Green Barley Leaves (*Hordium vulgare L. var. nudum Hook*). A polyethylene grid (1 m x 1 m x 1 cm) was placed on the surface of a nutrient solution supported by the rim of a container. Cheesecloth was stretched over the grid to provide enough moisture for growth. The cheesecloth was covered with a black plastic sheet with slits approximately 1 cm wide to prevent the growth of undesirable algae in the nutrient solution. The barley seeds were planted between the slits of the black plastic sheet about 5 mm apart.

An aerator constructed from a 60 cm glass tube and a Tygon tube was connected to the aeration manifold to control air flow. The conductivity and pH were kept above 0.25 mmol/cm and 4.5, respectively, and used as indicators of the availability of nutrients. These parameters were measured every other day. If one of the readings dropped below the optimum ranges, the solution was replaced. The seeds germinated two days after seeding.

Sample Preparations. The barley leaves were harvested two weeks after seeding when they grew to approximately 20 cm high. The stems were cut at the base and the leaves were freeze-dried for three days. The freeze-dried leaves were ground and the leaf powder was kept at -80°C until used.

The scheme of experimental procedures is shown in Figure 2. Freeze-dried barley leaf powder (20 g) was extracted twice with 500 ml *n*-hexane, followed by filtration and evaporation. The residue of the *n*-hexane extract was extracted twice with 500 ml 80% ethanol followed by filtration and evaporation. The residue of the ethanol extract was fractionated by a 40 cm x 4.5 cm i.d. glass column packed with an Amberlite XAD-2 nonionic polymeric adsorbent. The column was eluted stepwise with 1 l each of distilled water, 20% MeOH, 40% MeOH, 60% MeOH, 80% MeOH, and 100% MeOH, and 100% acetone at the rate of approximately 15 ml/min. The solvent from each eluate was evaporated and the residue was weighed.

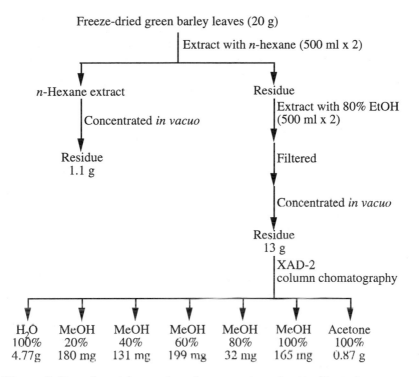

Freeze-dried green barley leaves (20 g)

Extract with *n*-hexane (500 ml x 2)

n-Hexane extract

Concentrated *in vacuo*

Residue
1.1 g

Residue

Extract with 80% EtOH
(500 ml x 2)

Filtered

Concentrated *in vacuo*

Residue
13 g

XAD-2
column chomatography

H₂O	MeOH	MeOH	MeOH	MeOH	MeOH	Acetone
100%	20%	40%	60%	80%	100%	100%
4.77g	180 mg	131 mg	199 mg	32 mg	165 mg	0.87 g

Figure 2. Experimental procedure for separation of antioxidants from green barley leaves.

Antioxidative Activity Testing. Antioxidative activity of the samples was measured using the thiobarbituric acid (TBA) assay according to the method of Ohkawa *et al.* (*13*). A lipid peroxidation system containing test sample was induced by Fenton's reagent as described by Tamura *et al.* (*14*). Each test sample (0.5 mg) was added to 5 ml 0.25 mmol Trizma HCl/0.75 mmol potassium chloride buffer (pH = 7.4) containing 10 μl ethyl linoleate, 0.2% sodium dodecyl sulfate (SDS), 1 μmol ferrous chloride, and 0.5 μmol hydrogen peroxide. The known natural antioxidant, α-tocopherol, was used as a comparative standard to evaluate the antioxidative activity of each fraction. The mixture was constantly stirred at 37°C for 16 h. Each reaction mixture (0.2 ml) was transferred into a separate test tube, followed by addition of 50 μl 4% butylated hydroxytoluene (BHT) in ethanol solution to prevent further oxidation.

Standard stock solution of MA was prepared by hydrolyzing 220 mg malonaldehyde bis(diethyl acetal) in 100 mL aqueous 1% H₂SO₄ and stirring for 2 h at room temperature. The stock solution was diluted with distilled water in various concentrations (10, 25, 50, 75, and 100 μM) in separate volumetric flasks. The assay procedures were also applied to the MA standard solutions. A calibration curve for TBA assay was prepared by plotting concentrations of standard solutions versus absorbance at 532 nm. The calibration curve was linear in the range of the standard solutions.

Isolation and Identification of the Active Compound. The most active fraction (60% MeOH eluate) was precipitated by adding cold methanol. The yellow precipitate was filtered and then washed with 1 ml cold methanol twice. The precipitate was further purified by HPLC equipped with a 25 cm x 4.6 mm i.d. column packed with Develosil-ODS 10 (Nomura Chemical Co., Ltd.). The elution solvent was methanol with a solvent flow rate of 1 ml/min. Over 99.9% pure compound was obtained.

Further Studies on Antioxidative Activities. An aqueous solution (5 ml) containing arachidonic acid (1.5 mg/ml), 0.25 mmol Trizma buffer (pH 7.4), 0.75 mmol potassium chloride, SDS (0.2%), 0.2 mmol ferrous chloride, and 0.1 mmol hydrogen peroxide was stirred with 100 µg/mL 2-O-GIV or 100 µg/mL α-tocopherol at 37°C for 16 h.

N-Methylhydrazine (NMH, 40 µl) was added to the oxidized solutions of arachidonic acid and then stirred for 1 h at 25°C. The reaction solution was extracted with 8 ml dichloromethane using a liquid-liquid continuous extractor for 3 h. The solution was saturated with NaCl prior to extraction to prevent emulsion formation. The volume of the extract was adjusted to 10 mL by adding dichloromethane. After 100 µl 2-methylpyrazine was added as a GC internal standard, the solution was analyzed by GC with a NPD.

A Hewlett-Packard (HP) Model 5880 gas chromatograph equipped with a nitrogen-phosphorous detector and a 30 m x 0.25 mm i.d. DBWAX bonded-phase fused silica capillary column (J & W Scientific, Folsom, CA) was used for routine analysis. GC peak areas were integrated using an HP-3390A integrator. The oven temperature was held at 35°C for 1 min, programmed to 190°C at 4°C/min, and held until all components were eluted. The helium carrier gas flow was 30 cm/sec. The injector and detector temperatures were 270°C and 300°C, respectively.

Results and Discussion

Figure 3 shows the results of antioxidative activity on the samples. The results obtained from antioxidative activity testing with TBA showed that a 60% methanol eluate inhibited lipid peroxidation by over 90%, which was almost equivalent to the activity of α-tocopherol. The fractions obtained near 60% methanol eluate exhibited moderate antioxidative activities. For example, the fraction with 80% methanol eluate suppressed approximately 70% of oxidation. On the other hand, the hexane eluate and water eluate had prooxidative activities.

Identification of 2-O-GIV in 60% Methanol Eluate. The fraction obtained from 60% methanol eluate exhibited an HPLC major peak at retention time 2.79 min with a purity of 96%. The compound collected from this major peak with a preparative column was analyzed by spectroscopic means.

The IR spectrum of the compound showed a strong broad band at 3422 cm^{-1} due to the OH stretching mode of the alcohol or the phenol. Bands at 3000 and 1660 cm^{-1} suggested the presence of an aromatic ring. A strong band at 1620 cm^{-1} is due to the C=O stretching of a carbonyl group. A strong broad band at 1100 cm^{-1} is indicative of a C-O bond.

The FAB-MS of the compound gave an (M^{+1}) ion peak at 595, suggesting a molecular ion peak (M^+) of 594.

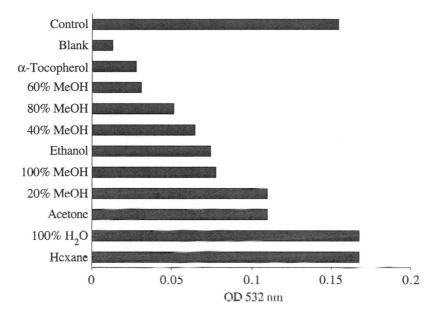

Figure 3. Results of TBA test on the extracts from green barley leaves.

The ^{1}H-NMR spectrum of the compound, dissolved in DMSO-d_6 indicated the aromatic protons of the B ring at δ 6.92 (H-3' and H-5' d, J = 10 Hz) and δ 7.97 (H-2' and H-6' d, J = 10 Hz). The aromatic protons of the A ring appeared at δ 6.88 (H-8 s) and δ 6.96 (H-3, s).

The ^{13}C-NMR of the compound, in DMSO-d_6, showed 27 signals of carbon atoms. A molecular formula of $C_{27}H_{30}O_{15}$ was suggested based on FAB-MS and ^{13}C-NMR spectral data. The chemical shifts of the compound have strong analogy with those of a known flavonoid, isovitexin (*15*). This suggested that the compound was a derivative of isovitexin. The chemical shifts of all the 27 carbon atoms were assigned by comparing those of isovitexin. The structure of compound was elucidated as 2″-O-glycosyl isovitexin (2-O-GIV) based on ^{13}C NMR, FAB-MS, UV, and IR spectra and is shown in Figure 4.

Effect of 2-O-GIV on Oxidation of Arachidonic Acid with Fenton's Reagent. The most commonly used method to monitor oxidation of lipids is the TBA assay. This method is very useful for measuring the relative degree of lipid oxidation because TBA reacts with many lipid peroxidation products. It is not possible, however, to monitor a single, specific product by the TBA method. Recently, we developed a new gas chromatographic method to monitor formation of reactive carbonyl compounds, such as MA and 4-hydroxynonenal (4-HN). MA and 4-HN were reacted with N-methylhydrazine (NMH) to yield 1-methylpyrazole and 5-(1-hydroxyhexyl)-1-methylpyrazoline, respectively, which were subsequently analyzed by a gas chromatograph equipped with a high-resolution, fused-silica capillary column and a nitrogen-phosphorus specific detector. The summary of products from reactive carbonyls and NMH is shown in Figure 5.

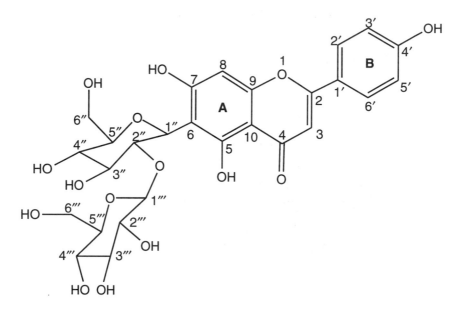

Figure 4. Structure of 2″-*O*-glycosyl isovitexin isolated from green barley leaves.

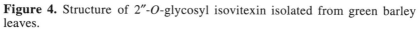

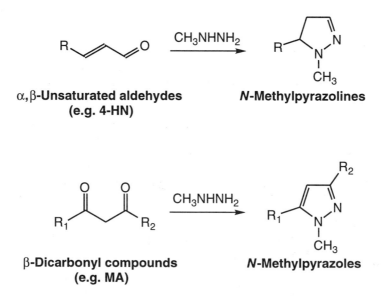

Figure 5. Derivatives used for analysis of reactive carbonyl compounds.

Figure 6 shows the gas chromatograms of lipid peroxidation products from arachidonic acid after reaction with NMH. Arachidonic acid (1 mg) produced 28 nmol MA and 7.5 nmol 4-HN. α-Tocopherol inhibited formation of MA by approximately 95%, whereas 2-*O*-GIV inhibited MA formation by over 99%. It should be noted that 2-*O*-GIV inhibited formation of 4-HN by approximately 80% in contrast to α-tocopherol, which did not inhibit its formation at all.

The antioxidative action of flavonoids has been known to be bi-modal. The phenolic moiety of flavonoids serves as a radical scavenger which is the antioxidative function of α-tocopherol. Flavonoids that possess the 3-hydroxy, 4-keto grouping and/or the 5-hydroxy, 4-keto grouping form complexes with metals and subsequently act as metal chelators (*16*). α-Tocopherol does not possess this function of chelating with metal ions. Both compounds scavenge free radicals formed by Fenton's reagent, but only 2-*O*-GIV can trap Fe^{2+} ions with its 5-hydroxy, 4-keto moiety. The formation of MA and 4-HN may proceed via different pathways.

Conclusions

Although natural antioxidants can be active by themselves, most of them become effective in the presence of synergists, such as α-tocopherol. Multiple purifications of crude extract of plant materials often result in failure to obtain a single strong antioxidant, due to the stability of the compound in the crude extract form or due to the presence of synergists (*15*). In addition, plants are complicated mixtures of numerous chemicals, and interactions with their components may affect the effectiveness of the antioxidant.

Although biological systems require oxygen, active oxygen species — such as hydroxy radical, superoxide, and singlet oxygen — may be involved in functional damage to organisms and may lead to pathological conditions. Besides intrinsic oxidation protecting enzyme systems, including superoxide dismutase (SOD), catalase, and glutathione peroxidase, food-derived substances such as tocopherols, ascorbic acid, and carotenes have been shown to diminish the adverse effects of oxidation in organisms. The effectiveness of flavonoids as *in vivo* antioxidants has been reported, but the metabolic pathway and action of naturally occurring antioxidative compounds is not clear. Their behavior and fate must be investigated *in vivo*.

Flavonoid compounds, which are widely distributed in the plant kingdom and occur in considerable quantities, show a wide range of pharmacological activities other than their antioxidative characteristics. These substances have been used to treat various pathological conditions including allergies, inflammation, and diabetes. In fact, the experimental data showing their high pharmacological activities such as antiviral and antitumor is accumulating; their therapeutic potential, however, has not been fully proven clinically. The investigation of the pharmacological potency of flavonoid compounds would provide evidence for their possible use in therapeutic treatments.

In the present studies, antioxidative activity of the flavonoid compound 2-*O*-glycosyl isovitexin present in young green barley leaves was demonstrated. Further work for application of this compound to actual pharmacological practice is in order.

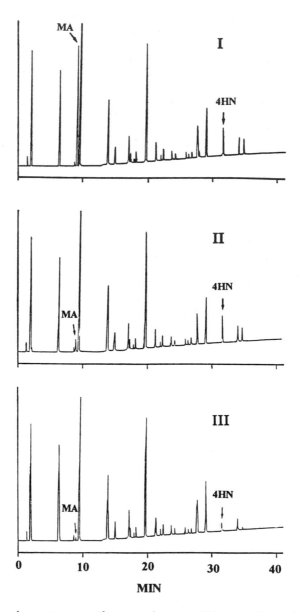

Figure 6. Gas chromatograms of extracts from arachidonic acid samples. I, control; II, sample with α-tocopherol; III, sample with 2-*O*-GIV.

Literature Cited

1. Inagaki, O.; Shoji, N.; Kaneko, K. *Eiyo To Shokuryo* **1986**, *33,* 309.
2. Paik, T. H.; Hoshino, T.; Kaneda, T. *Eiyo To Shokuryo* **1976**, *29*, 85.
3. Rice-Evans, C.; Hochstein, P. *Biochem. Biophys. Res. Comm.* **1981**, *100*, 1537.
4. Akimov, v. H.; Potapenko, A. Y.; Lashmanova, A. P.; Khorseva, N. I.; Polanskaya, N. P.; Bezdetnaya, L. N. *Studia biophysica 1988, 124*, 239.
5. Privett, O. S.; Cortesi, R. *Lipids* **1972**, *7*, 780.
6. Gardner, H. W. *J. Agric. Food Chem.* **1979**, *27*, 220.
7. Nielsen, H. K.; Loliger, J.; Hurrell, R. F. *Brit. J. Nutr.* **1985**, *53*, 61.
8. Reiss, U.; Tappel, A. L.; Chio, K. S. *Biochem. Biophys. Res. Comm.* **1972**, *48*, 921.
9. Tappel, A. L. *Fed. Proc.* **1973**, *32*, 1870.
10. Matsuoka, Y.; Seki, H.; Kubota, K.; Ohtake, H.; Hagiwara, Y. *Enshou* **1983**, *3,* 9.
11. Ohtake, H.; Yuasa, H.; Komura, C.; Miyauchi, T.; Hagiwara, Y.; Kubota, K. *Yakugaku Zasshi* **1985**, *105*, 1046.
12. Ohtake, H.; Nonaka, S.; Sawada, Y.; Hagiwara, Y.; Hagiwara, H.; Kubota, K. *Yakugaku Zasshi* **1985**, *105*, 1052.
13. Ohkawa, H.; Ohishi, N.; Yagi, K. *Anal. Biochem.* **1979**, *95*, 351.
14. Tamura, H.; Kitta, K.; Shibamoto, T. *J. Agric. Food Chem.* **1991**, *39*, 439.
15. Ramarathnam, N.; Osawa, T.; Namiki, M.; Kawakishi, S. *J. Agric. Food Chem.* **1988**, *36*, 732.
16. Pratt, D. E. In *Phenolic, Sulfur, and Nitrogen Compounds in Food*; Cody, V.; Middleton, E., Jr.; Harborne, J. B., Eds.; Alan R. Liss, Inc.: New York, 1976; pp 125–140.

RECEIVED June 22, 1993

Chapter 18

Antioxidative Compounds from Marine Organisms

K. Sakata[1], K. Yamamoto[2], and N. Watanabe[2]

[1]Research Laboratory of Marine Biological Science, Shizuoka University, Mochimune, Shizuoka 421–01, Japan
[2]Department of Applied Biological Chemistry, Shizuoka University, 836 Ohya, Shizuoka 422, Japan

Antioxidants from marine organisms are reviewed and our recent study on new antioxidants from marine organisms is described in detail. Among the extracts of viscera of various kinds of marine fish and bivalves, the extract of the short-necked clam, *Ruditapes philippinarum*, showed a very low peroxide value and mutagenicity by the rec assay. Several new compounds related to chlorophyll a — chlorophyllone a (**1**), chlorophyllonelactone a (**2**), chlorophyllonic acid a methyl ester (**3**) and 13^2-oxo-pyropheophorbide a (**8**) — as well as the known compounds pyropheophorbide a (**4**), purpurin 18 (**6**) and purpurin 18 methyl ester (**7**), were isolated as the main antioxidants from the edible parts of the short-necked clam. 13^2-*epi*-Chlorophyllone a (**5**), an epimeric isomer of **1**, together with **2**, **3** and **4** were also isolated from the viscera of the scallop, *Patinopecten yessoensis*. The presence of **1**, **2** and **3** in the viscera of the oyster *Crassostrea* sp. and **1** in the mixture of attached diatoms (*Fragilaria oceanica, F. cylindus, Nitzschia closterium, N. seriata, Cocconeis pseudomarginata, Hyalodiscus stelliger* are predominant) cultured for seedling production of juvenile abalone was also confirmed.

Organisms are well known to have defense mechanisms against oxidation, because peroxides of polyunsaturated fatty acids have shown many kinds of toxicity including mutagenicity (*1,2*). Many efforts have been made to identify new antioxidants for food processing and medical use. Most of the antioxidative compounds isolated so far are from terrestrial plants and only several from animal or marine organisms.

Antioxidants from Marine Organisms

Antioxidants from Algae. Marine edible algae are well known to be stable against oxidation during storage after drying in spite of their high content of highly unsaturated fatty acids. Several studies have been conducted to isolate antioxidants from algae (*3–9*). Nishibori *et al.* tested algal extracts for antioxidative activity and

0097–6156/94/0547–0164$06.00/0

found that the lipid fractions of the green alga *Enteromorpha* sp. and the brown alga *Undaria pinnatifida* showed potent antioxidative activity in the conventional test (*3*). They identified pheophytin a, one of the degradative products of chlorophyll a, as an active principle (*4*). Fujimoto *et al.* also screened 21 species of marine algal extracts for antioxidants and reported that more than half of them exhibited antioxidant activity to some extent. The active principles of the lipid fractions from the brown alga *Eisenia bycyclis* were ascribed to the phospholipid fraction (*5*), but they failed to identify the antioxidative components of the lipid fraction from *U. pinnatifida*. They reported that the active fraction contained several antioxidants or synergists, because the activity of the crude fractions declined as their fractionation proceeded. Four kinds of bromophenols were isolated as active principles from the most active fraction from the red alga *Polysiphonia ulceolate* (*6*). Furologlucinol related compounds were also identified as antioxidants from brown algae such as *Eisenia bycyclis* (*7*).

Le Tutour recently examined the antioxidative activity of several algal extracts and found that the extracts of brown algae like *Laminaria digtata* and *Himanthalia elongata* showed the strongest antioxidative activity and synergy with α-tocopherol (*8*). The active principles, however, have not been clarified yet.

Antioxidants From Other Marine Organisms. The antioxidative activity found in the lipid fraction of the krill *Euphausia suprba* was ascribed to the synergistic effect of α-tocopherol with phospholipids (*9*). The lipid fractions of shrimps and squids are known to contain antioxidants. Now the potent antioxidative activity is considered to be due to the synergistic effect of tocopherols with trimethylamine oxide (*10*). Homarine, which is found in many marine animals is also reported to show antioxidant activity (*11*). Recently, carotenoids were found to scavenge singlet oxygen. Astaxanthin, contained in crustaceans like krill, is the most active among the carotenoids (*12,13*).

Antioxidative Components from the Short-necked Clam, *Ruditapes philippinarum*

Antioxidative Activity and Mutagenicity of the Viscera Extracts of Fish and Clams. We focused our research interests on marine organisms containing a large amount of polyunsaturated fatty acids and measured peroxide value (POV) by the conventional antioxidant test and mutagenicity by the rec assay (*15*) of organic solvent extracts of their viscera as follows.

Extraction and Fractionation of Fish and Bivalve Viscera. Fresh viscera was chopped finely on ice and subjected to the solvent extraction and fractionation procedure as shown in Figure 1.

POV Measurement. Each sample solution (equivalent to 5 mg of the extract) was absorbed on a paper disk (φ8 x 1.5 mm). The disk was dried in a desiccator and then subjected to a modified American Oil Chemists' Society POV determination method.

Mutagenicity. The paper disk prepared in the same way as mentioned above (sample equivalent to 3 mg of each extract was applied) was subjected to the rec assay developed by Kada *et al.* (*15*) to measure mutagenicity.

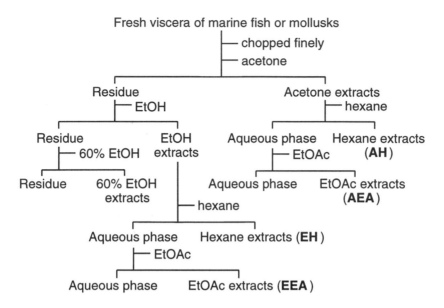

Figure 1. Extraction and fractionation of fish and bivalve viscera.

Each extract showed quite different activity. Positive correlation was observed between POV and mutagenicity of their extracts (Figure 2). Among the extracts of viscera of various kinds of marine fish and bivalves, clam extracts, especially from the short-necked clam *R. philippinarum*, showed a very low POV and mutagenicity, although they contained a lot of highly unsaturated fatty acids.

The extract of the short-necked clam was mixed with methyl linolate, then absorbed on a paper disk as described above. The disk was exposed to the air at 37°C in the dark for 62 hrs and then subjected to the POV measurement. It showed very low POV, indicating the presence of a potent antioxidant(s) in the extract. HPLC analysis, however, revealed that this extract contained too few tocopherols to account for the potent antioxidant activity. These observations encouraged us to isolate new antioxidants responsible for this low POV in the extract.

Isolation of Antioxidative Constituents from the Short-necked Clam *R. philippinarum*. The edible parts of the clam *R. philippinarum* (3.5 kg), collected at the Hamana lake, Shizuoka Prefecture, Japan, were finely chopped on ice and extracted with a mixture of CHCl$_3$-MeOH (2:3). The MeOH layer obtained by partitioning the concentrated extract between hexane and MeOH was chromatographed on a Sephadex LH-20 column (MeOH) to give the antioxidative fractions GPX and GPY. Isolation was guided by the antioxidant assay described in the following section. The active fraction GPY was further chromatographed on silica gel to yield four active compounds (**1, 2, 3** and **4** in Figure 3). Purification of fraction GPX in a similar manner gave compounds **6, 7** and **8** in Figure 4.

Antioxidant Assay. The thiocyanate method by Fukuda *et al.* (*14*) was modified as follows. Each sample, dissolved in appropriate solvent (*ca.* 30 μL),

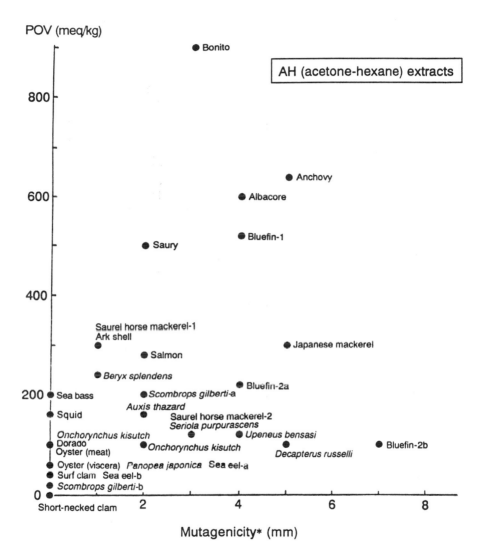

Figure 2. Peroxide value (POV) and mutagenicity of AH fraction of fish and bivalves. Each extract (equivalent to 3 and 5 mg fresh viscera) was subjected to the POV and mutagenicity assays.
*Judged by the rec assay using *Bacillus subtilis* (*15*).

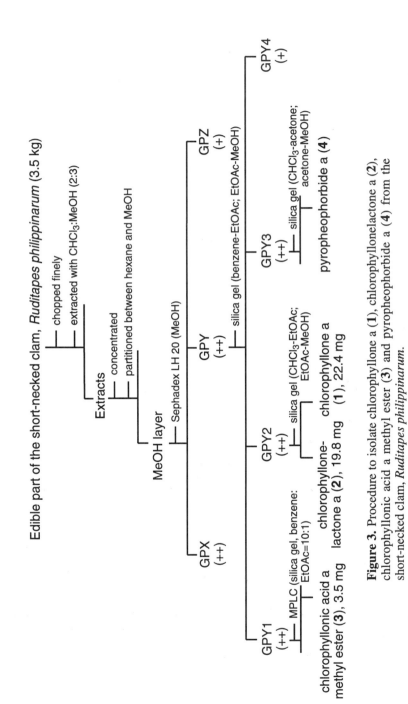

Figure 3. Procedure to isolate chlorophyllone a (**1**), chlorophyllonelactone a (**2**), chlorophyllonic acid a methyl ester (**3**) and pyropheophorbide a (**4**) from the short-necked clam, *Ruditapes philippinarum*.

was absorbed on a paper disk ($\phi 8 \times 1.5$ mm) in a 10 mL sample tube. After evaporation of the solvent, 200 μL EtOH, 200 μL 2.5% linoleic acid in EtOH, 400 μL 5×10^{-2} M phosphate buffer (pH 7.0) and 200 μL H_2O were added. The mixture, in a stoppered sample tube, was kept in a sonicator for a few minutes to dissolve the sample out of the paper disk and incubated at 40°C. At intervals during the incubation, 100 μL of the mixture was removed to a test tube, mixed with 3 mL 75% EtOH, 100 μL 30% ammonium thiocyanate and 100 μL 2×10^{-2} M ferric chloride. After 3 min the absorbance at 500 nm was measured.

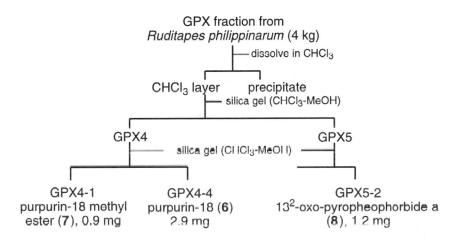

Figure 4. Procedure to isolate purpurin 18 (**6**), purpurin 18 methyl ester (**7**) and 13^2-oxo-pyropheophorbide a (**8**) from the short-necked clam, *Ruditapes philippinarum*.

Structure Determination of the Antioxidative Compounds from *R. philippinarum*. Compound **4** was identified as pyropheophorbide a by direct comparison of its spectroscopic data with those of an authentic specimen. The other compounds also showed characteristic UV and visible absorption spectra for chlorophyll a related compounds (Figure 5).

Compound **1**, a dark green solid ($C_{33}H_{32}N_4O_3$, $\nu_{c=o}$ (CHCl$_3$) 1722 cm^{-1}), was found to be a new chlorophyll a related compound and designated chlorophyllone a. Comprehensive ^{1}H- (Table I and Figure 6) and ^{13}C-NMR (Table II) spectroscopic analysis revealed the unique structure **1** of chlorophyllone a. The stereostructure of the new ring was deduced from NOE experimental results (Figure 7) (*16,17*).

Compound **2**, dark green crystals ($C_{33}H_{32}N_4O_4$, $\nu_{c=o}$ (CHCl$_3$) 1722 cm^{-1}), contained one more oxygen atom than **1**. Comprehensive ^{1}H-NMR analysis (Table I) revealed that **2** has a similar structure to that of **1**. In the HMBC spectrum of **2**, prominent crosspeaks from H-17$^{2b'}$ to C-13^2 and C-17^3 and H-17^{2b} to C-17^3 were observed. These observations revealed that C-13^2 and C-17^3 are within a three bond proximity to H-17^{2b} and H-17$^{2b'}$. Thus **2**, name chlorophyllonelactone a, has a structure in which ring V of **1** has been converted to a δ-lactone (*17,18*). The lower

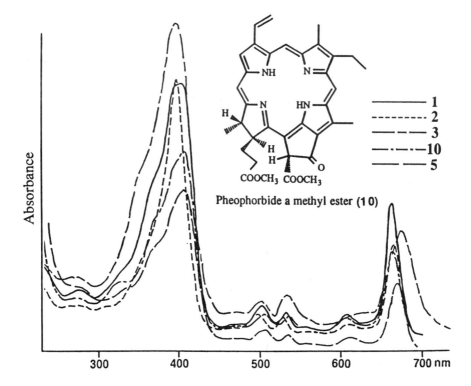

Figure 5. UV and visible spectra of chlorophyllone a (**1**), chlorophyllonelactone a (**2**), chlorophyllonic acid a methyl ester (**3**), 13^2-*epi*-chlorophyllone a (**5**) and pheophorbide a methyl ester (**10**).

Table I. ¹H-NMR Spectral Data (CDCl$_3$, 400 MHz)

proton	1*	2*	3*	5**	6*	7*	8*	10*
10	9.55 (s)	9.70 (s)	9.67 (s)	9.49 (s)	9.60 (s)	9.63 (s)	9.88 (s)	9.53 (s)
5	9.50 (s)	9.54 (s)	9.47 (s)	9.39 (s)	9.37 (s)	9.38 (s)	9.85 (s)	9.40 (s)
20	8.66 (s)	8.66 (s)	8.58 (s)	8.54 (s)	8.57 (s)	8.57 (s)	9.01 (s)	8.56 (s)
3¹	8.03 (dd)	8.02 (dd)	7.96 (dd)	7.98 (dd)	7.88 (dd)	7.90 (dd)	8.11 (dd)	8.00 (dd)
3²(E)	6.31 (dd)	6.34 (dd)	6.31 (dd)	6.28 (dd)	6.29 (dd)	6.31 (dd)	6.36 (dd)	6.30 (dd)
13²								6.25 (s)
3²(Z)	6.21 (dd)	6.20 (dd)	6.16 (dd)	6.19 (dd)	6.19 (dd)	6.20 (dd)	6.27 (dd)	6.19 (dd)
8¹	3.68 (q)	3.73 (q)	3.70 (q)	3.66 (q)	3.54 (q)	3.65 (q)	3.79 (q)	3.37 (q)
12¹	3.71 (s)	3.84 (s)	3.61 (s)	3.66 (s)	3.78 (s)	3.81 (s)	3.86 (s)	3.69 (s)
2¹	3.44 (s)	3.43 (s)	3.39 (s)	3.38 (s)	3.34 (s)	3.35 (s)	3.51 (s)	3.41 (s)
7¹	3.26 (s)	3.27 (s)	3.22 (s)	3.23 (s)	3.16 (s)	3.18 (s)	3.36 (s)	3.24 (s)
18¹	2.20 (d)	1.85 (d)	1.73 (d)	2.21 (d)	1.74 (d)	1.74 (d)	1.88 (d)	1.81 (d)
8²	1.70 (t)	1.71 (t)	1.69 (t)	1.69 (t)	1.66 (t)	1.68 (t)	1.75 (t)	1.70 (t)
18	4.36 (dq)	4.39 (dq)	4.41 (dq)	4.77 (dq)	4.39 (q)	4.39 (q)	4.69 (dq)	4.45 (dq)
17	4.91 (dt)	4.42 (ddd)	4.54 (ddd)	3.83 (ddd)	5.19 (dd)	5.20 (dd)	5.18 (ddd)	4.21 (ddd)
17¹a	2.26 (ddt)	2.22 (dddd)	2.39 (ddt)	3.70 (dddd)	1.99 (m)	2.00 (m)	2.83 (m)	2.64 (dddd)
17¹a'	2.90 (dddd)	2.88 (ddt)	2.91 (dddd)	2.67 (dddd)	2.5 (m)	2.5 (m)	2.44 (m)	2.32 (dddd)
17²b	2.81 (ddd)	3.54 (dt)	3.85 (dt)	3.86 (ddd)	2.5 (m)	2.5 (m)	2.75 (m)	2.52 (ddd)
17²b'	4.36 (ddd)	3.01 (ddd)	3.07 (ddd)	2.97 (ddd)	2.79 (m)	2.73 (m)	2.34 (m)	2.23 (dd)
NH	0.62 (br.s)	-0.8 (br.s)	-0.67 (br.s)	0.98 (br.s)	0.23 (br.s)	0.26 (br.s)	-2.35 (br.s)	0.56 (br.s)
NH	-1.90 (br.s)	-1.45 (br.s)		-1.50 (br.s)	-0.07 (br.s)	-0.05 (br.s)		-1.61 (br.s)
13²-OH	-1.90 (br.s)	5.86 (br.s)		4.16 (br.s)				
-OCH$_3$			4.04 (s)			3.60 (s)		3.57 (s)
-OCH$_3$								3.88 (s)
-OCH$_3$								

Concentration: *, 1 mg/0.6 ml; **, 0.6 mg/0.6 ml.
SOURCE: Reproduced with permission from reference 18. Copyright 1993.

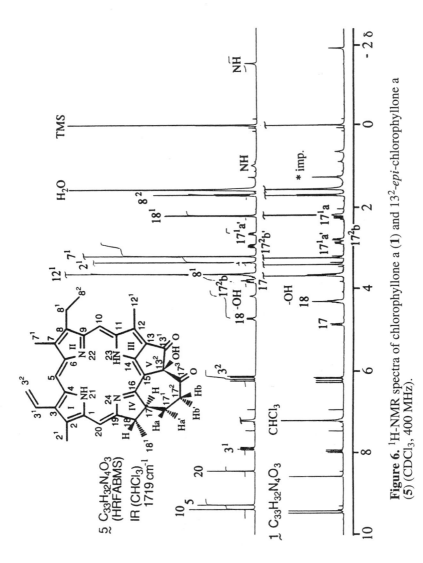

Figure 6. ^{1}H-NMR spectra of chlorophyllone a (**1**) and 13^2-*epi*-chlorophyllone a (**5**) (CDCl$_3$, 400 MHz).

Table II. ^{13}C-NMR Spectral Data (CDCl$_3$)

	1	2	3	5	6	10^{16}
17^3	208.10	203.39	196.9	206.1	177.5$^!$	173.34
13^1	195.62	162.26	166.9	193.5	177.2$^!$	189.59
19	172.65	173.00	173.1	172.2		172.15
16	163.19	163.55	164.1	162.7		161.23
6	154.43	155.60	155.2	154.6	156.2	155.58
9	150.81	150.07	149.7	150.8	150.1	150.95
14	147.66	133.97+	135.2	149.4	139.9	149.70
8	144.79	145.46	145.5	144.9	145.9	145.11
1	142.10	141.65+	142.1	142.3	144.1	142.04
11	138.09	131.36+	130.2	138.2	139.0$^&$	137.96
3	136.28	136.16	136.3	136.3	136.6$^&$	136.47
4	136.20	135.88+	136.3	136.3	136.6$^&$	136.16
7	135.68	136.31	135.7	135.8	137.8$^&$	136.16
2	131.53	131.33	130.8	131.5	131.5	131.79
12	129.04	138.41	138.4	129.0	129.2	129.02
3^1	129.04	128.86	128.9	129.0	128.4	129.02
13	127.71	111.49	121.5	129.5	122.6	129.02
3^2	122.78	122.79	122.7	122.7	123.6	122.64
15	105.38	100.23+	108.6	105.9	111.5	105.28
10	103.93	104.18	105.9	104.6	107.5	104.33
5	97.99	99.61	101.5	98.2	103.1	97.47
13^2	93.45	103.33+	192.0	92.7	176.4$^!$	64.77
20	92.84	93.32	93.4	91.6	95.0	93.09
17	51.89	50.06	50.3	53.7	55.0	51.21
18	51.51	51.13	51.2	50.3	49.2	50.17
17^2	40.12	34.81	36.4	43.2	32.2	29.92*
17^1	37.99	31.99	29.5	22.8	31.0	31.12*
18^1	22.35	23.50	23.6	16.9	23.8	23.08
8^1	19.18	19.27	19.5	19.3	19.3	19.34
8^2	17.27	17.41	17.5	17.3	17.3	17.32
12^1	12.16	12.11	12.5	12.2	12.2	12.01
2^1	12.07	12.11	11.9	11.9	11.9	12.01
7^1	11.11	11.17	11.2	11.1	10.9	11.08
OMe		52.3				
13^1						169.61
OMe						52.81**
OMe						51.64**
OMe						

1: 17.8 mg/0.6 ml, 100 MHz; 2: 18.8 mg/0.6 ml, 100 MHz;
3: 3.5 mg/0.1 ml, 22.5 MHz; 5: 1.8 mg/0.1 ml, 22.5 MHz;
6: 2.9 mg/0.1 ml, 22.5 MHz; 10: 2.0 mg/0.6 ml, 22.5 MHz.
*, **, !, &, $: interchangeable
+: deuterium shift 0.1–0.19 ppm

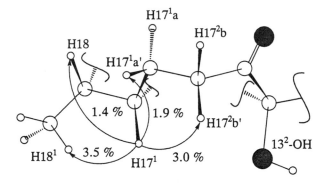

Figure 7. NOE experimental results of chlorophyllone a (**1**).

shift (δ 1.53) of the C-132 hydroxyl proton (δ 5.86) of **2** than that (δ 4.33) of **1** was reasonably ascribed to the hydrogen bonding between the hydroxyl proton and the C-173 ketone. NOE experiments confirmed the stereostructure.

Compound **3**, brown crystals ($C_{34}H_{34}N_4O_4$, $v_{c=o}$ (CHCl$_3$) 1731, 1693 cm^{-1}), was designated chlorophyllonic acid a methyl ester. The structure of **3**, with a sterically congested 1,2-diketone group, was deduced from ^{1}H- (Table I) and ^{13}C-NMR (Table II) spectroscopic data. The structure of **3** was confirmed by X-ray crystallographic analysis (*19*). The C-132 carbonyl was found to be oriented downward out of the plane of chlorin ring to reduce the steric hindrance around the carbonyl group.

In addition to compounds **1, 2, 3** and **4**, three other pyropheophorbide a related compounds — **6, 7** and **8** — were isolated in trace amounts. Compound **6**, a purplish red material, $C_{33}H_{32}N_4O_5$, was identified as purpurin 18 (*20, 21*).

Compound **7**, a purplish red solid ($C_{34}H_{34}N_4O_5$, FABMS (pos.) m/z 579 [MH]$^+$, FABMS (neg.) m/z 578 [M]$^-$, 577 [M-H]$^-$; $v_{c=o}$ (CHCl$_3$) 1750, 1718 cm^{-1}), was identified as a methyl ester of **6**. It seemed to be an artifact produced during MeOH extraction. This is, to our knowledge, the first report of isolation of naturally occurring **6** and **7**, although they have been already synthesized (*20, 21*).

Compound **8**, named 132-oxo pyropheophorbide a, a yellowish brown solid (FABMS (pos.) m/z 549 [MH]$^+$, FABMS (neg.) m/z 548[M]$^-$; $v_{c=o}$ (CHCl$_3$) 1706 cm^{-1}), was also found to be a pyropheophorbide a related compound from the ^{1}H-NMR (Table I) H-17 was found at δ 5.18, shifted 1 ppm down field from its position in pheophorbide a methyl ester (**10**). The lower field shift of this proton also occurs in the spectra of **6** and **7**, suggesting an anisotropic effect due to a carbonyl at C-13^2. Hence we propose the 1,2-diketone structure for **8**.

The structures of the chlorophyll a related compound are shown in Figure 8.

Origin of the New Chlorophyll a Related Compounds

Compounds **1, 2, 3** and **8** were new chlorophyll a related compounds and we were interested in their origin. These compounds seem to originate from chlorophylls in diatoms and detritus fed on by the bivalves. Chlorophyll a related compounds in the viscera extracts of the other plankton feeding bivalves, the scallop *Patinopecten yessoensis* and the oyster *Crassostrea* sp., as well as in the extract of attached diatoms mixture were analyzed as follows.

Scallop and Oyster.   Viscera (2.3 kg) of the scallop *P. yessoensis* was finely chopped on ice and subjected to the same extraction and isolation procedure (Figure 9) as in the case of the short-necked clam, yielding **2** (1.3 mg), **3** (0.4 mg), **4** (1.0 mg) and new pheophorbide a related compound **5**, 13^2-*epi*-chlorophyllone a (1.2 mg). Compound **1**, however, was not detected in the extract.

Compound **5**, a dark green solid ($C_{33}H_{32}N_4O_3$, $v_{c=o}$ (CHCl$_3$) 1719 cm^{-1}), has the same molecular formula as that of **1**. Comparative studies of their ^{1}H- (Figure 6, Table I) and ^{13}C-NMR (Table II) spectra with those of **1** revealed that **5** was 13^2-*epi*-chlorophyllone a as detailed.

Large coupling constants, $J_{H-17,17^1a}$=12.6 and $J_{H17^1a,17^2b'}$=12.1 Hz, and NOEs of H-17 (3.1%) and H-17^{1}a' (2.2%) by irradiation at H-17^{2}b' suggested their 1,2-trans diaxial relationships. The lower field shift of H-17^{1}a in **5** compared with that of **1** is ascribable to the anisotropic effect of 13^2-OH, which has the 1,4-diaxial conformation with H-17^{1}a. The upper field shifts of H-17 and H-17^{2}b' in **5** are due

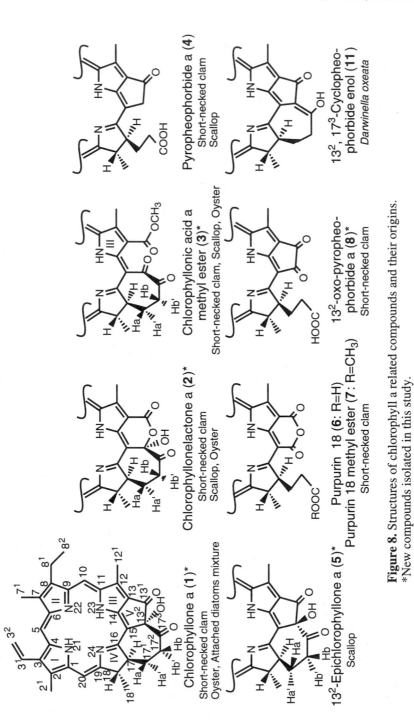

Figure 8. Structures of chlorophyll a related compounds and their origins.
*New compounds isolated in this study.

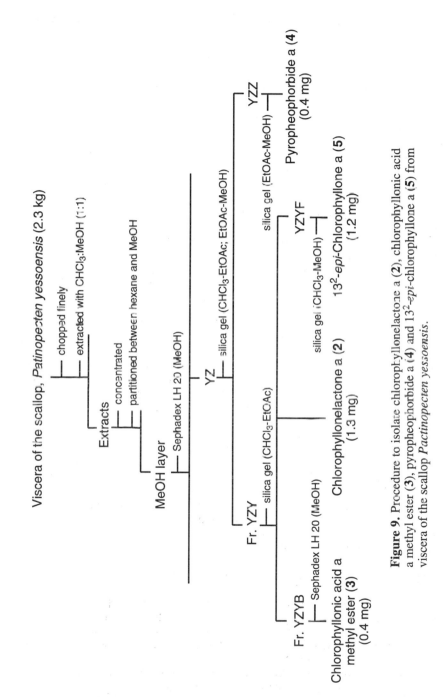

Figure 9. Procedure to isolate chlorophyllonelactone a (**2**), chlorophyllonic acid a methyl ester (**3**), pyropheophorbide a (**4**) and 13²-*epi*-chlorophyllone a (**5**) from viscera of the scallop *Pactinopecten yessoensis*.

to decreased anisotropic effect by 13^2-OH caused by the stereostructural change of **1** into **5**. Larger coupling constant ($J_{H\text{-}18,17}$=9.2 Hz) in **5** than that ($J_{H\text{-}18,17}$=1.6 Hz) in **10** indicated that methyl protons ($H_{3\text{-}18^1}$) of **5** were in the same plane of the chlorin ring, whereas $H_{3\text{-}18^1}$ of **10** are directed downward out of the chlorin ring. This conformational change of the ring VI increased the anisotropic effect from chlorin ring to $H_{3\text{-}18^1}$ to result in its lower field shift (δ= 0.40 ppm) in **5** than that of **10** (Table 1). The upper field shift (δ= -15.2 ppm) of C-17^1 of **5** from that of **1** is due to the steric compression effect caused by the conversion of 13^2-OH configuration. Thus the structure of **5** was unambiguously established to be an epimer of **1** at C-13^2 (*17,18*).

Similarly to scallop, 3.9 mg chlorophyllone a (**1**) was isolated from viscera (1.9 kg) of the oyster *Crassostrea* sp.

Attached Diatoms. Attached diatoms mixture (equivalent to 18 g dry weight) cultured on plastic plates for abalone seedling production at Shizuoka Prefecture Fish Farming Center was extracted with $CHCl_3$-MeOH (1:2). The mixture was found, by microscope, to consist mainly of *Fragilaria oceanica, F. cylindus, Nitzschia closterium, N. seriata, Cocconeis pseudomarginata* and *Hyalodiscus stelliger*. The EtOAc layer from the extract was subjected to Sephadex LH-20 (MeOH) and silica gel (CHCl3-MeOH) column chromatography to give **1** (0.6 mg), indicating that chlorophyllone a (**1**) was produced by the microalgae themselves.

Wafting Diatoms. Several species (*Nannochloropusis oculata, Chaetoceros gracilis, Pavlova* sp., *Phaeodactylum* sp., *Tetraselmis tetrathele, Dunaliala tatiolecta, Isochrysis galbana, Poryphyridium purupureum and Chaetoceros* sp.) of monocultured wafting diatoms were available and the presence of the new chlorophyll a related compounds in the microalgae was examined by silica gel HPLC [YMC packed A-012 (S-5 120A SIL; ϕ6 x 150 mm); $CHCl_3$]. Samples for HPLC analysis were prepared by collecting the fractions showing the same Rf value on silica gel TLC (hexane:benzene:acetone:MeOH:H_2O=150:70:50:20:2) as that of **1** (0.27) and **2** (0.25) from silica gel column ($CHCl_3$; $CHCl_3$:MeOH=50:2, 25:2). Each sample corresponding to 50 mg dry alga was injected for the analysis. A small peak corresponding to **1** was observed in the extract of *N. oculata, I. galbana* and *P. purupureum,* and that of **2** in *N. oculata, D. tatiolecta* and *Chaetoceros* sp. The peak of **2** in *D. tatiolecta* alone was big enough to be confirmed by cochromatography with authentic sample. Because of the shortage of the algal sample amount, the presence of the new chlorophyll a related compounds in the wafting diatoms was only chromatographically confirmed, suggesting the presence of the new degradation pathway of chlorophyll a (Figure 10).

A Hypothetical Degradation Pathway of Chlorophyll a.

Many degradation pathways of chlorophyll a have been reported (*22*). Enzymatic degradation of chlorophyll a to pyropheophorbide a has been confirmed in a mutant strain of the microalga, *Chlorella fusca* (*23*). The fact that **1, 2, 3, 5, 6, 7** and **8** have been isolated here and that **5** is an epimer of **1** at C-13^2 strongly suggests that both **1** and **5** are enzymatically produced.

Compound **4**, produced by decarboxylation of pheophorbide a, seems to be converted into $13^2,17^3$-cyclopheophorbide enol (**11**) by Claisen type condensation followed by oxidation to give **1** and **5** (Figure 10). Compound **11**, which was first

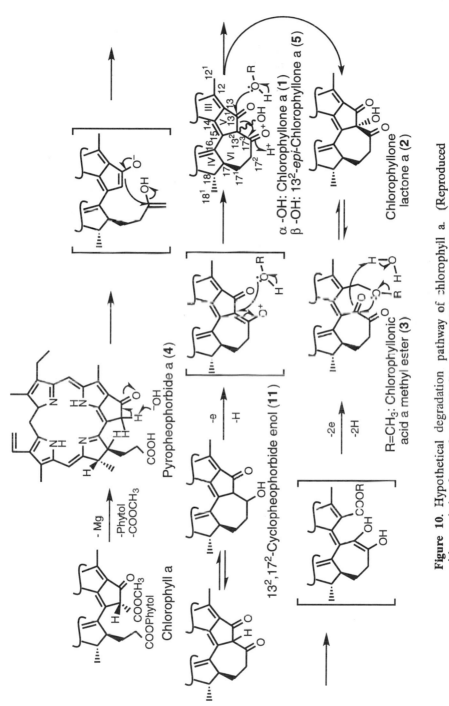

Figure 10. Hypothetical degradation pathway of chlorophyll a. (Reproduced with permission from reference 18. Copyright 1993.)

synthesized by Eschenmoser *et al.* (*24*), was recently isolated from the sponge *Darwinella oxeata* (*25*), and is also considered to originate from chlorophylls contained in the diatoms symbiotic with the sponge.

Compound **2** was assumed to be directly produced by Baeyer-Villiger type oxidation of **1** or to be derived from **1** via **3** by the loss of two electrons together with ring V opening under acidic conditions, followed by intramolecular cyclization. Compound **3** may be derived from **2** by ring V opening or directly produced by oxidation of **1** or **5** under acidic conditions.

Antioxidative Activity of the Chlorophyll a Related Compounds 1–8

Chlorophyll related compounds are well known to promote lipid oxidation in light (*26*), but to show antioxidative activity in the dark (*27*). Compounds **1–4** and **6–8** showed stronger antioxidative activity at a dose of 3 μg than that of 20 μg of α-tocopherol in the dark (Figures 11 and 12), suggesting that these compounds contribute to the antioxidative activity of the short-necked clam extract. Compound **5** from the scallop also showed antioxidative activity. The antioxidative mechanism of chlorophylls is considered to be due to radical scavenger effects (*28*).

Concluding Remarks

Chlorophyll related compounds are known to possess potent antioxidative activity in the dark, but to be photosensitizers in light. We are consuming a lot of chlorophylls from vegetables, but only a little is known about their catabolism and the action of the degradation products. Studies on these compounds must be carried out from the not only antioxidative but also anticarcinogenic points of view.

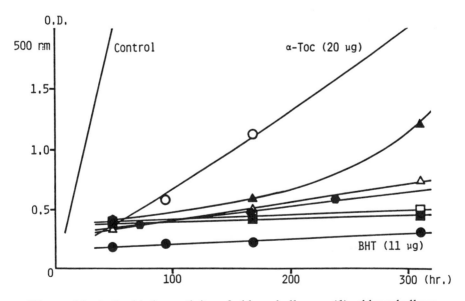

Figure 11. Antioxidative activity of chlorophyllone a (**1**), chlorophyllone-lactone a (**2**), chlorophyllonic acid a methyl ester (**3**), pyropheophorbide a (**4**) and 13^2-*epi*-chlorophyllone a (**5**).
▲ **1** (3 μg), ■ **2** (3 μg), □ **3** (3 μg), ● **4** (2 μg), △ **5** (3 μg)

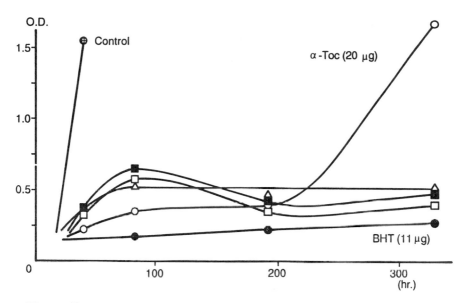

Figure 12. Antioxidative activity of purpurin 18 (**6**), purpurin 18 methyl ester (**7**) and 13²-oxo-pyropheophorbide a (**8**).
□ **6** (3 μg), ■ **7** (6 μg), Δ **8** (3 μg)

Acknowledgements. The authors are indebted to Mr. N. Kawashima (Shizuoka Prefecture Fish Farming Center) and Mr. S. Akera (Tasaki Institute for Marine Biological Research) for their generous gifts of the attached diatoms and monocultured wafting diatoms, respectively. The authors also thank the JEOL company and Taisho Pharmaceutical Co., Ltd. for HRFABMS and HMBC measurements, respectively.

Literature Cited

1. Uchiyama, M.; Matsuo, M.; Sagai, M. In *Kasanka Shishitsu to Seitai* Gakkai Shuppan Center: Tokyo, 1985; pp 255–287.
2. Niki, E; Kawakami, A; Saito, M; Yamamoto, Y; Tsuchiya, J.; Kamiya, Y. *J. Biol. Chem.* **1985**, *260*, 219–225.
3. Nishibori, S.; Namiki, K. *Kaseigaku Zasshi* **1985**, *36*, 845–850.
4. Nishibori, S.; Namiki, K. *Kaseigaku Zasshi* **1988**, *39*, 1173–1178.
5. Fujimoto, K.; Kaneda, T. *Bull. Japn. Soc. Sci. Fish.* **1980**, *46*, 1125–1130.
6. Fujimoto, K.; Ohmura, H.; Kaneda, T. *Bull. Japn. Soc. Sci. Fish.* **1985**, *51*, 1139–1143.
7. Maruyama, M.; Kimura, S.; Fujimoto, K.; Kida, N. *Japan Kokai Tokkyo Koho* **1990**, JP 02245087.
8. Le Tutour, B. *Phytochemistry* **1990**, *29*, 3759–3765.
9. Yanagimoto, M.; Sibasaki, M.; Umeda, K.; Kimura, S. *Nippon Shokuhin Kogyo Gakkaishi* **1977**, *24*, 20–25.

10. Ohta, S. *New Food Industry*, **1985**, *27*, 63–72.
11. Yamazaki, K.; Oyabu, D.; Ota, S. *Japan Kokai Tokkyo Koho*, **1975**, JP 50008783.
12. Miki, W. *Gekkan Food Chemical* **1990**, *No. 5*, 61–65.
13. Miki, W. *Pure & Appl. Chem.* **1991**, *63*, 141–146.
14. Fukuda, Y.; Osawa, T.; Namiki, M.; Ozaki, T. *Agric. Biol. Chem.* **1985**, *49*, 301–305.
15. Tajima, Y.; Kondo, S.; Kada, Y.; Sotomura, A. *Kankyo Hen'igen Jikkenho*, Kodansha: Tokyo, 1980; p 48.
16. Sakata, K.; Yamamoto, K.; Ishikawa, H.; Yagi, A.; Etoh, H.; Ina, K. *Tetrahedron Lett.* **1990**, *31*, 1165–1169.
17. Sakata, K.; Yamamoto, K.; Ishikawa, H.; Watanabe, N.; Etoh, H.; Yagi, A.; Ina, K. *32nd Symposium on The Chemistry of Natural Products Symposium Papers*, Organizing Committee of the 32nd Symposium on the Chemistry of Natural Products: Chiba, Japan, 1990; pp 57–64.
18. Watanabe, N.; Yamamoto, K.; Ishikawa, H.; Yagi, A.; Sakata, K. *J. Nat. Prod.* **1993**, *56*, 305–317.
19. Yamamoto, K.; Sakata, K.; Watanabe, N.; Yagi, A.; Brinen, L. S.; Clardy, J. *Tetrahedron Lett.* **1992**, *33*, 2587–2588.
20 Hoober, J. K.; Sery, T. W.; Yamamoto, N. *Photochem. Photobiol.* **1988**, *48*, 579–587.
21. Kenner, G. W.; McCombie, S. W.; Smith, K. M. *J. Chem. Soc.* Perkin I, **1973**, 2517–2523.
22. Rudiger, W.; Schoch, S. In *Plant Pigments* Goodwin, T. W. Ed.; Academic Press: London, 1988; p 48.
23. Ziegler, R.; Blaheta, A.; Guha, N.; Schonegge, B. *J. Plant Physiol.* **1988**, *132*, 327.
24. Falk, H.; Hoornaert, G.; Isenring, H. P.; Eschenmoser, A. *Helv. Chim. Acta* **1975**, *58*, 2347–2357.
25. Karuso, P.; Berguquist, P. R.; Buckleton, J. S.; Cambie, R. C.; Clark, G. R.; Rickard, C. E. F. *Tetrahedron Lett.* **1986**, *27*, 2177–2178.
26. Endo, Y.; Usuki, R.; Kaneda, T. *J. Am. Oil Chem. Soc.* **1984**, *61*, 781–784 .
27. Endo, Y.; Usuki, R.; Kaneda, T. *J. Am. Oil Chem. Soc.* **1985**, *62*, 1375–1378.
28. Endo, Y.; Usuki, R.; Kaneda, T. *J. Am. Oil Chem. Soc.* **1985**, *62*, 1387–1390.

RECEIVED July 6, 1993

Chapter 19

Chemistry and Antioxidative Mechanisms of β-Diketones

Toshihiko Osawa, Y. Sugiyama, M. Inayoshi, and Shunro Kawakishi

Department of Food Science and Technology, Nagoya University, Chikusa, Nagoya 464–01, Japan

In the course of an intensive search for novel plant antioxidants, two β-diketones, n tritriacontan-16,18-dione (TTAD) and 4-hydroxytri-triacontan-16,18-dione, were isolated as novel natural antioxidants from Eucalyptus leaf wax; we have also succeeded in obtaining chemically modified β-diketones, the tetrahydrocurcuminoids. Tetra-hydrocurcuminoids were shown to be strong heat-stable antioxi-dants. It was observed that the active methylene moiety between two carbonyls of β diketone derivatives plays an important role in the antioxidative mechanism.

We have been involved in the isolation and identification of novel types of antioxidative substances in plant materials (*1*), in particular, plant leaf waxes, and we have found two different types of lipid-soluble natural antioxidants — phenolic and β-diketone type antioxidative components. As for the former type of natural antioxidative substances, novel γ-tocopherol derivatives named prunusol A and B, and γ-tocopherol conjugates with *p*-coumaric acid as shown in Figure 1 were isolated from *Prunus* leaf wax (*2*). On the other hand, several β-diketone type natural antioxidative substances, *n*-tritriacontan-16,18-dione (TTAD) and 4-hydroxy-tritriacontan-16,18-dione (Figure 2), have been isolated from the leaf wax of Eucalyptus and identified as a novel class of natural antioxidants (*3–4*).

There are many reports on the antioxidative mechanisms of phenolic type antioxidants. In particular, tocopherols were reported to form several oxidative products such as tocopheryl quinones, tocopherol dimers and tocopherol conjugates (*5*) during oxidation by peroxy radicals formed in the process of lipid peroxidation (Figure 3). On the other hand, the antioxidative mechanism of β-diketone type anti-oxidants was not clear.

Recently, Hirose *et al.* (*6*) reported that TTAD strongly inhibited hepatic and pancreatic carcinogenesis. Several simple β-diketones such as 1,1,1-trifluoro-acetylacetone, acetylacetone, benzoylacetone and dibenzoylmethane were reported to inhibit the mutagenicity induced by 2-nitrofluorene using *Salmonella typhi-*

0097–6156/94/0547–0183$06.00/0

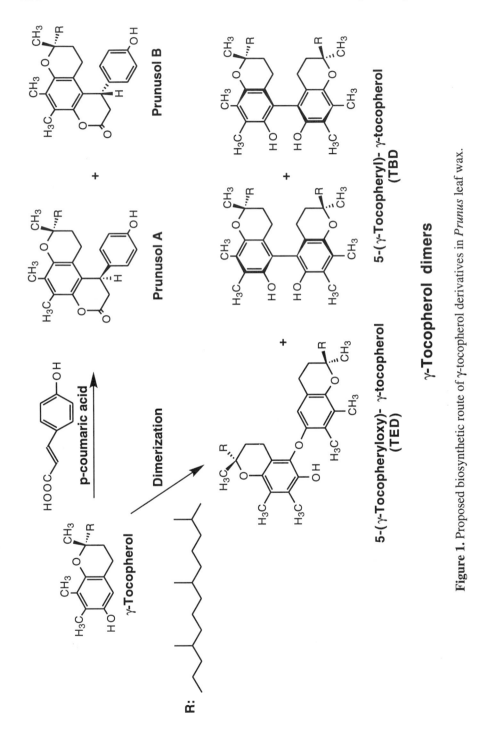

Figure 1. Proposed biosynthetic route of γ-tocopherol derivatives in *Prunus* leaf wax.

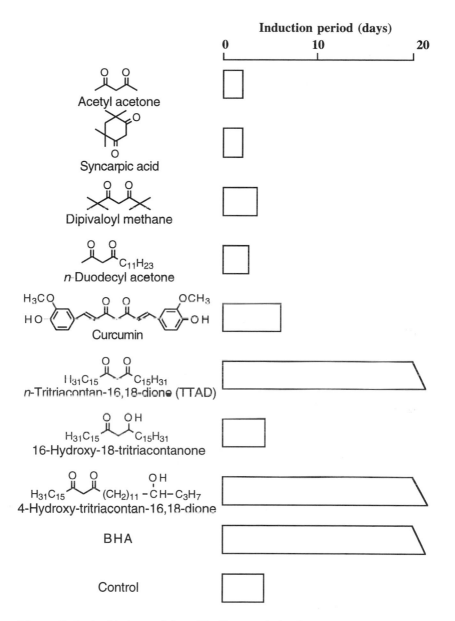

Figure 2. Antioxidative activity of β-diketone derivatives as determined by the thiocyanate method. Thiocyanate method was carried out by the method described in reference 4. Each sample (200 μg dissolved in 100 μl chloroform) was added to linoleic acid/99.0% ethanol/0.2 M phophate buffer and mixed in a conical flask. The solution was incubated at 40°C and the peroxide value was determined by reading the absorbance at 500 nm after coloring reaction with FeCl₂ and thiocyanate at intervals during the incubation.

murium (*7*), but this antimutagenicity does not seem to correlate with antioxidative activity, because acetylacetone was found to have no antioxidative activity (*4*).

Moreover, Pariza *et al.* (*8*) reported that conjugated dienoic derivatives of linoleic acid (CLA) are effective in inhibiting benzo[*a*]pyrene-induced forestomach neoplasia in mice, and also suppressing the process of tumor promotion in the mouse forestomach. One of the possible mechanisms for anticarcinogenicity of CLA is that an oxidized derivative of CLA must be the actual ultimate antioxidant form rather than CLA itself. The structure of the active form has not been determined yet. It has been speculated, however, that introduction of the β-diketone moiety into the CLA molecule is the most likely candidate (*9*). This background prompted us to investigate the antioxidative mechanism of β-diketone-type antioxidants.

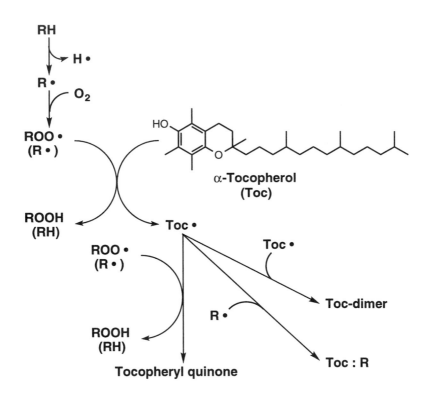

Figure 3. Proposed mechanism for oxidation of tocopherol by peroxy radicals.

Antioxidative Mechanism of TTAD

By the study of structure-activity relationships of β-diketone derivatives (*4*), it was found that long hydrocarbon side chains on both sides of β-diketones seemed to be essential for antioxidative activity although the detailed mechanisms are not clear (Figure 4).

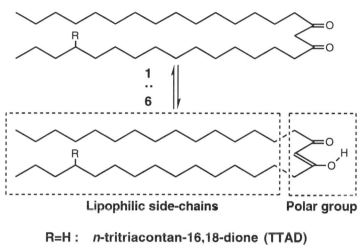

R=H : *n*-tritriacontan-16,18-dione (TTAD)
R=OH : 4-hydroxy-tritriacontan-16,18-dione

Figure 4. Keto-enol tautomerism of β-diketone type antioxidants.

In order to confirm the antioxidative mechanism of β-diketones, TTAD was incubated with 2,2′-azobis(2,4-dimethylvaleronitrile) (AMVN), a lipid-soluble radical initiator, in the presence or absence of linoleic acid methyl ester as shown in Table I.

After 30 hrs, the oxidative product of TTAD was detected by TLC analysis as shown in Figure 5. Lipid radicals formed from linoleic acid methyl ester were assumed to enhance the oxidation of TTAD, because the amount of oxidative product at Rf 0.43 on TLC was found to increase. Isolation and structure elucidation of the oxidative product have been carried out, and the product was identified as palmitic acid. From these data, an antioxidative mechanism of TTAD was proposed as shown in Figure 6. Both enol and keto forms of TTAD must be converted to the hydroperoxide after scavenging free radicals although this intermediate was not chemically identified. Then, it is speculated, the C-C bond at the position of the methylene carbon between the two carbonyls is cleaved.

Antioxidative Mechanism of Tetrahydrocurcumin

We recently succeeded in obtaining a strong heat-stable antioxidant, tetrahydrocurcumin (THC), which contains both phenolic hydroxy and β-diketone structures as functional groups (*10*). THC, derived from curcumin by hydrogenation, is supposed to be a more effective antioxidant than curcumin (Figure 7). Although curcumin has been used widely for many years in indigenous medicine for the treatment of sprains and inflammation, there is a limitation to utilizing curcumin for food and medicinal purposes because of its yellow color. THC, however, has no yellow color. Curcumin was reported to inhibit the microsome-mediated mutagenicity of benzo[*a*]pyrene and 7,12-dimethylbenz[*a*]anthracene as well as tumor promotion although the details are not clear (*11*). This background prompted us to investigate the antioxidative mechanism of THC.

**Table I. Reaction conditions of TTAD with AMVN
in the presence or absence of methyl linoleate**

a) 0.01 M TTAD + 0.01 M methyl linoleate + 0.2 M AMVN
b) 0.01 M TTAD + 0.2 M AMVN
c) 0.01 M methyl linoleate + 0.2 M AMVN
   Solvent: CH_2Cl_2, 3 ml reaction volume, 37°C incubation

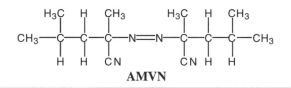

AMVN

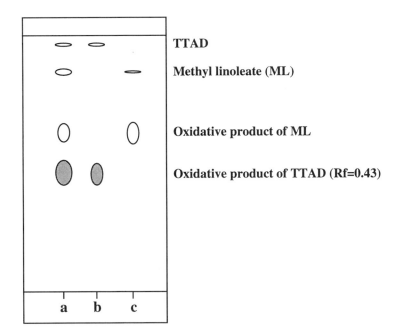

Figure 5. TLC chromatogram of oxidative products of TTAD after 30 hrs.
Solvent: Hexane:Ether:AcOH (8:7:0.1)
Detection: phosphomolybdic acid.

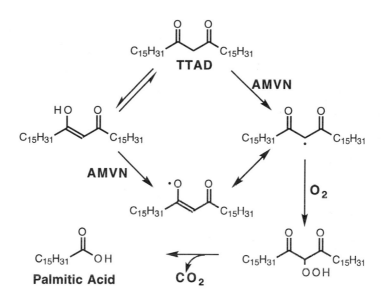

Figure 6. Proposed mechanism for oxidation of TTAD by AMVN.

In order to obtain preliminary information on which is more effective in scavenging lipid radicals, a phenolic hydroxy group or a β-diketone moiety, THC was incubated at room temperature in the presence of linoleic acid. By isolation and spectroscopic analysis of the oxidation products of THC, it was found that the phenolic hydroxy group of THC was involved in scavenging the free radicals at the early stage of lipid peroxidation, and then many degradation products were produced during the further oxidative process. The structure of the main oxidized product formed during the early stage was identified as the ketone-epoxide metabolite (Figure 8). Although the β-diketone moiety of THC must play an important role at the later steps of its antioxidative mechanism, too many reaction products were found to be formed.

In order to get more information on the involvement of the β-diketone moiety of THC, dimethyltetrahydrocurcumin (DMTHC) was prepared by the treatment of THC with diazomethane. DMTHC was also incubated with AMVN at 37°C, and four major products were detected by HPLC analysis as shown in Figure 9. Peaks 1, 2 and 3 were found to increase with time. Isolation and structure determination of these reaction products, peaks 1, 2 and 3, identified them as 3,4-dimethoxybenzoic acid, 3',4'-dimethoxy acetophenone and 3-(3,4-dimethoxy-phenyl)-propionic acid, respectively. Peak 4 decreased gradually after 40 hrs incubation, suggesting that it must be an unstable intermediate. In fact, ^{1}H-NMR data of the separated peak 4 suggested the possibility of the presence of hydro-peroxide, although the detailed structure determination of peak 4 is still in progress. These results suggested that C-C bond at the methylene carbon between two carbonyls in the β-diketone moiety is cleaved as summarized in Figure 10.

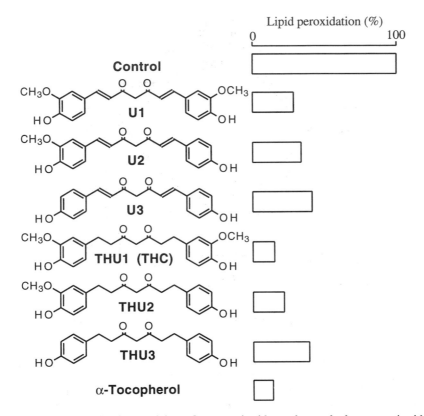

Figure 7. Antioxidative activity of curcuminoids and tetrahydrocurcuminoids. Antioxidative activity was determined by TBA method using rabbit erythrocyte membrane ghost after lipid peroxidation induction by *t*-butylhydroperoxide. (Adapted from reference 10.)

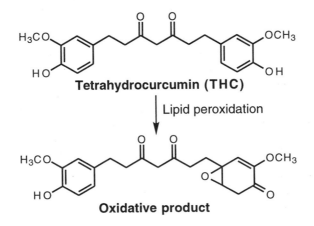

Figure 8. Structure of the oxidative product of THC.

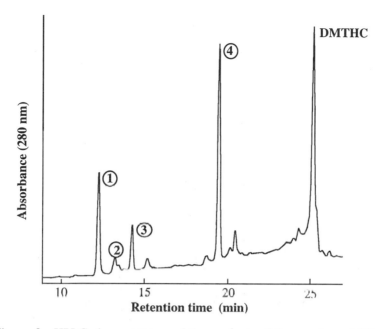

Figure 9. HPLC chromatogram of the products of the reaction of DMTHC with AMVN.
Column: Develosil ODS-5 (φ4.6 x 150 mm)
Flow Rate: 1.0 ml per min
Detection: UV 280 nm
Eluent: A) H_2O:MeOH (8:2, 0.1% TFA); B) MeOH (0.1% TFA)
               100% A → 100% B in 30 min

Potential for the Future Use of Tetrahydrocurcuminoids

These data strongly supported that β-diketones, the active methylene moiety between two carbonyls of β-diketone derivatives in particular, must play an important role in the antioxidative mechanism of THC. Many papers have reported on phenolic type natural antioxidants including vitamin E. This is the first report, however, to show the involvement of β-diketones in antioxidative mechanisms.

TTAD has the potential to be utilized for the purpose of cancer prevention, but it has poor solubility both in polar and lipophilic solvents. One approach to developing a useful β-diketone is the introduction of polar functional groups in the side chain of TTAD, such as 4-hydroxy-*n*-tritriacontan-16,18-dione. Another approach is the chemical modification of curcuminoids to develop a new class of antioxidant which has both phenolic and β-diketone functional groups.

Curcumin is the main yellow pigment present in the turmeric together with two minor yellow pigments, and these pigments are converted to tetrahydrocurcuminoids by hydrogenation. The antioxidative activities of these curcuminoids were evaluated using linoleic acid as the substrate in the ethanol/water system, by thiocyanate and TBA methods, and by an *in vitro* system using rabbit erythrocyte

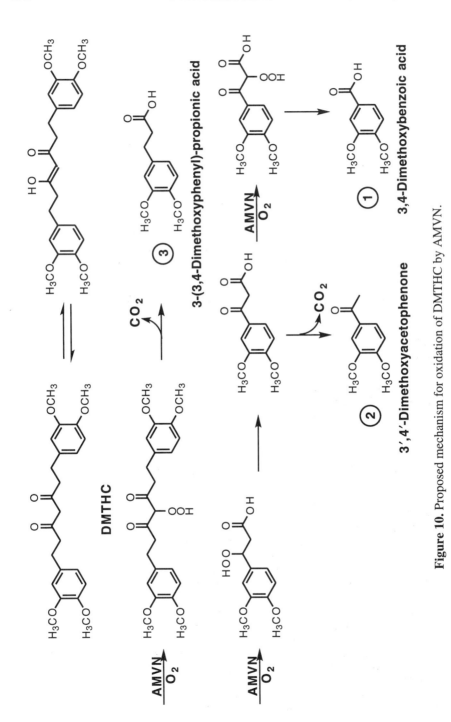

Figure 10. Proposed mechanism for oxidation of DMTHC by AMVN.

ghosts by determining TBARS formation after induction of lipid peroxidation by *t*-butylhydroperoxide. THC showed the strongest antioxidative activity, and the other two tetrahydrocurcuminoids also exhibited much more activity than the original curcuminoids. To our knowledge, there are no reports on the antioxidative properties of β-diketone derivatives, although recent works have indicated the importance of β-diketone structure not only for antioxidative activity but also for cancer prevention.

Literature Cited

1. Osawa, T.; Kawakishi, S.; Namiki, M. In *Antimutagenesis and Anticarcinogenesis Mechanisms II*; Kuroda, Y.; Shankel, D. M.; Waters, M. D., Eds.; Plenum: New York, 1990; pp 139-153.
2. Osawa, T.; Kumazawa, S.; Kawakishi, S. *Agric. Biol. Chem.* **1991**, *55*, 1727-1731.
3. Osawa, T.; Namiki, M. *Agric. Biol. Chem.* **1981**, *45*, 735-939.
4. Osawa, T.; Namiki, M. *J. Agric. Food Chem.* **1985**, *33*, 777-780.
5. Liebler, D. C.; Baker, P. F.; Kaysan, K. L. *J. Am. Chem. Soc.* **1990**, *112*, 6995-7000.
6. Hirose, M.; Ozaki, K.; Takabe, K.; Fukushima, S.; Shirai, T. *Carcinogenesis* **1991**, *12*, 1917-1921.
7. Wang, C. Y.; Lee, M. S.; Nagase, H.; Zukouski, K. *J. Natl. Cancer Inst.* **1989**, *81*, 1743-1747.
8. Pariza, M. W.; Ha, Y. L. In *Mutagens and Carcinogens in the Diet* Pariza, M. W.; Aeschbacher, H. U.; Felton, J. S.; Sato, S., Eds.; Wiley-Liss: New York, 1990; pp 217-221.
9. Ha, Y. L.; Storkson, J.; Pariza, M. W. *Cancer Res.* **1990**, *50*, 1097-1101.
10. Osawa, T.; Ramarathnam, N.; Kawakishi, S.; Namiki, M. In *Phenolic Compounds in Food and Their Effects on Health II : Antioxidants & Cancer Prevention*; Huang, M-T.; Ho, C-T.; Lee, C.Y.,Eds.; ACS, Washington, D.C., 1992; pp122-134.
11. Huang, M.-T.; Smart, R. S.; Wong, C. Q.; Conney, A. H. *Cancer Res.* **1988**, *48*, 5941-5946.

RECEIVED July 6, 1993

PHYTOCHEMICALS IN TURMERIC AND GINGER

Chapter 20

Molecular Mechanism of Action of Curcumin

Inhibition of 12-*O*-Tetradecanoylphorbol-13-acetate-Induced Responses Associated with Tumor Promotion

J. K. Lin, T. S. Huang, C. A. Shih, and J. Y. Liu

Institute of Biochemistry, College of Medicine, National Taiwan University, Number 1, Section 1, Jen-ai Road, Taipei, Taiwan, Republic of China

Curcumin is a potent inhibitor of tumor promotion by phorbol esters (Huang, M-T. *et al.*, 1988). Our recent studies on the molecular mechanism of this inhibition have demonstrated that suppression of c-Jun/AP-1 activation by curcumin is observed in mouse fibroblast cells (Huang, T. S. *et al.*, 1991). The functional activation of the transcription factor c-Jun/AP-1 is believed to play an important role in signal transduction of TPA-induced tumor promotion. Inhibition of c-Jun/AP-1 binding to its cognate motif by curcumin may be responsible for the inhibition of c-Jun/AP-1 mediated gene expression. The TPA-induced increase of protein kinase C in fibroblast cells is suppressed by curcumin. Furthermore, TPA-induced formation of 8-hydroxydeoxyguanosine in cellular DNA is significantly inhibited by curcumin. It appears that the processes of TPA-induced tumor promotion can be suppressed by curcumin through several biochemical pathways at different levels of genetic information.

Epidemiological investigations have demonstrated that environmental factors including dietary substances play a major role in the causation of cancer (*1,2*). Some dietary constituents have also received considerable attention in the chemoprevention of carcinogenesis (*3,4*). Many plant phenolics and carotenoids have been demonstrated to be effective blocking agents against chemical carcinogenesis (*4–6*). Curcumin is a typical plant phenolic and coloring matter from roots of *Curcuma longa* Linn., *Curcuma aromatica Salisb.*, and *Curcuma zedoaria Berg., Zingiberaceae*. It has been widely used as a spice and coloring agent in curry, mustard and other foods.

Curcumin has a wide range of biological and pharmacological activities including anti-thrombotic effect (*7*), antioxidant properties (*8*), antimutagenic effects *in vitro* (*9*), and in smokers (*10*), hypocholesterolemic effects in rats (*11*), and finally hypoglycemic effects in man (*12*). Recently, the inhibitory effects of curcumin and its structurally related compounds on chemical carcinogenesis in culture (*13*) and in mouse skin (*14*) have been investigated.

Effect of Curcumin on the Multiple Steps of Carcinogenesis

Carcinogenesis is a multiple step process that can be divided into initiation, promotion and progression (*15,16*). Curcumin inhibits tumor initiation by 7,12-dimethylbenz[*a*]anthracene (DMBA) and tumor promotion by 12-*O*-tetradecanoylphorbol-13-acetate (TPA) in female CD-1 mice (*14*). Topical application of curcumin also inhibits benzo[*a*]pyrene-mediated DNA adduct formation in the epidermis (*14*). Co-application of curcumin with TPA inhibits TPA-induced skin inflammation, TPA-induced increase in epidermal ornithine decarboxylase activity, TPA-induced epidermal thickness and leukocyte infiltration (*14*). Curcumin also inhibits arachidonic acid-induced edema of mouse ears *in vivo* and epidermal cyclooxygenase and lipoxygenase *in vitro* (*17*).

Effect of Curcumin on the Expression of Cellular Oncogenes

Neoplasia is a multistep process in which normal cells are converted to malignancy through a series of progressive changes. The role of oncogene activation and tumor suppressor gene inactivation in this process is clearly illustrated in several animal models as well as in some human neoplasms (*18*). In addition, cooperation of two or more distinct oncogenes has been demonstrated in a number of settings, and multiple steps in development of a malignancy could correspond to activation of multiple oncogenes in combination with loss of tumor suppressor genes (*15,18*).

Most oncogenes identified so far encode either extracellular growth factors or signal transducing molecules in the plasma membrane or cytoplasm. Ultimately, however, the behavior of a cell is governed by its pattern of gene expression, which is controlled in the nucleus. Some of the nuclear oncogenes such as c-jun, c-fos and c-myc have been clearly shown to function as transcriptional regulatory proteins, thereby indicating that abnormalities in the programmed control of gene expression can lead directly to neoplastic transformation (*19,20*).

Elevated expression of genes transcriptionally induced by TPA is among the events required for tumor promotion (*21*). Analysis of these genes reveals the highly conserved motif 5'-TGACTCAG-3' or 5'-TGAGTCAG-3' conferring TPA inducibility (*22*). This TPA-responsive element (TRE) also serves as the binding site for the AP-1 family of transcription factors. The proto-oncogene (cellular oncogene) product c-Jun/AP-1 represents one member of the AP-1 family. Enhanced binding of c-Jun/AP-1 to the TRE in TPA-stimulated cells has been associated with the increased transcription of different responsive genes (*23,24*). Therefore, activation of c-Jun/AP-1 is probably crucial in transmitting the tumor promoting signals from the extracellular environment to the nuclear transcriptional machinery.

Curcumin was reported to inhibit 98% of the TPA-induced tumor promotion on mouse skin (*14*). The molecular mechanism of this inhibition, however, remained to be explored. To study the mechanisms of the effect of curcumin on TPA, we examined c-jun gene expression. The c-jun gene is positively auto-regulated by its product c-Jun/AP-1 (*24*). When cells are treated with TPA, c-Jun/AP-1 activity is regulated at two levels: first, an immediate post-translational event leads to increased DNA-binding activity of pre-existing c-Jun/AP-1 protein, and second, the activated pre-existing c-Jun/AP-1 protein then binds to the TRE site in the c-jun promoter, activating transcription. In this study (*25*), we show that suppression of the TPA-induced c-jun mRNA correlates with the amount of

curcumin applied to mouse fibroblast cells. The increase in TRE-binding activity of
nuclear extract after TPA treatment can be blocked by curcumin. The reduction of
TPA-induced c-jun mRNA seems to be from attenuation of c-Jun/AP-1 TRE-
binding activity. We also observed that curcumin repressed the expression of c-
Jun/AP-1 responsive simian virus 40-chloramphenicol acetyltransferase (SV-40-
CAT). These results suggest that the suppression of TRE-binding activity is a
mechanism for terminating the elevated expression of genes transcriptionally
activated by TPA.

Curcumin Reduces the TPA-induced c-Jun mRNA Level

To examine the effect of curcumin on c-jun gene expression, Northern blot
hybridization was used to analyze the amount of c-jun mRNA in NIH 3T3 cells
treated with TPA in the absence or presence of curcumin. Total RNA from
confluent (quiescent) cells stimulated with TPA at 50 ng/ml was isolated at
different times and analyzed by Northern Blot Hybridization. Relative abundance
of c-jun mRNA was further quantified by liquid scintillation counting of the
excised spots of dot hybridization (Figure 1). The c-jun mRNA level of quiescent
cells increased steadily up to 45 min after TPA addition and reached maximum at
60 min. This induction profile differs from that of c-fos, which peaked at 30 min.
Thus, the maximal increase of c-jun mRNA after 60 min of TPA treatment was
used as the reference to study curcumin inhibition.
    Curcumin at 10, 15 or 20 µM together with TPA at 50 ng/ml were used to
treat quiescent cells for 60 min. The inhibitory effect of curcumin correlated
directly with curcumin dose. Dot hybridization indicated that 10, 15 or 20 µM of
curcumin inhibited the TPA-induced increase of c-jun mRNA by 21, 28 or 56%,
respectively (Figure 1). We also analyzed the effect of curcumin on the c-fos gene.
The product of c-fos is thought to be complexed and synergistic with c-Jun/AP-1
(23). In contrast to c-jun, however, c-fos mRNA was not affected by curcumin at
30 min after TPA treatment.

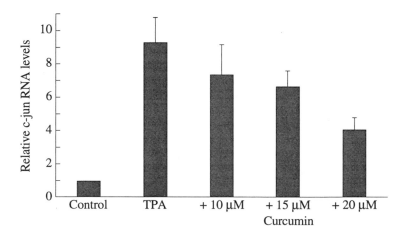

Figure 1. Effect of curcumin on the c-jun mRNA level in NIH 3T3 cells. The
abundance of c-jun mRNA was normalized to β-actin mRNA and represents
the mean of three independent experiments. (Adapted from reference 25.)

Curcumin Suppresses the TPA-induced TRE-binding Activity

The product of the c-jun gene, c-Jun/AP-1, is a transcriptional factor that functions by binding with a specific enhancer element TRE. To evaluate the effect of curcumin on the c-Jun/AP-1 protein, we examined the changes of TRE-binding activity in nuclear extract by using gel retardation assays. After TPA-treatment for 60 min with or without curcumin, crude nuclear extracts were isolated and incubated with a synthetic oligodeoxynucleotide:

5′-GATCG<u>TGACTCA</u>GCGCG-3′
3′-C<u>ACTGAGT</u>CGCGCCTAG-5′

containing a TRE site. We found that whenever 0.5 or 1 µg poly(dI-dC) was added, curcumin (20 µM) could inhibit more than 50% of the TPA-induced TRE-binding activity.

To confirm the suppressive effect of curcumin on TRE-binding activity, a time course experiment was done. NIH 3T3 cells were treated with TPA in the absence or the presence of 15 µM curcumin for 30, 60, 120 and 180 min. Nuclear extracts were isolated and TRE-binding activity was assayed. Each band corresponding to a DNA-protein complex was cut from the gel and the radioactivity was quantified by liquid scintillation counting (Figure 2). The data indicate a 3.8 fold increase of TRE-binding activity shortly after 30 min TPA treatment. This immediate induction was cycloheximide-resistant, suggesting that the induction was via a post translational modification (*24*).

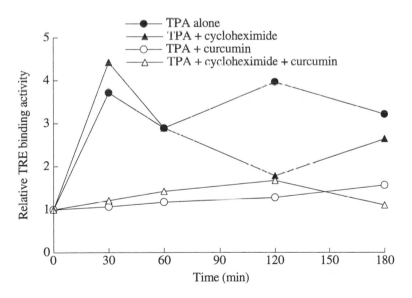

Figure 2. Effect of curcumin on the TRE-binding activity in the nuclear extracts of NIH 3T3 cells. The entire binding reaction mixture included 2 µg nuclear extract, 0.5 ng TRE probe and 0.5 µg poly (dI-dC). The TRE-binding activity of control was designated as 1. (Adapted from ref. 25.)

Curcumin Suppresses SV-40 CAT Expression

We tried to use a transient-expression assay to determine the effect of curcumin on TPA-induced transcriptional enhancing activity of the SV40-CAT (pSV2CAT) reporter gene. The pSV2CAT gene contains the SV40 promoter enhancer region located upstream of the CAT gene (22). Activation of c-Jun/AP-1 was indicated as important for conferring inducibility by TPA upon this promoter enhancer region (24). It has also been demonstrated that the nucleotide sequence of the TPA-inducible control region of SV40 contains a highly conserved sequence motif (31) similar to the TRE described above. As expected, 12-hr TPA-induction of NIH 3T3 cells transfected with pSV2CAT plasmids enhanced the level of CAT expression 2 to 3-fold relative to uninduced cells (Figure 3). The TPA-enhanced CAT level was abolished, however, when the transfected cells were treated with TPA plus 10 µM curcumin. The result showed the inhibitory effect of curcumin on TPA-induced *trans*-activating activity of c-Jun/AP-1.

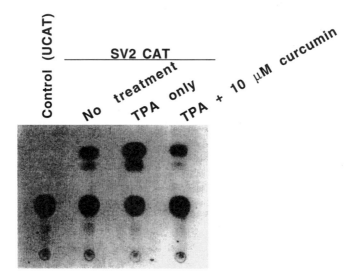

Figure 3. Effect of curcumin on TPA-induced c-Jun/AP-1 *trans*-activation activity. NIH 3T3 cells were transfected with plasmids bearing the CAT gene under control of SV40 promoter enhancer region (SV2CAT) or plasmids lacking the control region (UCAT). Cells were transfected by the calcium phosphate co-precipitation technique and exposed to the precipitate for 12 h. Fresh Dulbecco's modified Eagle's medium/0.5% fetal calf serum was added after 3-min glycerol shock; dishes were treated with TPA at 50 ng/ml in the presence or absence of 10 µM curcumin after 36 h and were harvested 12 h later. CAT assays were done as previously described (22) (Adapted from ref. 25.)

Curcumin Reduces the Activity of Protein Kinase C

Treatment of NIH 3T3 cells that were induced with 0.1 µM TPA with 15 or 20 µM curcumin for 15 min inhibited the TPA-induced protein kinase C (PKC) activity in the particulate fraction by 26 or 60%, respectively (Liu and Lin, unpublished data). It appears that curcumin can not affect the level of PKC as indicated by Western blot analysis. The inhibitory effect of curcumin can be reduced by the presence of mercaptoethanol (20 mM) or dithiothreitol (1 mM).

Curcumin Inhibits the Formation of 8-Hydroxydeoxyguanosine

Recent investigations in our laboratory have demonstrated that the hydroxylation of deoxyguanosine in a hydroxyl free radical generating system is significantly inhibited by curcumin (10 μM). Furthermore, the TPA-induced formation of 8-hydroxydeoxyguanosine and lipid peroxidation in NIH 3T3 cells are remarkably suppressed by the presence of curcumin (Shih and Lin, unpublished data). Prooxidant states and tumor promotion were described in many reports (26, 27). Biological systems are in approximate equilibrium between prooxidant state and antioxidant capacity of biological or dietary systems (28). Cellular prooxidant states appear to play a role in the promotion phase, presumably because tumor promoter treated cells overproduce activated forms of oxygen and/or are deficient in their ability to destroy them (29). Oxidative damage to DNA is considered an initial event in the molecular mechanism of mutagenesis and carcinogenesis (30). Therefore, the anti-promoting effect of curcumin may be attributed to its anti-oxidant and hydroxyl free radical scavenging action.

General Considerations on the Action Mechanism of Curcumin

Recent investigation (32) has demonstrated that the activity of AP-1 is stimulated by TPA and EGF in promotion-sensitive (P$^+$) but not in promotion-resistant (P$^-$) JB6 mouse epidermal cell lines. TPA and EGF also promote neoplastic transformation only in P$^+$ cells. Thus, it has been proposed that AP-1 dependent gene expression is involved in determining sensitivity to tumor promotion. Furthermore, the overall TPA- and EGF-induced levels of jun but not fos expression were higher in P$^+$ cells. This suggests that tumor-promoter regulated c-jun expression may contribute to the differential AP-1 activation observed in these cells and may be important in determining sensitivity to promotion of neoplastic transformation (33). It is of interest to note that mRNA of c-jun but not c-fos is selectively inhibited by curcumin in NIH 3T3 cells.

Intensive studies on the action of curcumin in various biological systems have indicated that this compound has engaged in multiple anti-tumor promoting pathways. It has been demonstrated that the TPA-induced tumor promotion is effectively inhibited by curcumin (14). TPA is a versatile biologically active agent which induces several biosynthetic processes, namely enhanced expression of cellular oncogenes such as c-jun, c-fos, and c-myc, induction of ornithine decarboxylase, elevation or translocation of PKC, induction of cyclooxygenase and lipoxygenase and others. It seems that all of these biochemical processes are required for anabolic pathways and cell proliferation. It is noteworthy that all of these processes have been effectively inhibited by the presence of curcumin (Table I).

It is apparent that the molecular mechanism of action of curcumin is quite complicated and dispersed. The targets of its action are varied from DNA level to RNA level and to protein level (enzyme level). The action of curcumin may proceed simultaneously or sequentially through these different levels. Accordingly, we propose the following pathways for the action of curcumin. The primary target of curcumin could be on the plasma membrane where the activity of PKC is first inhibited. Some PKC-mediated nuclear protein factors are then inhibited through various signal transduction mechanisms. The TRE binding activity of c-Jun/AP-1 is then repressed and finally the transcription of genes essential for cell proliferation are suppressed as indicated by the inhibition of related enzymes such as ornithine decarboxylase, PKC, cyclooxygenase and lipoxygenase. It is obvious that the proposed enzymes and biochemical pathways are far from complete and more enzyme systems should be considered and further investigated.

Table I. Biochemical Effects of Curcumin

Level of action	Parameter	Authors
DNA synthesis	Inhibition of thymidine incorporation into DNA	Huang, Smart, Wong & Conney (1988)
Transcription	Suppression of c-jun mRNA	Huang, Lee & Lin (1991)
Translation	Inhibition of TRE binding by c-Jun/AP-1 protein	Huang, Lee & Lin (1991)
Enzyme	(1) Inhibition of ornithine decarboxylase	Huang, Smart, Wong & Conney (1988)
	(2) Inhibition of cyclooxygenase and lipoxygenase	Huang, Lysz, Ferraro, Abidi Laskin & Conney (1991)
	(3) Inhibition of protein kinase C	Liu & Lin (1992) (unpublished observation)
Activated oxygen	Inhibition of the formation of 8-hydroxydeoxyguanosine	Shih & Lin (1992) (unpublished observation)

Acknowledgements

This study was supported by the National Science Council, NSC 81-0421-B002-05Z, Taipei, Taiwan, ROC.

Literature Cited

1. Doll, R.; Peto, R.*J., Natl. Cancer Inst.* **1981**, *66* 1191–1308.
2. Ames, B. N. *Science* **1983**, *221*, 1256–1264.
3. Wattenberg, L. W. *Cancer Res.* **1983**, *43 (suppl.)*, 2448–2453.
4. Newmark, H. *Nutr. Cancer* **1984**, *6*, 58–70.
5. Wang, C. J.; Chou, M. Y.; Lin, J. K. *Cancer Lett.* **1989**, *48*, 135–142.
6. Mehta, R. G.; Moon, R. C. *Anticancer Res.* **1991**, *11* , 593–596

7. Srivastava, R.; Dikshit, M.; Srimal, R. C.; Dhawan, B. N. *Throm. Res.* **1985**, *40*, 413–417.

8. Sharma, O. P. *Biochem. Pharmacol.* **1976**, *25*, 1811–1812.

9. Nagabhashan, M.; Amnokar, A. J.; Bhide, S. V. *Food Chem. Toxicol.* **1987**, *25*, 545–547.

10. Polasa, K.; Raghuram, T. C.; Krishna, J. P.; Krishnaswamy, K. *Mutagenesis* **1992**, *7(2)*, 107–109.

11. Subba Rao, D.; Chandrasekhara, N.; Satyanarayana, M. N.; Srinivasan, M. *J. Nutr.* **1970**, *100*, 1307–1315.

12. Srinivasan M. *Indian J. Med. Sci.* **1972**, *26*, 269–270.

13. Kuttan, P. R.; Bhanumathy, P.; Nirmala, K.; Georgwe, M. C. *Cancer Lett.*, **1985**, 29 197–202.

14. Huang, M-T.; Smart, R. C.; Wong, C-Q.; Conney, A. H. *Cancer Res.* **1988**, *48*, 5941–5946.

15. Fearon, E. R.; Vorgelstein, B. *Cell*, **1990**, *61*, 759–767.

16. Bertram, J. S.; Kolonel, L. N.; Meyskens, F. L., Jr. *Cancer Res.* **1987**, *47*, 3012–3031.

17. Huang, M-T.; Lysz, T.; Ferraro, T.; Abidi, T. F.; Laskin, J. D.; Conney, A. H. *Cancer Res.* **1991**, *51*, 813–819.

18. Cooper, G. M. In *Oncogenes*; Jones and Bartlett Publishers: Boston, Ma., 1990; pp 141–159.

19. Schuermann, M.; Neuberg, M.; Hunter, J. B.; Jenuwein, T.; Ryseck, R. P.; Bravo, R.; Muller, R. *Cell* **1989**, *56*, 507–516.

20. Cole, M. D. *Ann. Rev. Genet.* **1986**, *20*, 361–384.

21. Bernstein, L. R; Colburn, N. II. *Science*, **1989**, *244*, 566–569.

22. Lee, W.; Mitchell, P.; Tjian, R. *Cell* **1987**, *49*, 741–752.

23. Chiu, R.; Boyle, W. J.; Meek, J.; Smeal, T.; Hunter, T.; Karin, M. *Cell* **1988**, *54*, 541–552.

24. Angel, P.; Hattori, K.; Smeal, T.; Karin, M. *Cell* **1988**, *55*, 875–885.

25. Huang, T. S.; Lee, S. C.; Lin, J. K. *Proc. Natl. Acad. Sci. USA*, **1991**, *88*, 5292–5296.

26. Cerutti, P. A. *Science* **1985**, *227*, 375–381.

27. Cerutti, P. A. *Carcinogenesis* **1988**, *9*, 519–526.

28. Floyd, R. A. *FASEB J.*, 1990, 4, 2587–2597.

29. Amstad, P.; Cerutti, P. A. *Environ. Health Perspect.* **1990**, *8*, 77–82.

30. Breimer, L. H. *Mol. Carcinogenesis* **1990**, *3*, 188–197.

31. Imbra, R. J.; Karin, M.*Nature* **1986**, *323*, 555–558.

32. Bernstein, L. R.; Ben-Ari, E. T.; Simek, S. L.; Colburn, N. H. *Environ. Health Perspect.* **1991**, *93*, 111–119.

33. Ben-Ari, E. T.; Bernstein, L. R.; Colburn, N. H.*Mol. Carcinogenesis* **1992**, *5*, 62–74.

RECEIVED July 6, 1993

Chapter 21

Formation and Reactivity of Free Radicals in Curcuminoids

An Electron Paramagnetic Resonance Study

K. M. Schaich[1], C. Fisher[2], and R. King[3]

[1]Department of Food Science, Cook College, Rutgers, The State University of New Jersey, New Brunswick, NJ 08903
[2]Department of Food Science, University of Delaware, Newark, DE 19716
[3]Kalsec, Inc., P.O. Box 511, Kalamazoo, MI 49005

Intense interest in potential physiological roles of antioxidants and concern over the toxicity of chemical antioxidants commonly added to foods have led to evaluation of phenolic compounds from plants as natural antioxidants. Antioxidant activity by curcumin, demethoxycurcumin and bisdemethoxycurcumin from turmeric has been documented in a wide variety of systems, but detrimental actions have also been noted. Electron paramagnetic resonance (EPR) electrolysis and spin-trapping studies designed to provide a chemical basis explaining contradictory actions of curcuminoids have shown that free radicals in curcuminoids are short-lived and unstable and are very sensitive to oxygen, water and curcumin concentrations. At least two species appear to be formed, phenoxyl radicals on the rings by hydrogen abstraction and alkyl radicals, most likely at the keto-enol double bond, by electron transfer. Curcuminoids reduce HO• production in Fenton/Tiperox reactions by mechanisms involving both HO• scavenging and metal complexation. Curcuminoids also increase HO• production under some circumstances. Important factors appear to be solvent, presence of oxygen, H_2O_2:curcuminoid ratio, and redox cycling of metal complexes. Curcuminoid radical scavenging of lipid radicals may be much more efficient than scavenging of HO•. The net effect of curcuminoids in biological systems should be determined by phase localization and presence/balance between all of the above factors. Several alternative reaction pathways that may be available to curcuminoids under different reaction and environmental conditions are outlined.

Concern over toxicity of synthetic phenolic antioxidants such as butylated hydroxytoluene (BHT) and butylated hydroxyanisole (BHA) has stimulated interest in identifying natural phenolic compounds with comparable antioxidant activity for medical and food applications. One source of phenolic compounds with interesting properties is the ground dried rhizome of the plant *Curcuma longa* Linn, which has

0097–6156/94/0547–0204$06.00/0

been used for centuries as a spice, food preservative, and food colorant. Dried and powdered it is known as turmeric, one of the major components of curry.

Three phenolic compounds — curcumin, demethoxycurcumin, and bis-demethoxycurcumin (Figure 1) — provide the yellow pigmentation of turmeric. After extraction and purification to remove the bitter oils, these curcuminoids are used commercially to color such diverse items as yellow cake mixes, cheese, pickles, and even fruit drinks.

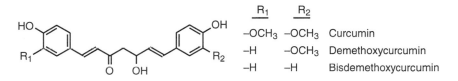

Figure 1. Structures of curcuminoids.

The lipid antioxidant action of these three compounds has been recognized for some time (*1,2*). Curcuminoids also have antiinflammatory (*3–5*) and anti-tumorigenic properties (*6*). They inhibit enzymes such as lipoxygenase (*7*), cyclo-oxygenase (*8*), and ornithine decarboxylase (*6*), which are involved in oxidative damage to cells. Nevertheless, there have been other reports of *opposite* effects, i.e. cytotoxic (*9*), phototoxic (*9,10*), and bactericidal (*11*) effects of curcuminoids, often unconnected to their ability to inhibit lipid oxidation.

While much attention has been given to determining the actions of turmeric pigments in a variety of systems, there is relatively little definitive information about the fundamental chemical properties which may underly the multiple activities of curcuminoids. It is clear from their structures that free radical scavenging capability should be expected, most likely at the phenoxyl groups. Curcuminoids are also metal complexers, with metal chelation occuring at the central keto-enol group (*12,13*) as quadridentate, tridentate, bidentate, or unidentate complexes.

Thus, the very structure of curcuminoids provides potential both for direct anti- and pro-oxidative action and indirect effects via competition with other cellular metal binders, particularly metalloenzymes. The factors that will govern overall action of the individual compounds in specific circumstances are: (a) the ease of phenoxyl hydrogen donation, and the stability or reactivity of the resulting radicals; (b) the effects that solvent phases have on this process; (c) the redox activity of the curcuminoid-metal complexes; and (d) the binding constants of the curcuminoid-metal complexes relative to other metal complexes (including proteins) in cells or other biological systems.

We report here electron paramagnetic resonance studies aimed at iden-tifying curcuminoid free radical species and elucidating the free radical and metal activity behaviors of these turmeric pigments.

Curcuminoid Radicals

Conventional theory holds that efficient antioxidants donate hydrogens readily and, in the process, form free radicals which are relatively stable and unreactive, or which readily convert to unreactive products. To test how curcuminoids behave vis-à-vis this theory, we first utilized an electrochemical cell with a platinum gauze electrode to generate curcuminoid radicals *in situ* in the EPR cavity in order to identify the specific radical species produced in curcumin and the two demethoxy homologues.

The electrochemical experiments were run in buffered aqueous systems as well as in protic (ethanol) and aprotic (acetonitrile and tetrahydrofuran) organic solvents using lithium chloride, lithium perchlorate, and tetraethylammonium perchlorate (TEAP) as electrolytes, as appropriate for the solvent. Current was supplied with either positive or negative potential using a voltage power supply (Power Design); applied voltage was monitored with a voltammeter and electrolysis current was measured with a microammeter.

No curcuminoid radicals could be identified directly by this technique under either oxidative or reductive conditions at any applied voltage level because (1) the curcuminoid radicals are short-lived and unstable (in contrast to BHT, e.g.) and (2) the curcuminoid compounds, even at low concentrations and in organic solvents, adhere to surfaces and rapidly foul the electrodes, impeding electron flow.

We are certain that free radicals were being formed because transient signals were detected a few times with curcumin and demethoxycurcumin using fresh electrode surfaces and thorough argon sparging. The signals decayed within seconds, however, before a single scan could be recorded. We also observed a transient signal with oxidation by iridium (no electrolysis).

Weak free radical signals also were obtained at pH 10 with electrolysis and during exposure to bright light *in situ* in the EPR cavity (Figure 2). These EPR signals <u>required</u> active current or light for generation and disappeared as soon as these sources were removed. No EPR signals were detected with bisdemethoxycurcumin. The spectra from both curcumin and demethoxycurcumin were broad singlets which lacked structural information. This is characteristic of matrix delocalization of free electrons. It is likely that they were observable at high pH because these solutions were very viscous with aggregation of the curcuminoids after exposure to light, thus immobilizing the electrons in the molecular matrices. Solutions were prepared immediately before analysis, and neither aggregation nor free radical formation occurred in controls without light or current. Thus, the radicals did not arise from hydrolysis, which is known to occur rapidly in alkaline media (*14,15*), and, furthermore, the aggregation was most likely mediated by free radicals.

Two EPR techniques that can be used to study free radicals which are very short-lived are continuous flow and spin trapping. Flow experiments are presently in progress. In spin trapping, nitrone or nitroso compounds react with unstable free radicals to form more stable nitroxide radicals whose spectra can provide information about the original radical.

Figure 3 shows the typical signal pattern observed when the spin trap α-(4-pyridyl 1-oxide)-*N-tert*-butylnitrone (4-POBN), reacted with curcuminoid radicals generated electrochemically in tetrahydrafuran (THF) with LiCl as the electrolyte. First signals appeared in curcumin and demethoxycurcumin at 0.2 V applied and increased in intensity as applied voltage levels increased. No signals were detected in the absence of either curcuminoid compound or current. We estimate that redox potentials for these two curcuminoids are a little lower that that of BHT, based on the applied potentials at which radicals were detected in our system. We have not been able yet to detect adducts arising from bisdemethoxycurcumin.

GENERATED BY LIGHT

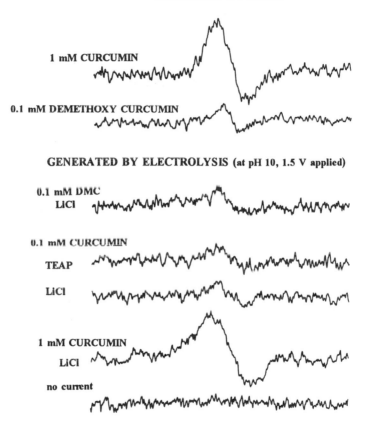

1 mM CURCUMIN

0.1 mM DEMETHOXY CURCUMIN

GENERATED BY ELECTROLYSIS (at pH 10, 1.5 V applied)

0.1 mM DMC
LiCl

0.1 mM CURCUMIN

TEAP

LiCl

1 mM CURCUMIN

LiCl

no current

Figure 2. EPR signals generated in curcumin and demethoxycurcumin by (+) electrolysis (at pH 10) and by light (at pH 7, 0.01 mM PO$_4$). Electrolytes in the electrolyses were lithium chloride (LiCl) or tetraethylammonium perchlorate, as noted on the spectra. Typical instrumental conditions: 10 mW microwave power, 5 G modulation amplitude, 5000–10,000 gain, 0.01 sec time constant.

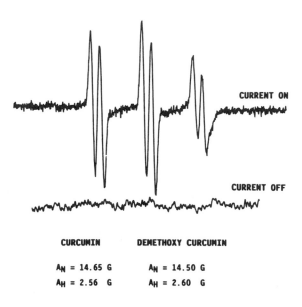

CURCUMIN	DEMETHOXY CURCUMIN
$A_N = 14.65$ G	$A_N = 14.50$ G
$A_H = 2.56$ G	$A_H = 2.60$ G

Figure 3. Typical EPR spectrum from 4-POBN trapping of curcumin radicals generated by (+)electrolysis in deaerated tetrahydrofuran with 0.1 M LiCl as the electrolyte. Reaction conditions: 0.1 mM curcuminoid, 10 mM 4-POBN, 0.5 V applied current. Instrument conditions: 5 mW microwave power, 0.5 G modulation amplitude, 5000 gain, 0.01 sec time constant.

Importantly, these experiments demonstrated clearly that the curcuminoid radicals are very sensitive to oxygen, water, and concentration. Detection of the curcumin adducts required deoxygenation of the samples on a vacuum manifold, use of dry hydrophobic solvent, an electrolyte without oxygen, and curcumin concentrations of 10^{-4} M or lower.

The splitting constants (curcumin: $a_N = 14.65$ G, $a_H = 2.56$ G; demethoxy-curcumin: $a_N = 14.50$ G, $a_H = 2.60$ G) are characteristic of alkyl radical adducts rather than the phenoxyl radicals expected (16). This carbon-centered radical formation by electron transfer during electrolysis (Figure 4) most likely occurs at the center double bond which is doubly activated by its conjugation and its keto-enol tautomerism. Such radicals are notoriously unstable in both oxygen, rapidly forming peroxyl radicals and hydroperoxides, and in water, which add hydrogens and either quenches the free radicals or induces hydrolysis.

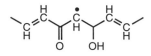

Figure 4. Curcumin and demethoxycurcumin alkyl radical.

That phenoxyl radicals were not observed in these studies may mean either that phenoxyl radicals are not formed at all or that they are formed but are not sufficiently stable for EPR detection. If phenoxyl radicals in curcuminoid compounds are formed only by hydrogen abstraction and not electron transfer, electrolysis would not have produced these radicals. Alternatively: (a) a two-electron process may be involved with comparable redox potentials for each step so that the phenoxyl free radical immediately converts to product, e.g. a quinone, (b) curcuminoid phenoxyl radicals may have sufficient reducing power to reduce oxygen to $O_2^-\bullet$, generating curcuminoid quinones in the process, or (c) hydrogen atoms from water rapidly regenerate the phenols. All of these possibilities are currently under investigation.

Mechanisms of Curcuminoid Action

Competitive spin trapping was used to provide information about relative reactivities of the three curcuminoid compounds. One of the mechanisms by which curcuminoids may exert antiinflammatory activity, in particular, is the scavenging of hydroxyl radicals generated by neutrophils (*1,3*).

To model this activity, we used Ti^{3+}/H_2O_2 (tiperox) and Fe^{2+}/H_2O_2 (Fenton) reactions to generate $HO\bullet$ from H_2O_2.

Tiperox Reaction $Ti^{3+} + H_2O_2 \rightarrow Ti^{4+} + HO\bullet + OH^-$
$$\downarrow DMPO$$
$$DMPO\text{-}adduct$$

Fenton Reaction $Fe^{2+} + H_2O_2 \rightarrow Fe^{3+} + HO\bullet + OH^-$
$$\downarrow DMPO$$
$$DMPO\text{-}adduct$$

The $HO\bullet$ radicals are then trapped with the nitrone 5,5-dimethyl-1-pyrroline *N*-oxide (DMPO) to produce a more stable nitroxide radical. The DMPO-HO• adduct signal will decrease if the curcuminoids react with HO• at competitive rates, about 10^9 $M^{-1}sec^{-1}$ (*17*). If curcuminoids bind Ti and Fe and limit their reducing activity, lower DMPO signals would also result because production of the source HO• radicals would decrease. If curcuminoid-metal complexes are themselves redox active or if curcuminoid radicals reduce oxygen, however, the net result could well be increased DMPO adduct signals. It appears that all of the above mechanisms were active, depending on reaction conditions, which may explain variable effects of curcuminoids in different test systems.

Fenton reactions were run in argon-sparged 2:1 water:acetonitrile. Curcumin, H_2O_2 and DMPO were mixed first. The iron was then added immediately before transferring the mixture to the EPR cavity and starting the analysis. Typical reactions contained 144 mM H_2O_2, 88 mM DMPO, 0.1 mM ferrous ammonium sulphate, and varying amounts of curcumin.

The EPR signals were monitored by averaging successive 30 second scans for up to 15 minutes. Results presented in Figure 5 are the cumulative averages

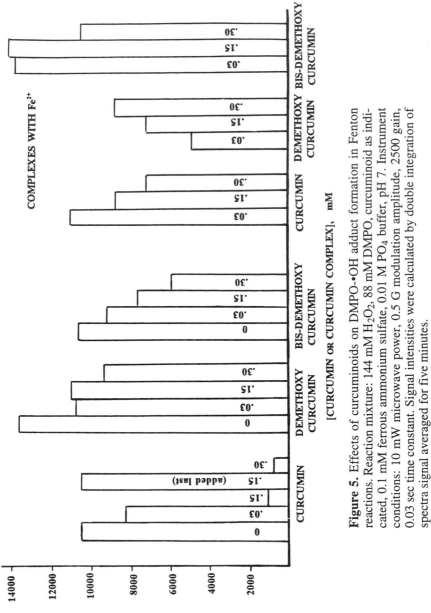

Figure 5. Effects of curcuminoids on DMPO-•OH adduct formation in Fenton reactions. Reaction mixture: 144 mM H_2O_2, 88 mM DMPO, curcuminoid as indicated, 0.1 mM ferrous ammonium sulfate, 0.01 M PO_4 buffer, pH 7. Instrument conditions: 10 mW microwave power, 0.5 G modulation amplitude, 2500 gain, 0.03 sec time constant. Signal intensities were calculated by double integration of spectra signal averaged for five minutes.

over the first five minutes, during which most of the signal intensity change occurred.

It can be seen that the three curcuminoids reduced the DMPO-HO• signal under the conditions of this experiment. In this sytem, curcumin was most effective. The two homologues were considerably less effective in reducing the DMPO-HO• signal. If curcumin was added last (after the iron), the signal diminution did not occur, so curcumin was not reacting with the nitroxide radical of DMPO.

A loss of HO• adduct could result either from radical scavenging, from complexing the iron and limiting the reduction of H_2O_2, or from increased reduction of the spin adduct to a non-radical species. To gain some insight into the role of iron complexation in curcuminoid activity, we ran the same reactions after incubation of the curcuminoids with Fe^{2+} under argon to form curcuminoid-Fe^{2+} complexes. Results, shown on the right side of Figure 5, were markedly different than in the first experiment.

HO• adduct signals were higher for curcumin-Fe^{2+} complexes than for curcumin. This would occur if the curcumin-metal complex was unable to scavenge HO• or if complexation by curcumin facilitated redox cycling of the iron by decreasing the iron redox potential. This suggests that metal complexation may well be an important mechanism involved in biphasic physiological and antioxidant actions of curcumin.

The iron complex of demethoxycurcumin, on the other hand, was more effective than bare demethoxycurcumin in decreasing HO• adducts, but the DMPO signal increased with demethoxycurcumin concentration, contrary to the pattern shown in other data sets.

Bisdemethoxycurcumin iron complexes showed a still different pattern, markedly increasing the DMPO signals substantially above control levels without bisdemethoxycurcumin. Such an increase could occur either if the complex had a lower redox potential than iron alone so that H_2O_2 reduction was facilitated, if there had been excess reducing power in the controls that limited the lifetime of the adducts, or if more HO• were being produced in the presence of the complex, e.g. by autoxidation of the iron. Which of these effects dominated could be determined by the redox poising of the system.

How does complexation by curcuminoid compounds affect iron? There have been reports in the literature that complexation of Fe^{3+} by curcuminoids reduces the iron, and thus curcuminoid complexation should facilitate redox cycling of iron (*18,19*). Increased redox cycling could explain increased HO• production in some curcuminoid reaction systems. Since ferric iron is paramagnetic but ferrous is not, EPR provides an excellent measure for determining whether valence changes are occurring during binding.

The top spectrum in Figure 6 is from an aqueous solution of $FeCl_3$, measured at -77°K. It shows the broad iron absorbance and some fine structure associated with the ligand environment. Incubation of the $FeCl_3$ with equimolar curcuminoids resulted in distinctive changes in the ligand fine structure for curcumin and demethoxycurcumin; little change was apparent with bisdemethoxycurcumin suggesting that binding was not occurring or was very weak. In none of the complexes was the signal envelope diminished, so Fe^{3+} was not being reduced to non-paramagnetic Fe^{2+}. Similarly, when equimolar amounts of Fe^{2+} and curcuminoids were mixed, there was no corresponding appearance of this ferric iron signal, as would have occurred if the Fe^{2+} was being oxidized. Thus, EPR provides no evidence for compulsory valence changes in either direction during what we presume is 1:1 complexation of iron by curcuminoids (based on optical spectral analyses and H^+ release during titration of curcuminoids with Fe^{3+}). This was further substantiated by a lack of appearance of the red ferrous complex when bipyridyl was present during Fe^{3+} complexation.

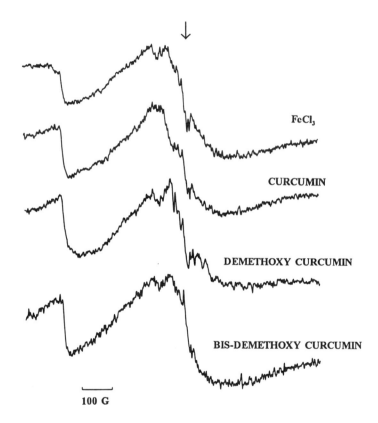

Figure 6. EPR spectra of 0.3 mM ferric iron solutions and 1:1 ferric-curcumin-oid complexes at -77 K. Instrument conditions: 5 mW microwave power, 2 G modulation amplitude, 10,000 gain, 0.03 sec time constant, 1000 G magnetic field scan. Arrow (↓) denotes region of hyperfine structure arising from metal ligands.

How complexation by curcuminoids may affect iron redox potentials, facilitating or inhibiting redox cycling of the iron, is another question. This issue will be discussed in greater detail in a subsequent section.

To be able to study radical trapping competition at lower concentrations of curcuminoids and H_2O_2 and also to avoid the complications of redox-cycling metal inherent in iron reactions, we used titanium-hydrogen peroxide (tiperox) systems for generating HO•. Results are presented in Table I.

At 36 mM H_2O_2, 0.2 mM curcumin and demethoxycurcumin reduced the amount of HO• adduct formed, while bisdemethoxycurcumin slightly increased the DMPO signal. Increasing the amount of curcumin increased the amount of HO• trapped.

Reducing the H_2O_2 to 15 mM H_2O_2 decreased the effectiveness of curcumin, but did not change the demethoxycurcumin reaction, while it markedly increased the effectiveness of bisdemethoxycurcumin in interfering with HO• trapping. At even lower H_2O_2 levels, curcumin increased the HO• adducts by about one-third, and, as in the first systems, doubling the amount of curcumin present increased HO• trapping still further. Bisdemethoxycurcumin doubled the HO•-adduct formation. Demethoxycurcumin, in contrast, reduced the levels HO• trapped by DMPO.

To differentiate effects of metal-binding from radical scavenging activity of curcuminoids, we formed inert barium complexes with all three compounds before incubating them with the tiperox reactions, reasoning that the complexes formed with Ba would be unable to bind Ti. Thus, any reduction in adduct concentrations should result from free radical scavenging reactions.

In the presence of Ba equimolar to the curcuminoids, DMPO-•OH adduct concentrations were reduced about 10% in curcumin and nearly 50% in bisdemethoxycurcumin relative to the uncomplexed reactions, although these levels were still higher than the tiperox control reaction without curcuminoids (Table I). The HO• adduct was dramatically increased with demethoxycurcumin. When Ba was present at only half the concentration of curcumin, the amount of HO• trapped was markedly increased. This could result from either diminished scavenging of HO• by curcumin or increased HO• production in that system.

We recorded optical spectra before and after the tiperox reactions to determine what changes in binding had occurred. Shoulders absorbing at about 460-480 nm are characteristic of barium complexes, and these were present in all three curcuminoid-Ba complexes before reaction with Ti (Figure 7). After the tiperox reaction, curcumin showed a strong absorbance at about 540 nm where titanium complexes have been reported to absorb, suggesting that Ti can either bind in addition to or displace Ba from the curcumin-Ba complex. Bisdemethoxycurcumin had a weak peak in this range, indicating some Ti complexation; Ba absorbance was weak also. This data, along with some additional optical studies of iron binding, EPR data, and some complexometric titrations lead us to feel at the present time that bisdemethoxycurcumin is a weaker metal binder than the other two compounds.

Demethoxycurcumin showed no absorbance peak above 500 nm, so little if any Ti complexation occurred. This means Ti was reacting in "free" form when it so dramatically increased HO• trapping, and suggests that curcuminoid complexation of Ti inhibits its redox activity. Until redox potentials are determined for these complexes, however, it is not possible to distinguish between increased reduction of H_2O_2 to HO• (decreased E^o), decreased reduction of the spin adducts (increased E^o), or increased autoxidation when Ti or other metal complexes of curcuminoids increase HO• trapping.

Why are the tiperox results so different from those with the Fenton reactions? Since the Fenton reactions were run in 33% acetonitrile in water and the tiperox reactions were run in 98% water, solvation and solvent effects may immediately be suspected. Bisdemethoxycurcumin is the most water-soluble of the

Table I. DMPO-•OH Signal Height from Tiperox Reactions: Effects of Curcuminoids

H$_2$O$_2$	DMPO	TiCl$_3$	Curcuminoid	Ba^{++}	Control	+ Curcumin	+ Demethoxy-curcumin	+ Bisdemethoxy-curcumin
36	44	2	2	—	3006	1905	1824	3366
36	44	2	20	—		2194		
15	18	0.9	2	—	1267	1152	822	792
15	18	0.9	4	—		1810		
1.5	18	0.9	2	—	1498	1919	1130	3063
1.5	18	0.9	4	—		2023		
* 1.5	18	0.9	2	—	1024	617	869	872
1.5	18	0.9	2	1		3632		
1.5	18	0.9	2	2		1753	2772	1659

Reaction mixture (in mM) — Signal height

* 50% acetonitrile

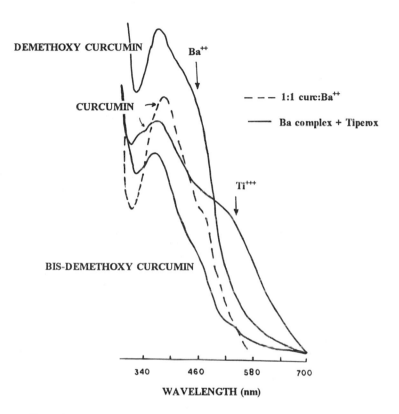

Figure 7. Optical spectra of Ba and Ti binding by curcuminoid complexes. 1:1 Curcumin-Ba complex showing typical Ba-curcuminoid absorption at 472 nm (– – –); spectra of curcuminoid-Ba complexes after reaction with Ti^{3+}-H_2O_2 (———). Shoulder at about 540 nm results from binding of Ti. Reaction conditions: 0.2 mM curcuminoid in argon-sparged 0.01 mM PO_4 buffer, pH 7, was mixed with 0.2 mM $BaCl_2$, equilibrated 15 min, then 15 mM H_2O_2 was added, followed by 0.9 mM $TiSO_4$ to start the binding competition reaction. Optical spectra presented were taken within two minutes after mixing, and did not change during subsequent scans.

three curcumin compounds, so it may have reactions and mechanisms possible in water that the two more hydrophobic compounds do not. When the low H_2O_2 tiperox reactions were run in 50% acetonitrile, the effects were comparable to those obtained in the Fenton reactions, namely all the curcuminoids reduced the HO• trapped, with curcumin being the most effective. This indicates that activity and/or reaction mechanisms of curcuminoid compounds may indeed be different in aqueous and organic environments.

Another possible source of differences is the presence of oxygen in the tiperox systems. The solutions used in the tiperox experiments were not sparged with argon because tiperox reactions are usually insensitive to oxygen. Oxygen could, however, alter the lifetime and fate of curcuminoid radicals or, if sufficient reducing power were present, could provide a source of additional H_2O_2 and HO• via O_2^{-}•. Both of these possibilities are presently under investigation.

DMPO is prone to formation of apparent HO• adducts by artifactual processes. To assess whether the observed HO• radical reactivity and production patterns are specific to the DMPO system, we have begun work with another spin trap, 4-POBN. Results with the 1.5 mM H_2O_2 tiperox system corroborate the DMPO observations of increased HO• in the presence of curcuminoids in aqueous media.

In designing these studies, HO• systems were deliberately chosen because HO• is known to be involved in inflammation and many of the conditions for which curcuminoids have shown a protective effect. Thus, we wanted specifically to test curcuminoid reactivity with HO•. When considering these results, however, we wondered whether HO• scavenging by curcuminoids might not be a minor pathway in cells and tissues, and whether reactivity with organic radicals may be preferred (faster). To address this possibility, a variation of the spin trap technique was used. Since the HO• signal of DMPO can be generated artifactually by non-HO• pathways, one way to show that authentic hydroxyl radicals have been trapped is to react them first with ethanol to form the abstraction radical on the carbon containing the alcohol. DMPO then traps this carbon-centered radical. We used this technique of adding ethanol to the tiperox reaction mixtures (1.5 mM H_2O_2, as in Table I) to determine whether curcuminoids could react with the ethanol radicals and reduce or eliminate DMPO-•OEt adduct formation.

As with HO• reactions, curcumin increased the intensity of the ethanol radical adduct signal detected (Figure 8). Initial levels were higher, and decay over time was slightly faster than when curcumin was not present. Demethoxycurcumin and bisdemethoxycurcumin similarly increased the DMPO-•OEt signal. These results indicate that curcuminoids cannot compete with DMPO for ethanol radicals, that is, the rate constant must be less than about 10^6 M^{-1} sec^{-1}. Since increased DMPO-•OEt adducts can only result from increased HO•, this study also corroborates our observations of apparent increases in HO• production in the presence of curcuminoids under some reaction conditions.

In contrast, when lipid (methyl linoleate) hydroperoxides were used in Fenton reactions in place of H_2O_2 in some preliminary investigations, curcumins completely eliminated the lipid adduct formation. This suggests that curcumins react more rapidly with lipid or other hydrophobic radical species than with HO•.

Summary and Conclusions

EPR and optical studies have demonstrated some patterns of reactivity that may explain how curcuminoids may be both pro- and antioxidants in biological systems and that indicate areas needing additional research.

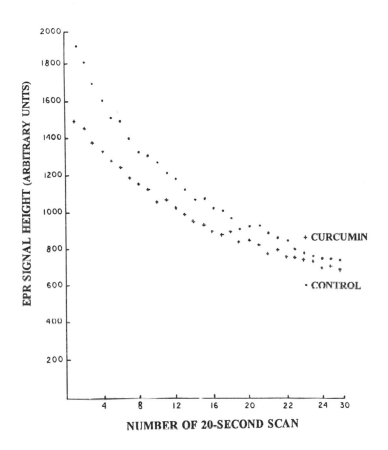

Figure 8. Effects of curcumin on ethanol radical adducts formed in Ti^{3+}/H_2O_2 reactions. Reaction conditions: 1.5 mM H_2O_2, 18 mM DMPO, 0.2 mM curcumin, 1.5 mM EtOH; reaction started by adding 0.9 mM $TiCl_3$. Instrument conditions: 5 mW microwave power, 1 G modulation amplitude, 2500 gain, 0.03 sec time constant. Signal intensities were determined by double integration.

a) Curcuminoid free radicals are short-lived and unstable. They are very sensitive to oxygen, water, and curcumin concentrations. At least two radical species appear to be formed, phenoxyl radicals on the rings by hydrogen abstraction and alkyl radicals, most likely at the keto-enol double bond, by electron transfer.

b) Curcuminoids reduce HO• production in Fenton/Tiperox reactions by mechanisms involving both HO• scavenging and metal complexation.

c) Curcuminoids increase HO• production under some circumstances. Important factors appear to be solvent, presence of oxygen, H_2O_2:curcumin ratio, and redox cycling of metal complexes.

d) Curcuminoid radical scavenging of lipid radicals may be much more efficient than scavenging of HO•.

e) The net effect of curcuminoids in biological systems should be determined by lipid/aqueous phase localization and the presence of, and balance between, all of the above.

The results of these studies suggest several alternative reaction pathways that may be available to curcuminoids under different reaction and environmental conditions. As depicted in Figure 9, curcuminoids may act by either free radical- or metal complex-mediated mechanisms. Phenoxyl radicals are formed by free radical scavenging reactions (hydrogen abstractions). The observed oxygen sensitivity of curcuminoid radicals may be explained by rapid oxidation of the phenoxyl radicals to quinones directly or via reduction of oxygen to O_2^-•. The latter has been claimed previously (10) and is supported by evidence of increased HO• production under some conditions in our studies. Increased reducing power under some conditions also suggests the possibility that curcuminoid radicals may reduce other radicals or metals. The former would have an antioxidant effect, while the latter would increase redox cycling of metals, enhancing HO• production and having a potentially pro-oxidant effect. The dependence of radical lifetime on curcuminoid concentration may be explained by dimerization of the phenoxyl radicals, a well-known phenomenon. The presence of water would regenerate the parent curcuminoid compound.

Electron transfer reactions — e.g., by electrolysis, light, oxidizing metals or by enzymes — yield alkyl radicals in the bridge which, in the presence of oxygen, may form hydroperoxides. These alkyl radicals or their secondary radical species are the most likely radicals formed during photodegradation and phototoxicity of curcumins. Little yet is known about the conditions under which this reaction dominates or about the stability of these hydroperoxides or the reactivity of its decomposition products, but subsequent secondary scission reactions would account for ferulic and vanillic acids and other photodegradation products of curcumins.

Complexation of metals by curcumins may have either pro- or antioxidant effects depending on how the redox potentials and ligand states of the metals are altered. If complexation moderately lowers the redox potential, e.g. of iron, increased redox cycling and production of H_2O_2 and oxygen radicals would be expected via both autoxidation and H_2O_2 reduction, respectively. If redox potentials are shifted to either extreme, redox activity of the metal would be inhibited by complexation.

Data of this study and others indicates that the ratio of curcuminoid to metal may be a critical determinant of the redox potential and reactivity of the complex. Crystal structures of curcuminoids suggest that unidentate (I), bidentate (II),

A. Radical reactions

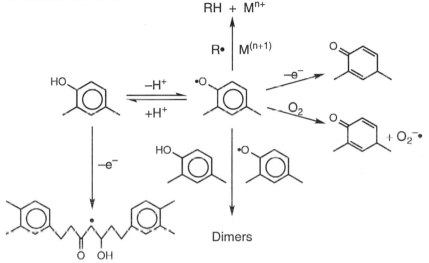

B. Redox activity of curcuminoid-metal complexes

Curcuminoid-M^{n+} + H_2O_2 → Curcuminoid-$M^{(n+1)+}$ + $HO\bullet$ + OH^-

Curcuminoid-M^{n+} + O_2 → Curcuminoid-$M^{(n+1)+}$ + $O_2^-\bullet$ →→ H_2O_2

Figure 9. Proposed alternative reaction pathways that may be available to curcuminoids under different reaction and environmental conditions.

tridentate (III), and quadridentate (IV) metal complexes of curcuminoids may be expected (Figure 10). The type of complex formed would depend on the relative concentrations of curcuminoid and metal present, and would determine the ligand structures and availability of metal ligand sites, factors known to have important influences on redox activity of metal complexes.

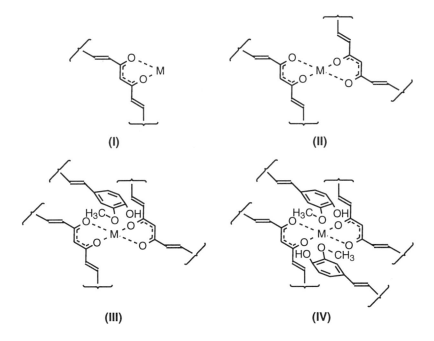

Figure 10. Proposed structures of unidentate (I), bidentate (II), tridentate (III), and quadridentate (IV) metal complexes of curcuminoids.

Thus, the overall effect of a curcuminoid compound in a biological system should be governed by a number of factors, including whether hydrogen abstraction or electron transfer processes are active, the stability or reactivity of the resulting radicals for a particular curcuminoid, the cellular localization — aqueous vs. lipid phase — of the curcuminoid, the redox activity of any metal complexes formed by the curcumin compound, and the binding constants of curcumin-metal complexes relative to other metal complexes (including proteins) in cells and biological tissues. Evidence from this study suggests that metal complex reactions dominate at low curcuminoid and peroxide concentrations and in aqueous media, while radical scavenging activity dominates at high curcuminoid concentrations and in organic phases. All of the reactions proposed in Figure 9 are currently under investigation to provide definitive information about the mechanisms of curcumin reactions.

Acknowledgements

The authors gratefully acknowledge the support of the Medical Department at Brookhaven National Laboratory where the EPR work was performed, New Jersey Agricultural Experiment Station Project 10116 (document 1-10116-92), and Kalsec Corporation.

Literature Cited

1. Sharma, O. P. *Biochem. Pharmacol.* **1976**, *25*, 1811–1812.
2. Toda, S.; Miyasi, T.; Arichi, H.; Tanizawa, H.; Takino, Y. *Chem. Pharm. Bull.* **1985**, *33(4)*, 1725–1728.
3. Srimal, R. C.; Dhawan, B. N. *J. Pharm. Pharmacol.* **1973**, *25*, 447–452.
4. Mukhupadhyay, A.; Basu, N.; Ghatak, N.; Gujral, P. K. *Agents Actions* **1982**, *12(4)*, 508–512.
5. Rao, T.; Basu, N.; Siddiqui, H. H. *Indian J. Med. Res.* **1982**, *75*, 574–578.
6. Huang, M.-T.; Smart, R. C.; Wong, C.-Q.; Conney, A. H. *Cancer Res.* **1988**, *48 (21)*, 5941–5946.
7. Flynn, D. L.; Rafferty, M. F.; Boctor, A. M. *Prostaglandins Leukotrienes Med.* **1986**, *24(2/3)*, 195–198.
8. Huang, M.-T.; Lysz, T.; Ferraro, T.; Abidi, T. F.; Laskin, J. D.; Conney, A. H. *Cancer Res.* **1991**, *51(3)*, 813–819.
9. Tønnesen, H. H.; de Vries, H.; Karlsen, J.; van Henegouwen, G. B. *J. Pharm. Sci.* **1987**, *76(5)*, 371–373.
10. Dahl, T. A.; McGowan, W. M.; Shand, M. A.; Srinivasan, V. S. *Arch. Microbiol.* **1989**, *151(2)*, 183–185.
11. Lutomski, V. J.; Kedzia, B.; Debska, W. *Planta Medica* **1974**, *26*, 9–19.
12. Jansen, A.; Gole, T. *Chromatographia* **1984**, *18(10)*, 546–549.
13. Sharma, K. K.; Chandra, S.; Basu, D. K. *Inorg. Chim. Acta* **1987**, *135*, 47–48.
14. Tønnesen, H. H.; Karlsen, J. Z. *Lebensm. Unters. Forsch.* **1985**, *180*, 132–134.
15. Tønnesen, H. H.; Karlsen, J. Z. *Lebensm. Unters. Forsch.* **1985**, *180*, 402–404.
16. Li, A. S. W.; de Haas, A. H.; Park, L. J.; Watson, M. S.; Chignell, C. F. *STDBII (4.5): A database for spin-trapping implemented on the IBM/compatible and Apple Macintosh personal computers* Library of Medicine, Biophysics, NIEHS: Research Triangle Park, NC, 1990.
17. Janzen, E. G. In *Free Radicals in Biology* Academic Press: New York, 1980, Vol. IV; 116–154.
18. Kunchandy, E.; Rao, M. N. A. *Int. J. Pharm.* **1989**, *57(2)*, 173–176.
19. Kunchandy, E.; Rao, M. N. A. *Int. J. Pharm.* **1990**, *58(3)*, 237–240.

RECEIVED August 11, 1993

Chapter 22

Anti-inflammatory and Cancer-Preventive Immunomodulation through Diet

Effects of Curcumin on T Lymphocytes

Marion Man-Ying Chan and Dunne Fong

Department of Biological Sciences and Bureau of Biological Research, Rutgers, The State University of New Jersey, Piscataway, NJ 08855–1059

Curcumin, found in turmeric and curry, is a cancer preventive anti-oxidant. Amongst its many functions, it inhibits cyclooxygenase and lipoxygenase dependent metabolism of arachidonic acid to prostaglandins and hydroxyeicosatetraenoic acids (HETEs), which have immuno-modulating activities. This study investigates the effects of curcumin on the T cell branch of the immune system: T helper-1 cells (Th1) which produce interleukin-2 (IL-2), T helper-2 cells (Th2) which produce interleukin-4 (IL-4) and cytotoxic T lymphocytes (CTL) which mediate tumor cell lysis. We found that commercial grade curcumin, at 10-60 μM, moderately enhanced IL-4 production, although IL-2 production remained unchanged. Moreover, the development of CTL was also unaffected.

The immunocompetence of the host is one of the major defenses against cancer development. In fact, the first experiment to define the role of lymphocytes in destroying cancer cells was performed as early as the 1950s (1,2). Tumors are probably outgrowths of transformed cells that have successfully escaped destruction by the immune system. Carcinogens, such as UV-irradiation, are immune suppressive (3,4), and some tumors produce immune suppressive substances; for example, mammary tumors produce prostaglandins (5,6).

Anti-tumor promoters have been identified in common food sources (7–10). They include dietary fiber, vitamins A, C and E, compounds in cruciferous vegetables, and some phenolic compounds in tea and spices. These compounds prevent cancer by various mechanisms; for example, reduction of oxygen radical formation, or enhancement of the immune response.

Curcumin

Curcumin (diferuloylmethane), obtained from rhizomes of the plant *Curcuma longa* Linn, is the major yellow pigment in turmeric and curry. It is a common spice, coloring agent, and herbal drug in Asia. The antitumorigenic effect of

0097–6156/94/0547–0222$06.00/0

curcumin is well established (*9–15*). It inhibits tumorigenesis in mouse skin, stomach and colon. The anti-tumorigenic effect may result from a combination of mechanisms. Curcumin is an antioxidant which reduces induction of ornithine decarboxylase activity by phorbol ester, reduces polyamine synthesis and blocks oxygen free radical formation. It inhibits cyclooxygenase-dependent metabolism of arachidonic acid to prostaglandins, and metabolism of arachidonic acid by lipoxygenase to 5-HETE and 8-HETE (*9*). These cyclooxygenase and lipoxygenase products have inhibitory effects on the immune system (*16*). Therefore, curcumin may rescue the immune system from suppression caused by tumors or tumor promoters.

In this chapter, we report on the effects of commercial grade curcumin on the T cell branch of the immune system. This crude extract contained 70% curcumin, together with 30% demethoxycurcumin and bisdemethoxycurcumin. T lymphocytes are divided into T helper cells, cytotoxic T lymphocytes and T suppressor cells. The T helper cells are further divided into Th1 and Th2 (*17*). Th1 cells produce IL-2 and γ-interferon (γ-INF), and are often referred to as the inflammatory T cells because they mediate delayed type hypersensitivity. IL-2 is important for the generation of CTL, which are tumoricidal. Th-2 cells produce IL-4, formerly called B cell growth factor (*18*). It has a unique function, namely the regulation of IgG$_1$ and IgE antibody production (*19*).

Effect of Curcumin on Lymphocyte Proliferation

First we tested whether curcumin is toxic to lymphocytes. Mice were infected with the parasitic protozoan *Leishmania mexicana amazonensis* subcutaneously. After one week, the draining lymph nodes were excised, lymph node cells were stimulated with *Leishmania*, and curcumin was added at various concentrations. The cells were grown in RPMI-1640 medium supplemented with 10% fetal calf serum, together with a final concentration of 0.5% acetone to aid curcumin solubility. After 3 days, the proliferation of the lymphocytes was determined by incorporation of ^{3}H-thymidine.

As shown in Table I, at concentrations up to 60 μM, curcumin did not inhibit proliferation of lymph node cells which contain T and B lymphocytes. We therefore chose to use a maximum of 60 μM when examining effects on T lymphocyte function. Only 5 μM curcumin (ED$_{50}$) is required to inhibit arachidonic acid metabolism *in vitro* and TPA induced tumor promotion (*11*).

Effect of Curcumin on IL-4 Production

We used the murine *Leishmania* infection model for testing lymphokine production because it is currently one of the few systems that clearly show dichotomy for the two branches of helper cell activity (*17*). After infection, in the resistant C57BL/6 mice, the Th1 cells, which produce IL-2 and γ-INF, are activated; whereas in the susceptible BALB/cJ mice, the Th2 cells, which produce IL-4, are activated (*20,21*). Therefore, in this system, the effect of the test compounds on these two compartments of the immune system can be measured independently without interference from each other.

To analyze the effect of curcumin on Th2 activity, lymphocytes were prepared from *Leishmania* infected BALB/c mice. After 5 days of re-stimulation *in vitro*, the amount of IL-4 secreted into the culture supernatants was assayed by the IL-4 specific cell line CT-4S, developed by Dr. William Paul of the National

Institutes of Health (22). The culture supernatants were serially diluted with complete RPMI (RPMI-1640 with 10% fetal calf serum) and CT-4S cells were added to each well. After an incubation of 48 hours, proliferation was determined by measurement of ^{3}H-thymidine uptake. Figure 1 is a representative experiment; relative degree of IL-4 production was determined by comparing the proliferation at a log slope of the curve.

Interestingly, at 10-60 μM, IL-4 production was enhanced, as indicated by a 2.3-fold increase in proliferation of CT-4S cells (Table II). Therefore, at a concentration that was effective for cancer prevention, curcumin enhances immune response *in vitro*. With respect to cancer immunity, Golumbek *et al.* (23) showed that introduction of IL-4 at the tumor site induces strong tumor specific CTL response and causes activation of macrophages. In the same study, they also described systemic protection from the tumor. IL-4 is synergistic with IL-2 in enhancing the *in vitro* and *in vivo* generation of CTL (24). IL-4 can also substitute for IL-2 to activate lymphokine activated killer (LAK) cells, and enhances antigen recognition by inducing major histocompatibility complex antigen (MHC) expression (25). It induces cytolytic activity of macrophages. In fact, the IL-4 induced killing is significantly higher than that of γ-INF (16).

Effect of Curcumin on IL-2 Production

Th1 cells secrete IL-2, a T cell growth factor that potentiates immunity in general. In cancer immunotherapy, IL-2 has been used to culture LAK and tumor-infiltrating lymphocytes (TIL) from excised tumors and has also been used directly with some degree of success, although IL-2 therapy produces severe adverse side effects (26, 27). Therefore, we investigated the effect of curcumin on IL-2 production.

In contrast to Th2, Th1 lymphocytes were prepared from *Leishmania* infected C57BL/6J mice, and the quantity of IL-2 in the culture supernatants was measured by IL-2 specific CT-2 cells from Dr. Frank Fitch of the University of Chicago (28). As shown in Table III, unlike its effect on IL-4, curcumin did not affect the production of IL-2. The proliferation index was 1.06 after addition of curcumin at concentrations that enhanced the production of IL-4.

Since both CT-2 and CT-4S are derived from the same parental line CTLL-2, comparison between the IL-2 and IL-4 experiments also indicated that it is highly unlikely that curcumin affects the indicator cells.

Effect of Curcumin on CTL

Cytotoxic T lymphocytes are responsible for lysis of cancer cells and IL-2 is essential for proliferation of CTL (29-31). Therefore, we studied the effect of curcumin on cytotoxic T lymphocyte generation. Allogeneic CTL were generated by culturing the spleen cells from two strains of mice, C57BL/6J (H-2^b) and BALB/cJ (H-2^d), for 5 days. The CTL activity was evaluated by cytolysis of ^{51}Cr-labeled P815 mastocytoma target cells (H-2^d). As shown in Table IV, curcumin, at 10-50 μM, did not affect the generation of allogeneic CTL.

Curcumin as a Potential Dietary Immune Modifier

The NCI has a program for the development of biological response modifiers that would modify a patient's response to fight tumors (32). Immunostimulants have

Table I. Toxicity of Curcumin on Lymphocytes

Experiment	Concentration of curcumin (µM)			
	0	30	60	125
1	55 ± 2[a]	58 ± 6	52 ± 9	3 ± 0.6
2	65 ± 12	64 ± 3	65 ± 8	52 ± 6
3	115 ± 10	126 ± 9	124 ± 4	112 ± 12

[a] Counts per minute (x 10^3, mean ± S. E.)

Lymph node cells (1×10^5) from mice that had been infected at the tail base with 4×10^6 *Leishmania* promastigotes were restimulated with promastigotes (1×10^4) *in vitro*, in 96 well plates. Curcumin was added to the cultures at the indicated concentrations. After 4 days, ^{3}H-thymidine was added and the cultures were incubated for 12-16 hours. The cells were then harvested by a cell harvestor and the amount of radioactivity incorporated was determined by liquid scintillation counting.

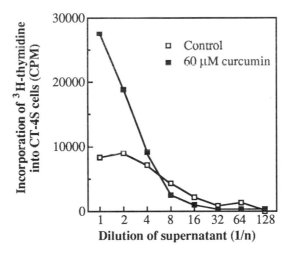

Figure 1. Example of an IL-4 assay. Culture supernatants were serially diluted at 1:2 ratio in 96 well plates, then equal volumes of thoroughly washed CT-4S cells, which had been grown in recombinant IL-4, were added at 5×10^4 cells/ml. The plates were incubated overnight, and then ^{3}H-thymidine was added. After another 12-14 hours, the cells were harvested by a cell harvestor and the samples were prepared for liquid scintillation counting.

Table II. Effect of Curcumin on IL-4 Production

Experiment	Curcumin (µM)	CPM	Proliferation index
1	0	8163[a]	
	60	26865	3.3[b]
2	0	12266	
	30	29009	2.1
	60	29196	2.4
3	0	2416	
	30	4148	1.7
	60	3249	1.3
4	0	13577	
	10	22521	1.7
	60	19621	1.5
5	0	3025	
	10	12266	4.2
	50	13034	4.3
6	0	27879	
	10	32940	1.2
	50	37839	1.4
		Average	2.3 ± 1.2[c]
		Range	1.1-3.5

[a] Counts per minute
[b] Proliferation index: CPM with curcumin/CPM without curcumin
[c] Mean ± S. E.

Lymph node cells from *Leishmania* infected C57BL/6J mice were restimulated *in vitro* in the presence of the indicated concentrations of curcumin as described in Table I. After 5 days of incubation at 37°C, the supernatants were collected and tested for IL-4 as described in Figure 1.

Table III. Effect of Curcumin on IL-2 Production

Experiment	Curcumin (μM)	CPM	Proliferation index
1	0	26672[a]	
	60	22150	0.83[b]
2	0	10384	
	30	12052	1.2
	60	19509	1.9
3	0	65558	
	10	67841	1.0
	30	67974	1.0
	60	34332	0.52
4	0	4899	
	10	5384	1.08
	30	4266	0.87
	60	4724	0.96
5	0	10931	
	10	10677	0.97
	50	13059	1.2
6	0	5772	
	10	6749	1.2
	50	6169	1.1
		Average	1.06 ± 0.30[c]
		Range	0.76-1.36

[a] Counts per minute
[b] Proliferation index: CPM with curcumin/CPM without curcumin
[c] Mean ± S. E.

Lymph node cells from *Leishmania* infected BALB/cJ mice were restimulated *in vitro* in the presence of the indicated concentrations of curcumin, as described in Table I. After 2 days of incubation at 37°C, the supernatants were collected and tested for IL-2 in a manner similar to that described in Figure 1, except CT-2 cells were used.

been successful in clinical trials. For example, bacillus Calmette Guerin (BCG) from mycobacterium, which enhances activation of T lymphocytes and macrophages, has protected patients against bladder carcinoma, acute nonlymphocytic leukemia, and lung cancer in clinical trials (33, 34). Levamisole is another immune enhancer which has been effectively used against mammary, hepatic, and colon carcinomas; it stimulates IL-2 synthesis, γ-INF production, and activates polymorphonuclear leukocytes and macrophages (34-36).

In this report, we identified curcumin, an antitumorigenic compound, as another potential immuno-modulatory compound. Since it is present in commonly comsumed food, this information will be useful for implementing cancer preventive diets. In addition, this information shows that avoiding curcumin in the diet may be helpful in some immune diseases, such as allergies and autoimmune diseases, because consumption of curcumin may enhance IL-4 and thus aggravate the symptoms.

Table IV. Effect of Curcumin on Cytotoxic T Lymphocyte Generation

Experiment	Concentration of curcumin (μM)		
	0	10	50
1	57[a]	65	76
2	66	56	57
3	74	72	84
Average	65 ± 8[b]	64 ± 8.3	72 ± 14

[a] Percent cytolysis
[b] Mean ± S. E.

Allogeneic CTL were generated as described in the text. The target cells were labeled with 0.1 mCi of $Na^{51}CrO_4$, for one hour at 37°C and washed to remove unbound label. The spleen cell preparation and target cells were then mixed. After incubation, an aliquot of the supernatant was removed for determination of the amount of radioactivity released. The percentage of ^{51}Cr released from the target cells due to CTL activity was calculated by the formula: Percent specific lysis = 100 x (^{51}Cr released - spontaneous ^{51}Cr released) / (maximum ^{51}Cr released - spontaneous ^{51}Cr released). Maximum ^{51}Cr release was determined by complete lysis of labeled target cells in 0.5 N HCl.

Acknowledgements

The authors thank Professors A. H. Conney and M-T. Huang for advice and supply of curcumin, and Professors W. E. Paul and F. Fitch for cell lines. This work was supported in part by grant 910024 from the World Health Organization to M.M.C., and grants BC695 from the American Cancer Society and CA49359 from the National Institutes of Health to D.F.

Literature Cited

1. Algire, G. H.; Weaver, J. M.; Prehn, R. T. *J. Natl. Cancer Inst.* **1954**, *15*, 493-507.
2. Penn, I. *Clin. Exp. Immunol.* **1981**, *46*, 459-475.
3. Kripke, M. L. *Immunol. Rev.* **1984**, *80*, 87-102.
4. Kripke, M. L.; Fisher, M. S. *J. Natl. Cancer Inst.* **1976**, *57*, 211-215.
5. Blomgren, H.; Rotstein, S.; Wasserman, J.; Petrini, B.; Hammarstrom, S. *Radiother. Oncol.* **1990**, *19*, 329-335.
6. Ip, C. *Am. J. Clin. Nutr.* **1987**, *45*, 218-224.
7. Good, R. A.; Lorenz, E.; Engleman, R.; Day, N. K. *Oncol. Tumor Pharmacother.* **1990**, *7*, 183-192.
8. Miller, A. B. *Diet and the Aetiology of Cancer*; Springer-Verlag: Berlin, 1989.
9. Huang, M-T.; Smart, R. C.; Wong, C.-Q.; Conney, A. H. *Cancer Res.* **1988**, *48*, 5941-5946.
10. Huang, M-T.; Deschner, E. E.; Newmark, H. L.; Wang, Z-Y.; Ferraro, T. A.; Conney, A. H. *Cancer Lett.* **1992**, *64*, 117-121.
11. Conney, A.; Lysz, T.; Ferraro, T.; Abidi, T. F.; Marchand, P. S.; Laskin, J. D.; Huang, M-T. *Adv. Enzyme Regulation* **1990**, *31*, 385-396.
12. Kuttan, R.; Sudheeran, P. C.; Joseph, C. D. *Tumori* **1987**, *73*, 29-31.
13. Kuttan, R.; Bhanumathy, P.; Nirmala, K., George, M. C. *Cancer Lett.* **1985**, *29*, 197-202.
14. Nagabhushan, M.; Amonkar, A. J.; Bhide, S. V. *Food Chem. Toxicol.* **1987**, *25*, 545-547.
15. Soudamini, K. K.; Kuttan, R. *J. Ethnopharmacol.* **1989**, *27*, 227-233.
16. Rose, D. P.; Connolly, J. M. *Med. Oncol. Tumor Pharmacother.* **1990**, *7*, 121-130.
17. Mosmann, T. R.; Coffman, R. L. *Immunol. Today* **1987**, *8*, 223-227.
18. Paul, W. E.; Ohara, J. *Ann. Rev. Immunol.* **1987**, *5*, 429-459.
19. Kuhn, R.; Rajewsky, K.; Muller, W. *Science* **1991**, *254*, 707-709.
20. Heinzel, F. P.; Sadick, M. D.; Holaday, B. J.; Coffman, R. L.; Locksley, R. M. *J. Exp. Med.* **1989**, *169*, 59-72.
21. Scott, P. *Exp. Parasitol.* **1989**, *68*, 369-372.
22. Hu-Li, J.; Ohara, J.; Watson, C.; Tsang, W.; Paul, W. E. *J. Immunol.* **1989**, *142*, 800-807.
23. Golumbek, P. T.; Lazenby, A. J.; Levitsky, H. I.; Jaffee, L. M.; Karasuyama, H.; Baker, M.; Pardoll, D. M. *Science* **1991**, *254*, 713-715.
24. Widmer, M.; Grabstein, K. *Nature* **1987**, *326*, 795-797.
25. Trenn, G.; Takayama, H.; Hu-Li, J.; Paul, W. E.; Sitkovsky, M. V. *J. Immunol.* **1988**, *140*, 1101-1106.
26. Rosenberg, S. A. In *Important Advances in Oncology 1988*; DeVita, V. T., Jr.; Hellman, S.; Rosenberg, S. A., Eds.; Lippincott: Philadelphia, 1988; pp 217-257.
27. Rosenberg, S. A. *Immunol. Today* **1988**, *9*, 58-62.
28. Chan, M. M. *Eur. J. Immunol.* **1993**, in press.
29. Rouse, B. T.; Röllinghoff, M.; Warner, N. L. *Nature New Biol.* **1972**, *238*, 116-117.
30. Tevethia, S. S.; Blasecki, J. W.; Waneck, G.; Goldstein, A. L. *J. Immunol.* **1974**, *113*, 1417-1423.
31. Henney, C. S. *J. Immunol.* **1971**, *107*, 1558-1566.

32. Mihich, E.; Sakurai U. *Biological Responses in Cancer. 3. Immunomodulation by Anti-Cancer Drugs*; Plenum Press: New York, 1985.
33. Pagano, F.; Bassi, P.; Milani, C.; Meneghini, A.; Maruzzi, D.; Garbeglio, A. *J. Urol.* **1991**, *146*, 32-35.
34. Klefstrom, P.; Nuortio, L. *Acta Oncol.* 1991, *30*, 347-352.
35. Symoens, J.; Decree, W. F.; Van Bever, M.; Janssen, P. A. J. In *Pharmacological and Biochemical Properties of Drug Substances. Vol 2*; Goldberg, M., Ed.; American Pharmaceutical Association: Washington DC, 1979; pp 407-464.
36. Wirth, A.; Green, M.; Zalcberg, J. R. *Aust. N. Z. J. Surg.* **1991**, *61*, 13-22.

RECEIVED May 17, 1993

Chapter 23

Analysis of Curcuminoids by High-Performance Liquid Chromatography

Thomas H. Cooper, Joseph G. Clark, and James A. Guzinski

Kalsec, Inc., P.O. Box 511, Kalamazoo, MI 49005

The effect of trace metal ions on the peak separation of curcuminoids using reverse phase HPLC has been demonstrated. At pH 3, using acetic acid buffer, severe peak tailing occurred after an amount of Fe(II) was added equal to 1000 times the background concentration found in the turmeric extract. The use of a citric acid buffer was found to be effective in separating curcuminoids without peak tailing. Calcium was found not to form a strong complex with curcumin under these conditions. Relative response factors for the three curcuminoids were found to be 1:1:1 at 422 nm.

The turmeric plant is used as a natural colorant in a variety of prepared foods. The yellow color it imparts is caused by curcumin (C) and its two related analogues, demethoxycurcumin (DMC) and bisdemethoxycurcumin (BDMC) (Figure 1). In addition, antimutagenic and pharmaceutical applications have been found for curcuminoids and compounds made from curcuminoids (1–3).

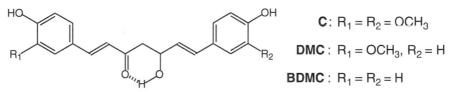

C: $R_1 = R_2 = OCH_3$

DMC : $R_1 = OCH_3, R_2 = H$

BDMC : $R_1 = R_2 = H$

Figure 1. Structures of curcumin, demethoxycurcumin and bisdemethoxycurcumin.

Chemistry

Curcumin decomposes under alkaline conditions to ferulic acid and feruloylmethane (4,5). These researchers found that the rate constant for hydrolysis

0097–6156/94/0547–0231$06.00/0

was pH-dependent and attributed this to various ionizable species existing in equilibrium in solution. This can also be inferred from the color changes from red to yellow to red as the pH is raised from <1 to 5 to >8. Because of these ionized species in solution, the presence of strong acids or bases can result in errors in the spectrophotometric analysis of curcuminoids. Therefore, a buffer is necessary for baseline separation of curcuminoids by reverse phase HPLC.

Curcumin fades rapidly in solution (6). We have measured an approximately 5% decrease in absorbance during the time for typical sample preparation when clear rather than amber glassware is used.

Coordination to Metals

It is well known that curcumin will react with borate to produce a characteristic red complex (7-9). This is due to the keto-enol structure which is also capable of forming a six membered ring around metals such as iron(II), copper(II), and nickel(II) (10,11). Plants require small amounts of many metals for normal growth (12). In view of this, it is reasonable to expect that some of the total curcuminoids found in turmeric roots are actually coordinated with metals.

In order to determine the approximate level of metal ion species in turmeric root, a sample from India was ground with a mortar and pestle and the curcuminoids extracted with acetone using a Soxhlet apparatus. Inductively Coupled Plasma emission spectroscopy was used to determine the concentrations of iron, calcium and magnesium. These results are shown in Table I.

There is approximately a 10% weight yield of turmeric extract obtained from turmeric root. Thus, these results represent an extraction recovery of 0.91% for iron, 0.54% for calcium and 0.04% for magnesium. This would seem to indicate a strong iron and curcumin complex followed by much weaker complexes being formed with calcium and magnesium. In addition, it can be seen that metal contamination from the packing material and metal fittings in an HPLC system is not significant compared to the amount present in the sample.

Table I. Metal Ion Concentration[a]

Sample	[Ca]	[Fe]	[Mg]
Turmeric Root	1690	197	2800
Turmeric Extract	91	18	12
HPLC Effluent[b]	0.5	0.12	<0.1

[a] in ppm
[b] 40% THF, 60% buffer

Metal Addition Study

Separation of curcuminoids is possible both by normal phase (amino or silica) and reverse phase (C_{18}) (13-15). Due to the very labile characteristics of curcuminoids, less reactive phases such as C_{18} are preferred for HPLC analysis (6). Of the three commonly used reverse phase solvents, methanol, acetonitrile and tetrahydrofuran (THF), methanol does not provide the necessary selectivity for the separation of curcuminoids. Acetonitrile shows adequate resolution of curcuminoids by accepted criteria (16). Using THF in place of acetonitrile as the organic modifier reverses the elution order of the curcuminoids with near baseline separation of C and DMC and baseline separation of DMC and BDMC (14,15).

With acetonitrile and THF, curcuminoids show excessive tailing in the presence of metal ions. The pH of the mobile phase can be reduced with acetic acid to provide better separation. The use of citric acid or other additives to the mobile phase has been found to give better separation of species which are capable of complexing with metal ions present in the sample (*17–19*).

In addition, it is possible that residual silanol groups on the C_{18} packing can produce non-specific interactions with curcuminoid keto-enolic groups. In this case, citric acid would act to compete with curcuminoids for these active sites and reduce tailing. This would be analogous to the addition of triethyl amine to reduce tailing when trying to separate amines which have strong affinity for residual silanol groups.

The purpose of this work was to investigate the effect of metal ions on chromatographic resolution. This effect was demonstrated by adding iron(II) in 10-fold increments of its initial concentration in the turmeric extract. These additions had no effect at the 1:10 and 1:100 level, but tailing peaks were observed at the 1:1000 and 1:10,000 levels. This is shown in Figure 2a and 2b.

When citric acid was used in place of acetic acid, iron(II) did not affect the chromatography until the 1:10,000 level. At this point, additional iron(II) actually caused decreased absorbance of the curcuminoids at 420 nm. This has been observed previously for iron(III) (*20*) and is possibly due to oxidation of curcuminoids. It is interesting to note that resolution was not affected by additional iron(II) when citric acid buffer was used (Figure 3a and 3b), although peak areas decreased at the highest level of iron(II).

A study using calcium in place of iron(II) showed no effect. This indicates very weak coordination with curcuminoids as compared to iron(II).

It is presumed that transition metal ions such as nickel(II) and copper(II) would show similar effects to those of iron(II) since they have been shown to form strong complexes with curcuminoids (*10*).

Relative Response Factors and Detection Limits

In a previously published curcuminoid HPLC method using an amine column and absorbance at 420 nm for detection (*13*), the relative response factors for curcuminoids were found to be 1:2.4:3.5 (C:DMC: BDMC). Using a reverse phase column with 40% THF and 60% water buffered to pH 3 with citric acid, calibration curves were generated for pure curcuminoids. Detection limits and determination limits were estimated from the intercepts of the calibration curves (*21*).

Table II. Detection Limits and Relative Response Factors

	Limits of Detection[a]	Limits of Determination[a]	Relative Responses
430 nm			
C	4	13	1.1
DMC	5	17	1.1
BDMC	4	13	1.0
422 nm			
C	4	13	1.0
DMC	6	20	1.0
BDMC	7	22	1.0

[a] in ppm

These are shown in Table II along with the relative response factors determined from the slopes of the curves.

These results seem to refute the claim made by Tønnensen *et al.* that BDMC has an absorption 3.5 times that of curcumin at 420–425 nm (*13*). In order

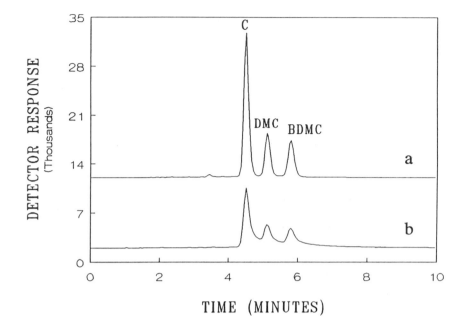

Figure 2. Chromatogram of turmeric extract with acetic acid buffer (a) and with 10 ppm added Fe (II) (1000 times background) (b).

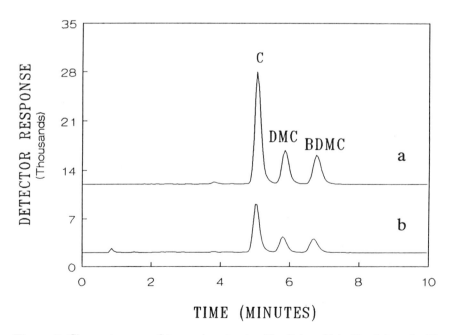

Figure 3. Chromatogram of turmeric extract with citric acid buffer (a) and with 100 ppm added Fe (II) (10,000 times background) (b).

to check this, molar absorptivities were determined in several different solvents, including the HPLC mobile phase. These are shown in Table III.

There is a nine nanometer difference in wavelength maxima between C and BDMC in all solvent systems. There is no single wavelength at which there is a three-fold increase in signal for BDMC over C in any solvent studied.

Table III. Molar Absorptivities of Curcuminoids in Various Solvents

Solvent	Curcuminoid	Wavelength	Logε
Methanol	C	423	4.770
"	DMC	419	4.706
"	BDMC	414	4.675
Acetone	C	421	4.799
"	DMC	416	4.775
"	BDMC	412	4.748
THF	C	423	4.770
"	DMC	419	4.650
"	BDMC	414	4.725
40% THF, 60% Buffer	C	429	4.780
"	DMC	425	4.806
"	BDMC	420	4.702
"	C	425	4.770
"	BDMC	425	4.690

HPLC Method

The following procedure proved satisfactory for the HPLC analysis of curcuminoids: 50 mg turmeric extract is dissolved in 10 ml THF. A 1 ml aliquot of this solution is diluted to 10 ml with a mixture of 40% THF/60% water. Five μl of the diluted sample solution is injected into the HPLC. Any high quality octadecylsilane column should be suitable for the analysis. The mobile phase is 40% THF and 60% water containing 1% citric acid adjusted to pH 3.0 with concentrated KOH solution. The system is run isocratically at a flow of 1.5 ml/min. Detection is at 420 ± 2 nm.

Summary

Baseline separation of curcuminoids was achieved using THF/water buffered with citric acid at pH 3.0 on a C_{18} column. This buffer system allowed quantitation of curcuminoids even in the presence of moderately high concentrations of iron(II). At 420 nm, the three naturally occurring curcuminoids had equal response factors.

Literature Cited

1. Huang, M-T.; Smart, R. C.; Wong,C-Q.; Conney, A. H. *Cancer Res.* **1988**, *48*, 5941.

2. Mulky, N.; Amonhar, A. J.; Bhide, S. V. *Indian Drugs* **1987,** *25*(3), 91.
3. Sharma, K. K.; Chandra, S.; Basu, D. K. *Inorg. Chim. Acta.* **1987,** *135*(1), 47.
4. Tønnesen, H. H.; Karlsen, J. *I. Lebensm. Unters. Forsch.* **1985,** *180,* 132.
5. Tønnesen, H. H.; Jarlsen, J. *I. Lebensm. Unters. Forsch.* **1985,** *180,* 402.
6. Khurana, A.; Ho, C. T. *J. Liquid Chromatogr.* **1988,** *11,* 2295.
7. Grotheer, E. W. *Analytical Chem.* **1979,** *51,* 2402.
8. Janßen, A.; Gole, T. *Chromatographia* **1984,** *18,* 546.
9. Quint, P.; Umland, F.; Sommer, H.D. *Z. Anal. Chem.* **1977,** *285,* 356.
10. Arrieta, A.; Dietze, F.; Mann, G.; Beyer, L.; Hartung, J. *J. Prakt. Chemie* **1988,** *330,* 111.
11. Pai, S.; Zhu, Y. *Rock and Mineral Analysis* **1991,** *10,* 178.
12. Miller, E. V. *The Chemistry of Plants*; Reinhold Publishing: New York, **1957;** pp 145-56.
13. Tønnesen, H. H.; Karlsen, J. *J. Chromatogr.* **1983,** *259,* 367-71.
14. Smith, R.; Witowska, B. *Analyst.* **1984,** *109,* 259.
15. Rouseff, R. L. *J. Food Sci.* **1988,** *53*(6), 1823.
16. Poole, C. F.; Schuette, S. A. *Contemporary Practice of Chromatography*; Elsevier: New York, **1984;** p 20.
17. Verzele, M.; Dewaele, C. *Chromatographia* **1984,** *18*(2), 84.
18. Verzele, M.; Dewaele, C. *J. Chromatogr.* **1981,** *217,* 377.
19. Nawrocki, J.; Buszewski, B.J. *J. Chromatogr.* **1988,** *449,* 1.
20. Wei, L.; Ziqing, D. *Food and Fermentation Industries* **1991,** *2,* 61.
21. Kieckhefer, T. J; Munroe, J. H. *J. Am. Soc. Brew. Chem.* **1984,** *42*(4), 173.

RECEIVED June 7, 1993

Chapter 24

Structure of Antioxidative Compounds in Ginger

Hiroe Kikuzaki, Yoko Kawasaki, and Nobuji Nakatani

Department of Food and Nutrition, Osaka City University, Sumiyoshi, Osaka 558, Japan

Fourteen gingerol related compounds and seven diarylheptanoids were isolated from ginger rhizomes. Their structures were elucidated on the basis of chemical and spectroscopic evidence. Most of the isolated compounds exhibited stronger antioxidative effect than α-tocopherol. The antioxidant activity depended upon substituents on the side chain and benzene ring, and the length of side chain.

The fact that ancient people empirically used spices and herbs as preservatives of foods suggests they have antioxidative or antimicrobial properties. Accordingly, many scientists have given their attention to spices and herbs as a source of natural antioxidants with the idea of finding safe and effective compounds.

The rhizome of ginger (*Zingiber officinale* Roscoe) is currently widely used as a spice and a food seasoning due to its sweet aroma and pungent taste. It is known from several investigators that ginger has antioxidative effects (*1–2*), but there have been few reports on its antioxidative constituents (*3–4*). Therefore, we studied the chemical constituents of ginger and their antioxidant activity.

Structures of the Constituents of Ginger

Extraction and Isolation of the Constituents of Ginger. Ground rhizomes of dried steamed ginger were successively extracted with dichloromethane and methanol. The former was separated by steam distillation to give a volatile and a nonvolatile fraction. The antioxidant effect of each extract and fraction was measured by the ferric thiocyanate method and thiobarbituric acid method (TBA) using linoleic acid as a substrate in ethanol-water system. The dichloromethane extract showed stronger activity than α-tocopherol, whereas the methanol extract exhibited the same activity as α-tocopherol. Between two fractions of the dichloromethane extract, only the nonvolatile one showed strong activity. This fraction was purified by column chromatography on silica gel and Sephadex LH-20, and by preparative HPLC to give fourteen gingerol related compounds (Figure 1, compounds **1–14**) and seven diarylheptanoids (**15–21**). We have already reported the structure

0097–6156/94/0547–0237$06.00/0

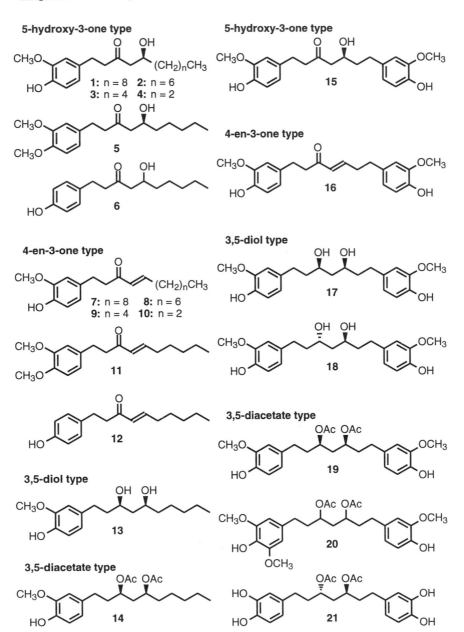

Figure 1. Structures of the isolated compounds of ginger.

elucidation of compounds **13–21** (*5–7*). In this paper we describe the structure determination of compounds **1–12** and their antioxidant activity.

Structure Elucidation of the Isolated Compounds. The structures of the isolated compounds were determined by the methods described below.

Compound 3. Compound **3** was obtained as one of the major components of ginger rhizome. The spectral data (IR, ^{1}H- and ^{13}C-NMR and MS) and specific rotation ([α]$_D^{25}$: +23.9°) showed this compound to be [6]-gingerol (*8*).

Compounds 1, 2 and 4. The spectral characteristics of compounds **1**, **2** and **4** were very similar to those of **3** except for their molecular ion peaks. A [M]$^+$ peak of compound **1** at m/z 350 was 56 mass larger than that of **3**, which indicated **1** to be [10]-gingerol, having a side chain composed of fourteen carbon atoms. Compounds **2** and **4** were determined to be [8]- and [4]-gingerol, respectively, from their mass fragmentation patterns (Figure 2) (*9*).

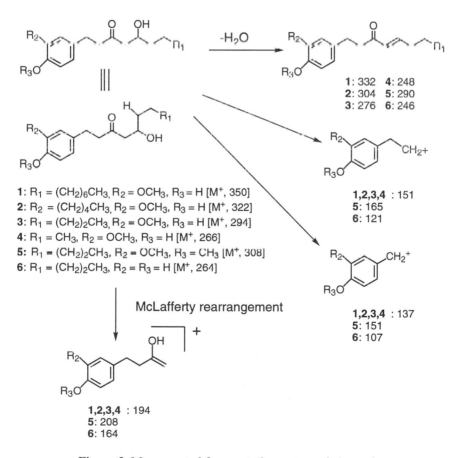

Figure 2. Mass spectral fragmentation pattern of gingerols.

Compound 5. Compound **5** was obtained as colorless needles. The spectral data were very close to those of [6]-gingerol. The EI mass spectrum exhibited a $[M]^+$ peak at m/z 308 and $[M-H_2O]^+$ at m/z 290, which are 14 mass larger than those of **3**. The presence of a 3,4-dimethoxyphenyl group was supported by aromatic proton signals [δ6.73 (2H, m) and 6.79 (1H, d, J=8.6 Hz)] accompanied with arylic methoxyl groups [δ3.85 (3H, s) and 3.86 (3H, s)] in the ^{1}H-NMR, aromatic carbon signals (δ148.6, 147.3, 133.9, 120.2, 111.7 and 111.4) in the ^{13}C-NMR spectrum and a typical fragment ion peak at m/z 151 $[CH_2C_6H_3(OCH_3)_2]^+$ in the mass spectrum. The ^{13}C-NMR spectrum revealed ten more carbons, corresponding to the straight chain carbons, in which a carbonyl (δ211.3) and a hydroxyl methine (δ67.7) carbons were found. In the ^{1}H-NMR spectrum, the signal at δ2.73 (2H, t), corresponding to methylene protons at C-2, indicated that C-3 was substituted by a carbonyl group. When the signal of a methine proton adjacent to a hydroxyl group at δ4.02 was irradiated, two double doublets of geminal methylene protons at δ2.48 and 2.57 changed into two doublets with a coupling constant of 17.4 Hz each. This fact showed a hydroxyl group to be at C-5. The substitution pattern of the decane was supported by the presence of a characteristic fragment ion peak at m/z 208 derived from McLafferty rearrangement between C-4 and C-5 (Figure 2).

These data indicated compound **5** to be a methylated derivative of [6]-gingerol. To confirm the structure of **5**, [6]-gingerol was methylated with methyl iodide in the presence of dry K_2CO_3. All spectral data and specific rotation of **5** were identical with those of the synthetic compound. Consequently, the structure of **5** was determined to be 5-hydroxy-1-(3,4-dimethoxyphenyl)decan-3-one. The presence of this compound was speculated to occur only by GC-MS and HPLC in ginger (*10,11*), and has not been isolated previously.

Compound 6. Compound **6** gave a $[M]^+$ peak at m/z 264.1703 in the HR mass spectrum, corresponding to a molecular formula of $C_{16}H_{24}O_3$. The spectral characteristics indicated that compound **6** had an analogous structure to [6]-gingerol. In comparison with the ^{1}H-NMR spectrum of [6]-gingerol, a pair of two *ortho*-coupling protons were observed at δ6.74 and 7.03 in **6** in place of the signals due to a 4-hydroxy-3-methoxyphenyl group. This finding denoted compound **6** possessed a 1,4-disubstituted aromatic ring. In addition, the presence of aromatic carbon signals [δ154.0, 132.9, 129.4 (2C) and 115.4 (2C)] in the ^{13}C-NMR spectrum and a stable ion peak at m/z 107 $[CH_2C_6H_4(OH)]^+$ confirmed that a 4-hydroxybenzyl moiety was present in the molecule. This fact indicated that compound **6** is a demethoxylated derivative of [6]-gingerol, which was supported by a typical mass ion peak at m/z 164 $[C_{10}H_{12}O_2]^+$ (Figure 2). Consequently, compound **6** is concluded to be 1-(4-hydroxyphenyl)-5-hydroxydecan-3-one, which was isolated for the first time.

Compounds 7–10. Compounds **7–10** were identical with homologues of shogaol from their spectral data and were determined to be [10]-, [8]-, [6]- and [4]-shogaol, respectively. Among them, compound **10**, [4]-shogaol, was isolated for the first time, though it had been detected in ginger by GC-MS before (*12,13*).

Compound 11. Compound **11** had a molecular formula of $C_{18}H_{26}O_3$ ($[M]^+$, m/z 290.1880). The spectral data were very close to those of [6]-shogaol, except for the presence of another methoxyl group instead of the hydroxyl one. In the IR spectrum, no hydroxyl absorption band was observed in **11**. The ^{1}H-NMR spectrum showed the signals of two methoxyl groups as each 3H singlet at δ3.85 and 3.86. The pattern of signals due to an aromatic ring was similar to that of compound **5** in the ^{1}H- and ^{13}C-NMR spectra. The signals at δ6.09 (1H, brd, J=15.9 Hz) and 6.82 (1H, td, J=7.0, 15.9 Hz) in the ^{1}H-NMR spectrum showed the presence of an α,β-*trans*-unsaturated carbonyl group, which was supported by the signals at δ199.7, 147.8 and 130.3 in the ^{13}C-NMR spectrum. A fragment ion at m/z 219 $[C_{13}H_{15}O_3]^+$ produced by cleavage at α position of double bond indicated that a carbonyl group was located on C-3 and a conjugated double bond on C-4. Furthermore, the methylated compound of [6]-shogaol with methyl iodide gave the same compound as **11**. Accordingly, compound **11** was determined to be 1-(3,4-di-methoxyphenyl)-4-decen-3-one, which is newly isolated from a natural source.

Compound 12. Compound **12** gave a $[M]^+$ peak at m/z 246.1608, corresponding to a molecular formula of $C_{16}H_{22}O_2$. The signals due to an aromatic ring of **12** resembled to those of compound **6** in the ^{1}H- and ^{13}C-NMR spectra, which indicated compound **12** to have a 4-hydroxyphenyl group. The presence of an α,β-*trans*-unsaturated double bond was supported by the signals at δ6.09 (1H, td, J=1.2, 15.9 Hz) and 6.82 (1H, td, J=6.7, 15.9 Hz) in the ^{1}H-NMR and at δ200.0, 148.0, 130.3 in the ^{13}C-NMR spectrum. Furthermore, the mass spectrum exhibited characteristic fragment ion peaks at m/z 175, produced by a cleavage at α-position of a double bond, and at m/z 107, corresponding to a benzylic cation. Consequently, compound **12** was concluded to be 1-(4-hydroxyphenyl)-4-decen-3-one. This is the first time that compound **12** was isolated from a natural source.

Structures of the Isolated Compounds of Ginger. Gingerol related compounds and diarylheptanoids which we have isolated until now can be classified into four groups according to the substitution pattern of side chain as follows: (1) 5-hydroxy-3-one type, (2) 4-en-3-one type, (3) 3,5-diol type and (4) 3,5-diacetate type as shown in Figure 1. Taking note of the substituents on benzene ring, compounds having five types of phenyl group such as 4-hydroxyphenyl, 3,4-dihydroxyphenyl, 4-hydroxy-3-methoxyphenyl, 3,4-dimethoxyphenyl and 4-hydroxy-3,5-dimethoxy-phenyl group were found in ginger. Among them, there were the greatest number of compounds with 4-hydroxy-3-methoxyphenyl group in the isolated ones. In addition, many gingerol related compounds had a side chain with ten carbons.

Antioxidant Activity of the Isolated Compounds

The antioxidant activity of the isolated compounds was examined by the ferric thiocyanate method. Gingerol related compounds with the same length of side chain and with a 4-hydroxy-3-methoxyphenyl group (**3**, **9**, **13** and **14**) had a higher activity than α-tocopherol when they were added at a concentration of 200 μM to a substrate solution. The activity increased in the order **3** (type 1), **9** (type 2), **13** (type 3), **14** (type 4). Diarylheptanoids having two 4-hydroxy-3-methoxyphenyl groups (**15–19**) were also more active than α-tocopherol at the same concentration. The activity increased in the order **18**, **15**, **16**, **17**, **19**, which indicated the similar result at gingerol related compounds. Furthermore, these five compounds showed a stronger activity than curcumin, a known antioxidative compound from turmeric (*Curcuma domestica*). These results suggested that the substitution pattern of side chain has an influence on the antioxidant activity.

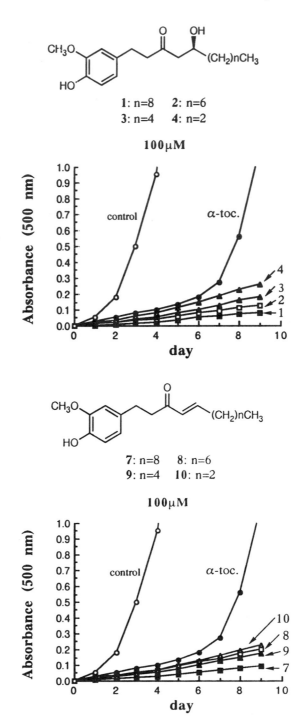

Figure 3. Antioxidant activity of gingerols and shogaols with different length side chains (ferric thiocyanate method: concentration 100 μM).

Concerning the substitution pattern on the benzene ring, compound **3** showed a remarkable activity, while compound **5** had no activity and **6** exhibited slight activity. Similar results were obtained in the case of shogaols (**9, 11** and **12**). In addition, compound **19** had a higher activity than **20** or **21**. These results indicated that the substitution pattern of the benzene ring definitely played an important role in the antioxidant activity.

Figure 3 shows the effects on the activity of gingerol related compounds with different length side chains when they were added at a concentration of 100 µM. All these compounds were effective and higher activity is correlated with longer side chains. Accordingly, it is suggested that antioxidant activity also depends upon the length of side chain.

Literature Cited

1. Hirosue, T.; Kawai, H.; Hosogai, Y. *Nippon Shokuhin Kogyo Gakkaishi* **1978**, *25*, 691.
2. Lee, C. Y.; Chiou, J. W.; Chang, W. H. *J. Chinese Agric. Chem. Soc.* **1982**, *20*, 61.
3. Fujio, H.; Hiyoshi, A.; Asari, T.; Suminoe, K. *Nippon Shokuhin Kogyo Gakkaishi* **1969**, *16*, 241.
4. Lee, I. K.; Ahn, S. Y. *Korean J. Food Sci. Technol.* **1985**, *17*, 55.
5. Kikuzaki, H.; Usuguchi, J.; Nakatani, N. *Chem. Pharm. Bull.* **1991**, *39*, 120.
6. Kikuzaki, H.; Kobayashi, M.; Nakatani, N. *Phytochemistry* **1991**, *30*, 3647.
7. Kikuzaki, H.; Tsai, S. M.; Nakatani, N. *Phytochemistry* **1992**, *31*, 1783.
8. Connell, D. W. ; Sctherland, M. D. *Aust. J. Chem.* **1969**, *221*, 1033.
9. Shoji, N.; Iwasa, A.; Takemoto, T.; Ishida, Y.; Ohizumi, Y. *J. Pharm. Sci.* **1982**, *71*, 1174.
10. Chen, C. C.; Ho, C. T. *J. Chromatogr.* **1986**, *360*, 163.
11. Masada, Y.; Inoue, T.; Hashimoto, K.; Fujioka, M.; Uchino,C. *Yakugaku Zasshi* **1974**, *94*, 735.
12. Masada, Y.; Inoue,T.; Hashimoto, K.; Fujioka, M.; Siraki, K. *Yakugaku Zasshi* **1973**, *93*, 318.
13. Harvcy, D. J. *J. Chromatogr.* **1981**, *212*, 75.

RECEIVED April 4, 1993

Chapter 25

Chemistry of Ginger Components and Inhibitory Factors of the Arachidonic Acid Cascade

Shunro Kawakishi, Y. Morimitsu[1], and Toshihiko Osawa

Department of Food Science and Technology, Nagoya University, Chikusa, Nagoya 464–01, Japan

Ginger extracts exhibited a strong inhibitory effect on human platelet aggregation; the inhibitory factors were isolated and studied in detail. Six kinds of gingerol analogues from n-hexane extracts of ginger were identified as the inhibitors of platelet aggregation, but their activities were very weak compared with that of eugenol analogues found in several spice species. Since the inhibitory activity of ginger could not be accounted for by the six gingerols, new inhibitors from the n-hexane extracts of ginger were investigated. Two labdane-type diterpene dialdehydes isolated from the extracts strongly inhibited the platelet aggregation as much as indomethacin, but these compounds did not suppress the activity of prostaglandin endoperoxide (PGH) synthase in the arachidonic acid cascade. On the other hand, these diterpene dialdehydes also inhibited human 5-lipoxygenase as strongly as the α-sulfinyl disulfides found in onion.

Ginger (*Zingiber officinale* Roscoe) is widely used as a spice because of fragrant and pungent principles among its constituents and is well known as a crude drug with several pharmacological functions. Aqueous extract of ginger has exhibited inhibitory effects against the biosynthesis of thromboxane (TX) and prostaglandin (PG) (*1,2*) and the inhibitory principles have been reported to be gingerol analogues (*3*).

Nutmeg oil also inhibited PG biosynthesis (*4*); its main inhibitor is eugenol, a related effect of which is the suppression of TXA_2 formation in the arachidonic acid cascade (*5*). The methanol extracts of several other spices have been studied to characterize the inhibitors of human platelet aggregation, and the extracts of clove and allspice exhibited the strongest activities. It was determined that their active factors were also eugenol and its analogues, and o-methoxyphenol-containing components are essential for their activity. There are many o-methoxyphenol-containing compounds in ginger such as the gingerol analogues.

[1]Current address: School of Food and Nutritional Sciences, University of Shizuoka, Yada 52–1, Shizuoka 422, Japan

0097–6156/94/0547–0244$06.00/0

This review paper concerns the re-evaluation of ginger components as the inhibitors of platelet aggregation and their characterization. We have also isolated and identified two labdane type diterpenes as strong inhibitors of both platelet aggregation and 5-lipoxygenase of the arachidonic acid cascade.

Inhibitory Effects of Ginger Extracts Against Human Platelet Aggregation

Sliced ginger (560 g) was extracted successively with *n*-hexane, chloroform, ethyl acetate and water at room temperature. The inhibitory activity of each fraction against human platelet aggregation was measured by the human platelet-rich plasma method described in (6). The results (Table I) showed the n-hexane extract, which gave four spots in TLC, to be the most active.

Table I. Inhibitory Effects of Ginger Extracts Against Human Platelet Aggregation

Dose[a] (μg)	*n*-Hexane extract (683 mg)[b]	CHCl$_3$ extract (105 mg)[b]	EtOAc extract (29 mg)[b]	Aqueous extract (3100 mg)[b]
100				±
40	+++	++	±	
20	++	+	±	
10	++	+		

[a] Amount added to 200 μl platelet rich plasma.
[b] Yield from 560 g sliced ginger.

From GC-MS analysis of the three higher Rf spots, the major products were identified as zingiberene and α,β,γ-bisabolene, but these components did not exhibit any activity against platelet aggregation. The lowest Rf spot was partially purified by silica gel column chromatography followed by preparative HPLC. Nine components were isolated from the HPLC (Figure 1), and among them, components 1–5 and 7 were all gingerol analogues. Component 3 was identified as newly isolated 5-methoxy-[6]-gingerol.

The activities (IC$_{50}$ values) and structures of six gingerol analogues are shown in Table II. The inhibitory activities of [6]-gingerol, [6]-gingerdione and 5-methoxy-[6]-gingerol were medium level, but the others were very weak. Therefore, the strong activity of *n*-hexane extract could not be accounted for by only gingerol analogues, and the further investigations were undertaken to isolate other potent components from *n*-hexane extracts.

Isolation and Characterization of New Inhibitors

The components which were not gingerol analogues, 6, 8 and 9 in the HPLC chromatogram in Figure 1, were further isolated and purified by HPLC to give three pure materials termed compound 1, 2 and 3. Further investigation of this fraction identified galanolactone which has been already found and characterized as an anti-5-hydroxytryptamine (serotonin) factor from ginger (7). Compounds 1 and 2 have

**Table II. Structures and Activity Against Human Platelet Aggregation
of Gingerol Analogues**

Structure	IC_{50} Value
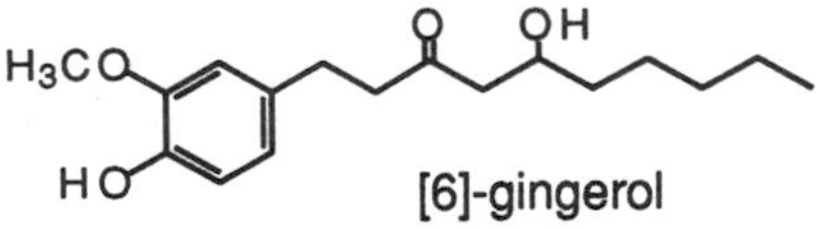 [6]-gingerol	97.4 μM
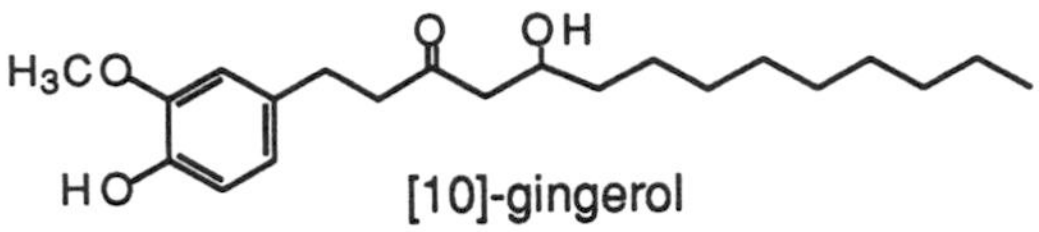 [10]-gingerol	311.6 μM
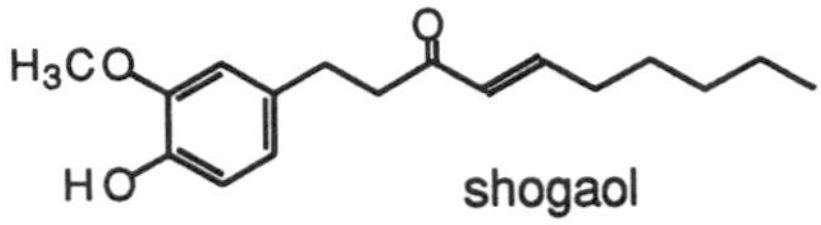 shogaol	124.8 μM
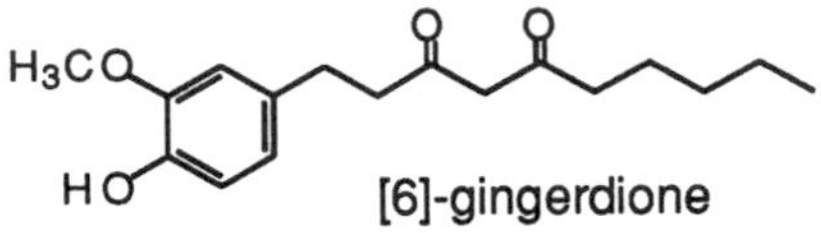 [6]-gingerdione	98.1 μM
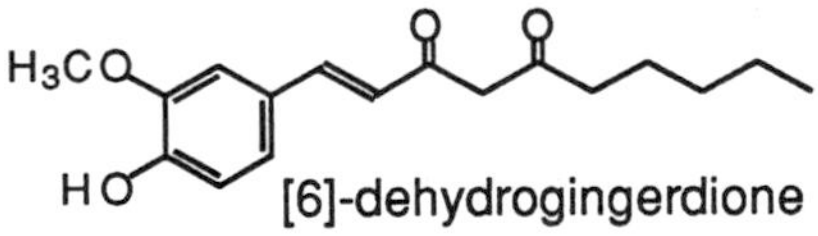 [6]-dehydrogingerdione	223.3 μM
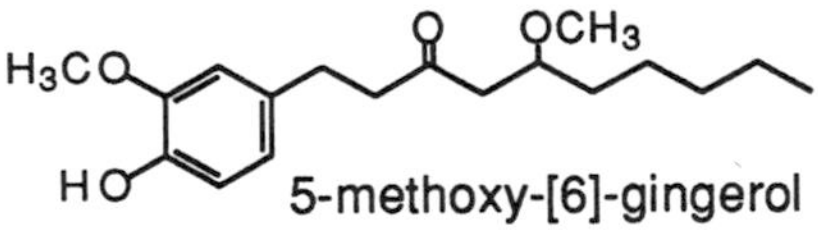 5-methoxy-[6]-gingerol	81.9 μM

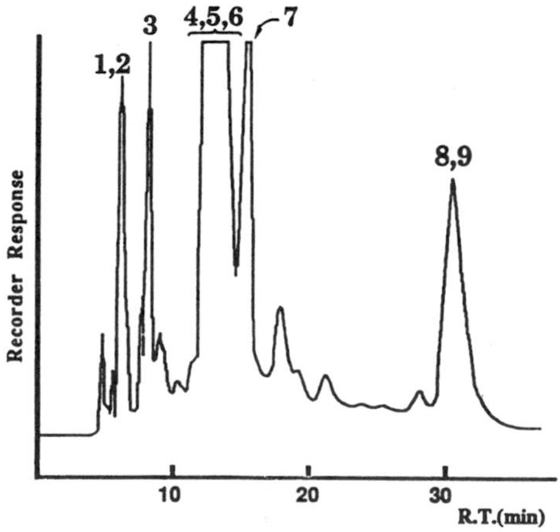

Figure 1. HPLC of *n*-hexane extracts of ginger. Conditions: Develosil ODS-5 column (8x250 mm); MeOH/H₂O/AcOH (80/20/0.1, v/v) mobile phase at a flow rate of 3 ml/min; detection at 254 nm.

been also isolated from Alpina galanga (*Zingiberaceae*) as antifungal diterpenes (*8*) and compound **3** was newly isolated from ginger by us. Compounds **1, 2, 3** and galanolactone were all labdane-type diterpenes and their spectroscopic data agreed with that of the references (*7,8*). The chemical structures of compounds **1** and **2** were determined as (*E*)-8(17),12-labd-diene-15,16-dial and (*E*)-8β(17)-epoxylabd-12-ene-15,16-dial, respectively. Compound **3** was homologous to **2** and easily determined from its spectroscopic data to be 15-hydroxy-(*E*)-8β(17)-epoxylabd-12-ene-16-al. The chemical structures of these four labdane-type diterpenes are shown in Figure 2.

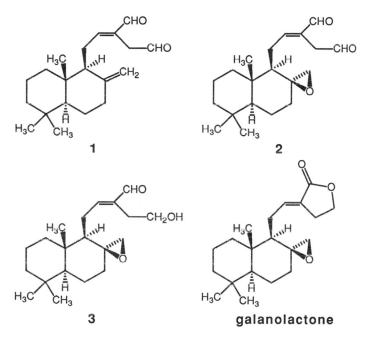

Figure 2. Structures of labdane-type diterpenes isolated from ginger.

The inhibitory activities of these compounds against platelet aggregation are compared to indomethacin in Table III. Compounds **1** and **2** exhibited strong activity, similar to indomethacin, but the activity of **3** was weak and galanolactone had none. These results suggested that the dialdehyde structure in labdane-type diterpenes is required for the developement of activity. This was also supported by the finding that when compound **2** is reduced with $NaBH_4$ in THF from dialdehyde to dicarbinol, the activity of compound **2** is completely destroyed.

Inhibitory Effects of Ginger Constituents on the Arachidonic Acid Cascade

The gingerol analogues we isolated had weak activities against platelet aggregation as previously described, so studies on whether gingerol inhibits the biosynthesis of TXA_2 in the arachidonic acid cascade of platelets were performed. When TXA_2 is formed from PGH_2 in arachidonic acid metabolism, PGH_2 is simultaneously

degraded to 12-(S)-hydroxy-5,8,10-heptadecatrienoic acid (HHT) and malondialde-
hyde (MDA) (9,10). The suppressive effect of [6]-gingerol on TXA_2 biosynthesis
was examined by the determination of MDA. As shown in Table IV, gingerol
exhibited weak inhibitory activity against PG biosynthesis compared with
indomethacin. Other gingerol analogues might also suppress the formation of
TXA_2 like [6]-gingerol.

Since labdane-type diterpene dialdehyde 1 and 2 strongly inhibited platelet
aggregation, their suppressive actions against TXA_2 biosynthesis and 5-lipoxy-
genase activity, which catalyzes leukotriene formation from arachidonic acid in
leukocytes, were studied by using rabbit renal microsomes and human 5-lipoxy-
genase, respectively (6). Table III shows the unexpected result that PG biosynthesis
was not suppressed by the dialdehydes while 5-lipoxygenase was strongly
inhibited. These results suggest that compounds 1 and 2 do not inhibit platelet
aggregation by suppression of the arachidonic acid cascade.

**Table III. Inhibitory Activities of Labdane-type Diterpenes Against
Human Platelet Aggregation, Prostaglandin Biosynthesis and 5-Lipoxygenase**

Compound	IC_{50} (µM)		
	Platelet aggregation	PG biosynthesis	5-lipoxygenase
1	3.2	>100	18.9
2	3.0	>100	4.0
3	90.4	—	>100
Galanolactone	>1000	—	>100
Indomethacin	2.1	0.75	—
AA861	—	—	0.3

— : not measured

**Table IV. Inhibitory Activity of [6]-Gingerol Against
Prostaglandin Biosynthesis**

		MDA (nmol /10^9 platelets)	Inhibition (%)
Control		5.75	—
[6]-Gingerol	200 µM	4.26	25.9
	1 mM	2.62	60.7
Indomethacin	50 µM	3.07	46.6
	200 µM	1.31	75.2

MDA values are proportional to PGH_2 and TXA_2.

Proposed Mechanism for Inhibition of Platelet Aggregation by Labdane-type Diperpenes

Since compounds 1 and 2 did not affect PG biosynthesis, the effects of many kinds of inducers on platelet aggregation were studied to make clear their mechanism of action. These compounds strongly inhibited the platelet aggregation induced by ADP, and moreover, the primary aggregation arising from low concentrations of ADP was also suppressed by compound 2. These results suggest that compounds 1 and 2 may block the ADP receptor site on platelets like *o*-phthalaldehyde which is well known to inhibit ADP binding on platelets by reaction with thiol and amino groups of its binding site (*11,12*). This speculation is also supported by the fact that UV absorption maximum at 232 nm of compound 2 rapidly disappeared after the addition of platelets. This UV disappearance suggests that the chemical changes at the α,β-unsaturated carbonyl group of compound 2 occur according to the above reaction. Moreover, the inhibitory activity of compound 3 containing mono-aldehyde was low (Table III) and compound 2 completely lost its activity when the dialdehyde was reduced to dicarbinol.

Conclusion

The *n*-hexane extracts of ginger exhibited strong inhibitory action against human platelet aggregation. Among them, gingerol analogues, major pungent principles, had only a weak inhibitory activity against platelet aggregation depending on the formation of TXA_2. Isolation of more active components identified two labdane-type diterpene dialdehydes having inhibitory activity against platelet aggregation and 5-lipoxygenase in leukocytes. It is noteworthy that these labdane dialdehydes differed from the active components of onion and garlic in that they inhibited platelet aggregation without inhibition of the arachidonic acid cascade. It is postulated that the ADP binding site on platelets may be masked by these labdane dialdehydes and as a result, the activation of platelets is suppressed, inhibiting platelet aggregation.

Acknowledgments

The authors wish to thank to Drs. T. Matsuzaki and T. Matsumoto of Japan Tobacco Inc. for kindly supplying rabbit renal microsomes and human 5-lipoxygenase and for their helpful discussion.

Literature Cited

1. Srivastava, K. C. *Prostaglandins Leukotrienes Med.* **1984**, *13*, 227.
2. Srivastava, K. C. *Prostaglandins Leukotrienes Med.* **1986**, *25*, 187.
3. Kiuchi, F.; Shibuya, M.; Sankawa, U. *Chem. Pharm. Bull.* **1982**, *30*, 754.
4. Misra, V.; Misra, R. N.; Unger, W. G. *Indian J.Med.Res.* **1978**, *67*, 482.
5. Rasheed, A.; Laekeman, G.; Totte, J.; Vlietinck, A. J.; Herman, A. G. *New Engl.J.Med.* **1984**, *310*, 50.
6. Kawakishi, S.; Morimitsu, Y. This Proceeding.
7. Huang, Q.; Iwamoto, M.; Aoki, S.; Tanaka, N.; Tajima, K.; Yamahara, J.; Takaishi, Y.; Yoshida, M.; Tomimatsu, T.; Tamai, Y. *Chem. Pharm. Bull.* **1991**, *39*, 397.

8. Morita, H.; Itokawa, H. *Planta Med.* **1988**, *54*, 117.
9. Lassmann, G.; Odenwaller, R.; Curtis, J. F.; DeGray, J. A.; Mason, R. P.; Marnett, L. J.; Eling, T. E. *J. Biol. Chem.* **1991**, *266*, 20045.
10. Okuma, M.; Steiner, M.; Baldini, M. *J. Lab. Clin. Med.* **1970**, *75*, 283.
11. Puri, R. N.; Colman, R. W. *Arch. Biochem. Biophys.* **1991**, *286*, 419.
12. Puri, R. N.; Roskoski, R., Jr. *Anal. Biochem.* **1988**, *173*, 26.

RECEIVED April 14, 1993

Chapter 26

Antitumor Promoters from Edible Plants

H. Ohigashi, A. Murakami, and K. Koshimizu

Department of Food Science and Technology, Faculty of Agriculture, Kyoto University, Kyoto 606, Japan

Possible inhibitory properties against tumor promotion (anti-tumor promoting activity) food items as well as their active constituents have been studied by an *in vitro* assay detecting inhibition of tumor promoter-induced Epstein-Barr virus (EBV) activation. In screening tests of 121 edible plants, 12% of the methanol extracts exhibited strong activity, and 6 and 10% showed moderate and weak activities, respectively. Oleanolic acid (OA) from a green perilla, mokko lactone (ML) and arctic acid from an edible burdock, and gingerol from a ginger were isolated as inhibitors of EBV activation in Raji cells. OA and ML were proven to be anti-tumor promoters by *in vivo* tests using a mouse skin model. Similar screening tests of marine algae showed that the algae in the genus *Phaeophyta*, which contains many edible species such as wakame seaweed and sea tangles, in particular possess significant activities. An *in vivo* test of an extract of wakame seaweed, an important daily food in Japan, indicated strong anti-tumor promoting activity. Further screening tests of tropical plants used as condiments and occasionally medicines suggested that the Zingiberaceae plants possess highly effective anti-tumor promotion activities. (1'S)-1'-Acetoxychavicol acetate (ACA), isolated from *Languas galanga*, inhibited EBV activation greater than 10 times more potently than other inhibitors from edible plants. On the basis of this investigation, the combination of food constituents is suggested to be a significant source of anti-tumor promoting activity.

The two step model of carcinogenesis involving initiation and promotion, first proposed by Berenblum (*1,2*), has recently been accepted as occurring in a variety of organs (*1–8*). Thus, chemical inhibition of either process would result in prevention of cancer. While inhibition of tumor initiation has hitherto been extensively studied (*9*), less attention has been given to the inhibitors of the promotion process, called anti-tumor promoters. This may provide a challenging

0097–6156/94/0547–0251$06.00/0

new area for cancer chemoprevention. We have investigated naturally occurring anti-tumor promoters, especially in medicinal (*10–15*) and edible plants (*16,17*) by using a test of inhibition of Epstein-Barr virus (EBV) activation in Raji cells (*10*). Here we report on part of our research on the anti-tumor promoting properties of edible plants as well as their active constituents.

Materials and Methods

Chemicals.   TPA (12-*O*-tetradecanoylphorbol-13-acetate; Funakoshi Chemicals, Tokyo) was used as a promoter for *in vitro* and *in vivo* assays. HPA (12-*O*-hexadecanoylphorbol-13-acetate) and teleocidin B-4, also used as promoters, were isolated from the plant *Sapium sebiferum* (*18*) and an actinomycete *Streptoverticillium blastmyceticum* (*19*) in our laboratory. Because of sample availability, HPA and teleocidin B-4 were used for screening tests in place of TPA. Teleocidin B-4 was also used for the determination of the final *in vitro* activities of the isolated compounds. 3-Oxoursolic acid, which was used as a positive control for the inhibition of EBV activation, was derived from ursolic acid as previously reported (*10*). 7,12-Dimethylbenz[*a*]anthracene (DMBA; Nakarai Chemicals, Kyoto) was used as an initiator in the *in vivo* anti-tumor promotion tests.

Inhibition of EBV Activation.   Except for the promoters used and their amounts, the test for inhibition of EBV activation in Raji cells was conducted by the standard method reported previously (*10*). When either TPA or HPA (40 ng/ml) was used as a promoter, 4 mM (352 µg/ml) sodium *n*-butyrate induced the highest level of EBV activation-positive cells, and in tests using teleocidin B-4 (20 ng/ml), 3 mM (264 µg/ml) sodium *n*-butyrate was best.

Screening Tests for *In Vitro* Anti-tumor Promoting Activity.   Fresh edible plants were extracted with MeOH. Each extract (200 µg) was assayed in 1 ml medium containing HPA (40 ng) as a promoter. Some inactive MeOH extracts were partitioned between EtOAc and water, and both parts (40 µg/ml) were assayed again. The inhibitory activity was classified into four ranks; strongly active (+++): inhibition rate (IR) at 70% or more, moderately active (++): IR at 50–69%, weakly active (+): IR at 30–49%, and inactive (–): IR at 29% or less. In the screening tests of tropical Zingiberaceae plants, each EtOAc soluble fraction (10 µg/ml) of the MeOH extract was used in the assay with teleocidin B-4 (20 ng/ml), and the activity was evaluated in the same way as the case of the screening test of edible plants. In the case of marine algae, MeOH extracts of non-edible fresh algae were re-extracted with EtOAc, and each EtOAc extract (4 µg/ml) was submitted to the assay with teleocidin B-4 (20 ng/ml) as a promoter. Dried edible algae (commercially obtained) were extracted with dichloromethane, and the extract (4 µg/ml) was used for the assay with teleocidin B-4 (20 ng/ml). In every test group, the inhibitory activity of 3-oxoursolic acid (2 µg/ml), a potent inhibitor of EBV activation, was measured and the relative inhibitory activity (RIA) of each marine algae extract to the activity of 3-oxoursolic acid was determined. The activity was divided into four ranks; strongly active (+++): RIA at 0.9 or more, moderately active (++): RIA at 0.7-0.89, weakly active: RIA at 0.5-0.69 and inactive (–): RIA at 0.49 or less.

In Vivo **Anti-tumor Promotion Tests.**   Except for the amounts of chemicals, *in vivo* activities were determined by the standard initiation (DMBA)-promotion (TPA) protocol using ICR mouse skin. Each experiment was performed on 15 female ICR mice. One week after initiation with DMBA, each mouse was repetitively promoted by twice weekly application of TPA. The inhibitors were applied one hour before each TPA treatment. The anti-tumor promoting activities were evaluated by both the ratio of tumor-bearing mice and the average number of tumors (more than 1 mm in diameter) per mouse after 15–20 weeks.

Results and Discussion

TPA, diterpene esters and teleocidins are known to cause several biological and biochemical responses, called pleiotropic effects (*20*). Some of these responses have been utilized as short-term detection methods for tumor promoters, and anti-tumor promoters as well (*21*). Among them, we have used inhibition of tumor promoter-induced EBV activation in Raji cells to detect anti-tumor promoting activity *in vitro* (*10*). This assay is based on the finding by Ito that EBV latently infecting B-lympho-blastoid (Raji) cells is highly activated by the combination of TPA and *n*-butyrate (*22*), while the activation by TPA alone is fairly weak (*23*). EBV is closely associated with Burkitt's lymphoma and nasopharyngeal carcinoma (*24*), but the Raji cell is an EBV non-producer cell line, and the use of the highly sensitive Raji cells has great advantages in experimental safety, rapidity, and convenience.

Anti-tumor Promoting Properties of Edible Plants, the Active Constituents, and Their Activities.   Possible anti-tumor promoting properties of edible vegetables and fruits, including herbage, roots and fruits, were explored using the EBV activation assay (*16*). A total of 121 plant species were extracted with methanol to be assayed (Figure 1). Of the plants tested, 14 (12% of the total tested) species showed strong activity, and 7 (6%) and 12 (10%) species exhibited moderate and weak activities, respectively. Strong activity was mostly found in the herbage-vegetables (7 out of 47 species tested) and fruits (6 of 29 species).
     Further screening tests indicated that some interference factors might frequently co-occur in the methanol extracts. When the methanol extracts of 26 inactive species were partitioned between ethyl acetate and water, strong activities appeared in seven (*e.g.*, a garland chrysanthemum, Japanese ginger and taro), and moderate activities in three species (a celery, edible burdock and carrot), particularly in the ethyl acetate soluble parts. Moreover, preliminary experiments to purify the inhibitors from the strongly active methanol extracts did not result in recovery of the inhibitory activities of EBV activation in a specified fraction. Hence, the anti-tumor promoting activity in the crude extracts may be enhanced or reduced with co-occurring factors acting additively, synergistically or antagon-istically.
     Oleanolic acid (OA), mokko lactone (ML) and arctic acid, and gingerol (Figure 2) have been isolated as *in vitro* active compounds from a green perilla, an edible burdock, and a ginger, respectively. OA completely inhibited both TPA and teleocidin B-4-induced EBV activation at 1000 times the molar concentration of TPA or teleocidin B-4 (Figure 3) (*10*). These activities were comparable to those of the well known anti-tumor promoter retinoic acid (*10*). The activities of other compounds were almost equal to those of OA.

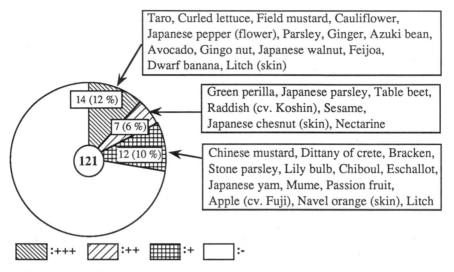

Taro, Curled lettuce, Field mustard, Cauliflower, Japanese pepper (flower), Parsley, Ginger, Azuki bean, Avocado, Gingo nut, Japanese walnut, Feijoa, Dwarf banana, Litch (skin)

Green perilla, Japanese parsley, Table beet, Raddish (cv. Koshin), Sesame, Japanese chesnut (skin), Nectarine

Chinese mustard, Dittany of crete, Bracken, Stone parsley, Lily bulb, Chiboul, Eschallot, Japanese yam, Mume, Passion fruit, Apple (cv. Fuji), Navel orange (skin), Litch

:+++   :++   :+   :-

Figure 1. Possible anti-tumor properties of edible plants. See reference 16 for names of inactive plants not shown here.

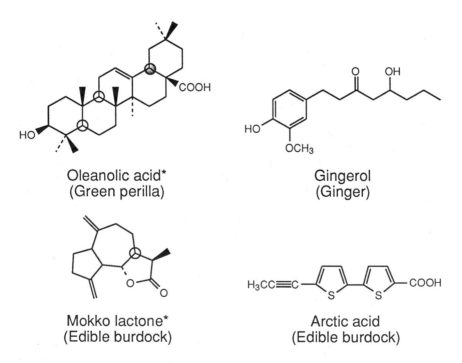

Oleanolic acid*
(Green perilla)

Gingerol
(Ginger)

Mokko lactone*
(Edible burdock)

Arctic acid
(Edible burdock)

Figure 2. Inhibitors of EBV activation from edible plants. *Anti-tumor promoting activity has been confirmed.

A previous study showed an anti-tumor promoting effect of OA (Figure 4) (*11*). Application of OA at 10 times the molar concentration of TPA reduced the incidence of tumor-bearing mice and the average number of tumors by 60% and 45%, respectively, compared to TPA controls. ML was also indicated to possess the effect (Table I), but it seems weaker than OA.

Table I. *In Vivo* Anti-tumor Promoting Activities of
ML, ACA, and Wakame Seaweed Extract (W. S. Ext.)

	ML (371 µg)[a]		ACA (37.4 µg)[b]		(374 ng)[b]		W. S. Ext.(1 mg)[b]	
	%[c]	No.[d]	%	No.	%	No.	%	No.
TPA alone	60	8.7	81	13.3	81	13.3	50	3.8
Sample+TPA	60	2.4	25	0.7	59	5.8	7	0.2

[a] Fifty µg DMBA and 1 µg TPA after 20 weeks. The dose of ML was 10 molar equivalents of TPA.

[b] Fifty µg DMBA and 1 µg TPA after 15 weeks. The doses of ACA were 100 molar equivalents (37.4 µg) and equimolar (374 ng) of TPA.

[c] Percent of mice bearing tumors.

[d] Average number of tumors per mouse.

Anti-tumor Promoting Properties of Marine Algae, and *In Vivo* Activity of A Wakame Seaweed Extract. Screening tests for possible anti-tumor promoting properties were also conducted in marine algae (*17*), some of which are important foods in Japan. In the tests, the activity was measured at reduced concentrations of both tumor promoter (teleocidin B-4 at 20 ng/ml) and test extracts (4 µg/ml), because most extracts were toxic to the Raji cells at high doses. The condition used here might be stricter than that in the case of edible plants. Interestingly, strong activity was found only in the algae Phaeophyta which includes several edible species such as wakame seaweed and sea tangles (Figure 5).

The dichloromethane extract of wakame seaweed was then tested for *in vivo* anti-tumor promoting activity in mouse skin, and the data are shown in Table I. Application of 1 mg of the crude extract remarkably inhibited both the incidence of tumor-bearing mice and the number of tumors per mouse after 15 weeks (*17*). An effort to purify the active compounds was not successful, however, because the strong inhibitory activity of the crude extract against EBV activation was not fully recovered in a specified fraction as mentioned above. Thus, the significance of a combination of multiple compounds for anti-tumor promotion is again suggested.

Anti-tumor Promoting Properties Of Tropical Zingiberaceae Plants, and the Active Constituents, ACA and Curcumin. Recently, high incidence of plants with strong activity *in vitro* was found in tropical Zingiberaceae plants used as condiments and occasionally also as local medicines. The result of the screening tests is summarized in Table II. The activity of *Languas galanga* and *Gingiber*

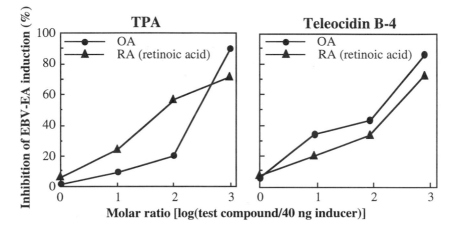

Figure 3. Inhibitory activity of oleanolic acid against EBV activation in Raji cells. TPA (left) and teleocidin B-4 (right) were used as inducers at a concentration of 40 ng/ml.

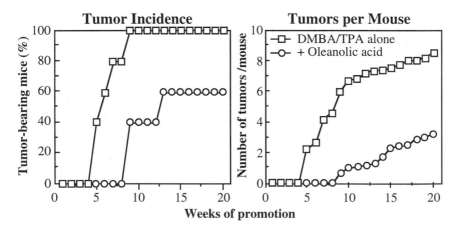

Figure 4. Anti-tumor promoting activity of oleanolic acid. Dosages of DMBA, TPA and OA were 100 μg, 2.5 μg and 19 μg (10 molar equivalent of TPA), respectively.

Table II. *In Vitro* Anti-tumor Promotion Property of
Tropical Zingiberaceae Plants

Plant	Part	Use	Activity[a]
Languas galanga	rhizhome	condiment, medicine	+++
Zingiber cassumunar	rhizhome	condiment, medicine	+++
Zingiber zerumbet	rhizhome	condiment, medicine	+++
Boesenbergia pandurata	rhizhome	condiment	+
Amomum krarvanh	rhizhome	condiment, medicine	-
Kaemphera galanga	rhizhome	condiment, medicine	+
Curcuma mangga	rhizhome	condiment, medicine	++
Nicolata elatior	rhizhome	medicine	+
	leaves, stems	vegetable	++

Sample: EtOAc soluble part (10 μg/ml) of MeOH extract.
Promoter: teleocidin B-4 (20 ng/ml).
[a] Inhibition of promoter-induced EBV activation in Raji cells.
  +++: inhibition > 70%   +: 50% > inhibition > 30%
  ++: 70% > inhibition > 50%   -: 30% > inhibition

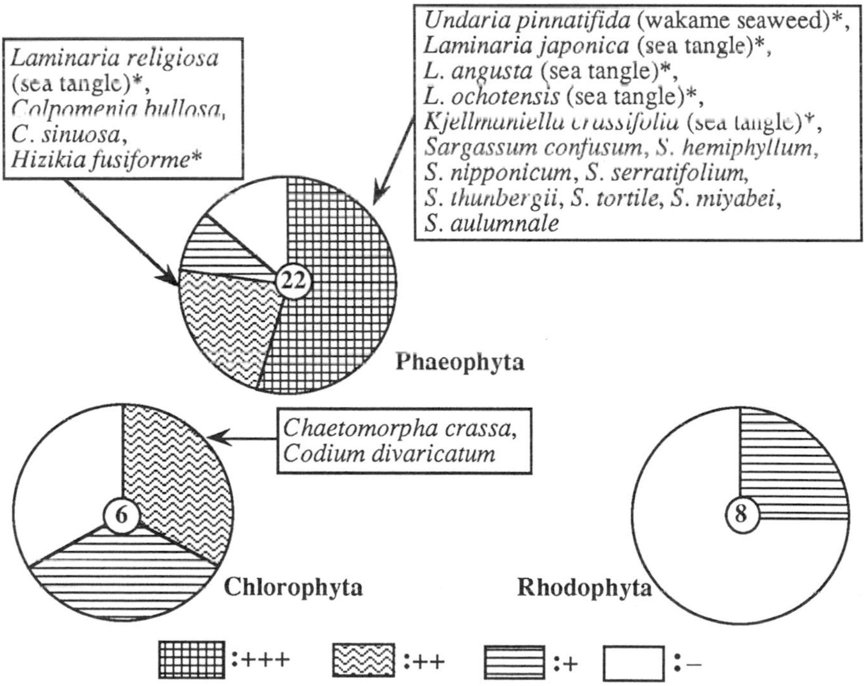

Figure 5. Possible anti-tumor promoting activity of marine algae. See reference
17 for the names of weakly active and inactive algae not shown here.

cassumunar were particularly remarkable. Both methanol extracts exhibited strong inhibitory activities at a concentration of 10 μg/ml against EBV activation by teleocidin B-4 at 20 ng/ml.

(1'*S*)-1'-Acetoxychavicol acetate (ACA) was isolated as an inhibitor of EBV activation from *L. galanga*. ACA completely inhibited EBV activation at a dose equivalent to 100 times the molar concentration of teleocidin B-4 (Figure 6); this activity was greater than 10 times more potent than that of inhibitors thus far obtained from the edible plants described above.

As shown in Table I, the strong anti-tumor promoting effect of ACA was confirmed. ACA at a dose equivalent to 100 times the molar concentration of TPA strongly inhibited the promoting activity of TPA, and remarkable activity was still observed even at a dose equimolar with TPA.

None of the ACA derivatives shown in Figure 7 exhibited remarkable inhibition of EBV activation, suggesting that both the acylated phenolic and acylated benzylic hydroxyls are important for the activity. The role of the terminal methylene for the activity is undetermined yet.

ACA is known to be an inhibitor of xanthine oxidase (*25*), which generates superoxide anions from the substrates, xanthine or hypoxanthine. Generation of the superoxide anion is the one of the biological responses of TPA (*26*). Furthermore, xanthine oxidase activity was shown to be stimulated in epidermis of SENCAR mouse by treatment with TPA (*27*), indicating the close association of the enzyme, and hence the anion, with tumor promotion. The anti-tumor promoting activity of ACA may be partly due to inhibition of the anion generation by xanthine oxidase.

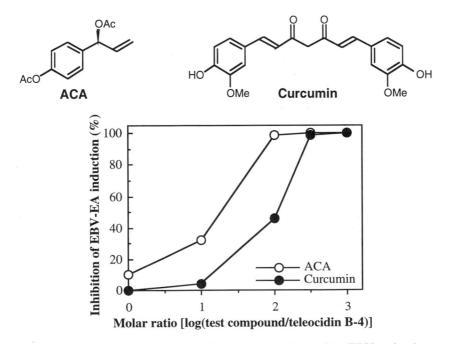

Figure 6. Inhibitory activities of ACA and curcumin against EBV activation. Teleocidin B-4 was used as a promoter at a concentration of 20 ng/ml.

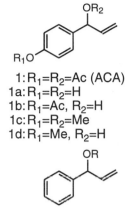

1: R_1=R_2=Ac (ACA)
1a: R_1=R_2=H
1b: R_1=Ac, R_2=H
1c: R_1=R_2=Me
1d: R_1=Me, R_2=H

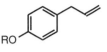

2: R=H (Chavicol)
2a: R=Ac
2b: R=Me

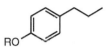

3: R=H (1-Phenyl-2-propene-1-ol)
3a: R=Ac
3b: R=Me

4: R=H (p-Propylphenol)
4a: R=Ac

Figure 7. Structures of ACA and ACA-derivatives.

Fine structure-activity relationships on inhibitions of not only EBV activation but also xanthine oxidase activity are now in progress.

Curcumin, which has anti-tumor promoting activity (28,29), was detected in an active fraction from *Z. cassumunar* by HPLC analysis. The inhibitory activity of curcumin against EBV activation seemed to be slightly weaker than that of ACA (Figure 6). Furthermore, a partially purified fraction, in which curcumin was removed by preparative HPLC, still possessed strong activity, indicating the presence of additional active constituents. The tropical Zingiberaceae plants may be one of the most promising sources of anti-tumor promoting food constituents.

Conclusions

The results obtained in these investigations may provide part of a chemical basis for epidemiological surveys on the daily intake of vegetables and a reduction of the risk of cancer. Furthermore, detailed chemical investigation of the indication that the anti-tumor promoting action could be enhanced by the combination of multiple compounds would be necessary.

Acknowledgements

We thank Dr. Harukuni Tokuda, Mrs. Akira Kondo, Yasushi Sakai and Shin Ohura, and Miss Kanako Yamaguchi for their technical assistance. This study was supported by a grant from the Ministry of Health and Welfare, by a Grant-in-Aid for Scientific Research on Priority Areas from the Ministry of Education, Science and Culture of Japan and by a subsidy from the Iyaku-Shigen Shinkokai Foundation.

Literature Cited

1. Berenblum, I. *Cancer Res.* **1941**, *1*, 44–48.
2. Berenblum, I. *Cancer Res.* **1941**, *1*, 807–814.
3. Peraino, C.; Fry, R. J. M.; Staffeldt, E; Christofer, J. P. *Cancer Res.* **1975**, *35*, 2884–2890.
4. Weisburger, J. H.; Madison, R. M.; Ward, J. M.; Viguera, C.; Weisburger, E. K. *J. Nat. Cancer Inst.* **1975**, *54*, 1185–1188.
5. Hicks, R. M.; Wakefield, J. St. J.; Chowaniec, J. *Chem. Biol. Interactions.* **1975**, *11*, 225–233.
6. Reddy, B. S.; Watanabe, K.; Weisburger, J. H.; Wynder, E. L. *Cancer Res.* **1977**, *37*, 3238–3242.
7. Wynder, E. L.; Hoffman, D.; McCoy, G. D.; Cohen L. A.; Reddy, B. S. In *Mechanisms of Tumor Promotion and Cocarcinogenesis*; Slaga, T. J.; Sivak, A.; Boutwell, R. K., Eds.; Carcinogenesis-A Comprehensive Survey; Raven Press: New York, 1978; Vol. 2, pp 11–48.
8. Jick, H.; Watkins, R. N.; Hunter, J. R.; Dinan, B. J.; Madsen, S.; Rothman, K. J.; Walker, A. M. *New Engl. J. Med.* **1979**, *300*, 218–222.
9. Wattenberg, L. W. *Cancer Res.* **1985**, *45*, 1–8.
10. Ohigashi, H.; Takamura, H.; Koshimizu, K.; Tokuda, H.; Ito, Y. *Cancer Lett.* **1986**, *30*, 143–151.
11. Tokuda, H.; Ohigashi, H.; Koshimizu, K.; Ito, Y.; *Cancer Lett.* **1986**, *33*, 279–285.
12. Murakami, A.; Ohigashi, H.; Nozaki, H.; Tada, T.; Kaji, M.; Koshimizu, K.; *Agric. Biol. Chem.* **1991**, *55*, 1151–1153.
13. Murakami, A.; Ohigashi, H.; Jisaka, M.; Hirota, M.; Irie, R.; Koshimizu, K. *Cancer Lett.* **1991**, *58*, 101–106.
14. Murakami, A.; Tanaka, S.; Ohigashi, H.; Hirota, M.; Irie, R.; Takeda, N.; Tatematsu, A.; Koshimizu, K. *Biosci. Biotech. Biochem.* **1992**, *56*, 769–772.
15. Murakami, A.; Tanaka, S.; Ohigashi, H.; Hirota, M.; Irie, R.; Takeda, N.; Tatematsu, A.; Koshimizu, K. *Phytochemistry* **1992**, *31*, 2689–2693.
16. Koshimizu, K.; Ohigashi, H.; Tokuda, H.; Kondo. A.; Yamaguchi, K.; *Cancer Lett.* **1988**, *39*, 247–257.
17. Ohigashi, H.; Sakai, Y.; Yamaguchi, K.; Umezaki, I.; Koshimizu, K. *Biosci. Biotech. Biochem.* **1992**, *56*, 994–995.
18. Ohigashi, H.; Ohtsuka, T.; Hirota, M.; Koshimizu, K.; Tokuda, H; Ito, Y. *Agric. Biol. Chem.* **1983**, *47*, 1617–1772.
19. Irie, K.; Hirota, M.; Hagiwara, N.; Koshimizu, K.; Hayashi, H.; Murao, S.; Tokuda, H.; Ito, Y.; *Agric. Biol. Chem.* **1984**, *48*, 1269–1274.
20. Slaga, T. J.; Fischer, S. M.; Weeks, C. E.; Nelson, K.; Mamrack, M.; Klein-Szanto, A. J. P. In *Cocarcinogenesis and Biological Effects of Tumor Promoters*; Hecker, E.; Fusenig, N. E.; Kunz, W.; Marks, F.; Thielmann, H. W., Eds.; Carcinogenesis-A Comprehensive Survey; Raven Press: New York, 1982; Vol. 7, pp 19–34.
21. Muto, Y.; Ninomiya, M.; Fujiki, H.; *Jpn. J. Clin. Oncol.* **1990**, *20*, 219–224.
22. Ito, Y.; Yanase, S.; Fujita, T.; Harayama, T.; Imanaka, H. *Cancer Lett.* **1981**, *13*, 29–37.
23. zur Hausen, H.; Bornkamm, G. W.; Schmidt, R.; Hecker, E. *Proc. Natl. Acad. Sci. USA.* **1979**, *76*, 782–785.

24. Henle, W.; Henle, G. In *Epstein Barr Virus*; Epstein, M. A.; Achong, B. G., Eds.; Springer-Verlag: Berlin, 1979; pp 1–22.
25. Noro, T.; Sekiya, T.; Katoh (nee Abe), M.; Oda, Y.; Miyase, T.; Kuroyanagi, M.; Ueno, A.; Fukushima, S. *Chem. Pharm. Bull.* **1988**, *36*, 244–248.
26. Kensler, T. W.; Egner, P. A.; Taffe, B. G.; Trush, M. A. In *Skin Carcinogenesis Mechanisms and Human Relevance*; Slaga, T. J.; Klein-Szanto, A. J. P.; Boutwell, R. K.; Stevenson, D. E.; Spitzer, H. L.; D'Motto, B., Eds., Progress in Chemical and Biological Research; Alan R. Liss, Inc.: New York, 1989; Vol. 298, pp 233–258.
27. Reiners, J. J.; Pence, B. C.; Barcus, C. S.; Cantu, A. R. *Cancer Res.* **1987**, *47*, 1775–1779.
28. Nishino, H.; Nishino, A.; Takayasu, J.; Hasegawa, T. *J. Kyoto Pref. Univ. Med.* **1987**, *96*, 725–728.
29. Huang, M. T.; Smart, R. C.; Wong, C. Q.; Conney, A. H. *Cancer Res.* **1988**, *48*, 5941–5946.

RECEIVED May 4, 1993

LIGNANS

Chapter 27

Chemistry of Lignan Antioxidants in Sesame Seed and Oil

Y. Fukuda[1], Toshihiko Osawa[2], Shunro Kawakishi[2], and M. Namiki[3]

**[1]Ichimuragakuen College, Inuyama-shi Aichi 484, Japan
[2]Department of Food Science and Technology, Nagoya University, Chikusa,
Nagoya 464-01, Japan
[3]Department of Brewing and Fermentation, Tokyo University
of Agriculture, Sakuragaoka, Setagaya-ku, Tokyo 156, Japan**

Sesame has been considered to be an important oil seed from ancient
times, not only because of its high oil content, but also because of its
resistance to oxidative deterioration and its medicinal effects. The
authors' studies on the antioxidative constituents in sesame seed and
oil revealed four lignanphenols. Among them, sesamolinol and sesaminol were novel antioxidants. Sesaminol was also found in high
concentration in unroasted sesame oil, due to the high yield conversion of sesamolin to sesaminol by intermolecular group transfer
catalyzed by the acid clay used for decolorization. The reason for the
especially high stability of the oil from roasted sesame has not been
clarified fully, but γ-tocopherol and minor amounts of sesamol as
well as the presence of products from the Maillard reaction by
roasting may be responsible. In this connection, it has been found
that large amounts of sesamol are produced from sesamolin during
frying process and contribute to the stability of fried food.

Sesame is one of the oldest oilseeds known to man and is considered to have not
only nutritional value, but also some medicinal effects (*1*). It is written in some
ancient Chinese medicinal books (*2,3*) that sesame increases energy and prevents
aging. The seed contains unusual minor components, including sesamin and
sesamolin, which are lignan compounds, and the seed oil exhibits more unusual
chemical and physiological properties than any other edible oil (*4*). Besides, sesame
oil has been used as a domestic Ayurvedic remedy (*5*) in India. The high resistance
of sesame oil to oxidative deterioration compared with other vegetable oils has long
been known, but the reason has not been scientifically clarified. The phenolic
compound sesamol, which is produced from sesamolin, has been considered to be
responsible (*6,7*). Our preliminary examination showed that sesamol is a trace
component in both the seed and its oil (*8*). Our group has been conducting a series
of studies on sesame antioxidants.

Nowadays great attention is being paid to the antioxidants because they prevent lipid peroxidation, which is proposed to be closely related to aging, mutation, cancer and other diseases (*9,10*).

Antioxidants in Sesame Seed

Four lignanphenol (Figure 1) and carboxyphenol antioxidants have been newly isolated and identified from the acetone extract of sesame seed after sample preparative and instrumental analysis (*11*). Among them were sesamolinol and sesaminol, which were found to be novel compounds by stereochemical study with X-ray analysis (*12,13*). It was also revealed that the lignanphenols and carboxyphenols exist as their glucosides by β-glucosidase treatment of defatted sesame flour (*11*). Recently two novel glycosides of pinoresinol have been identified (*14*).

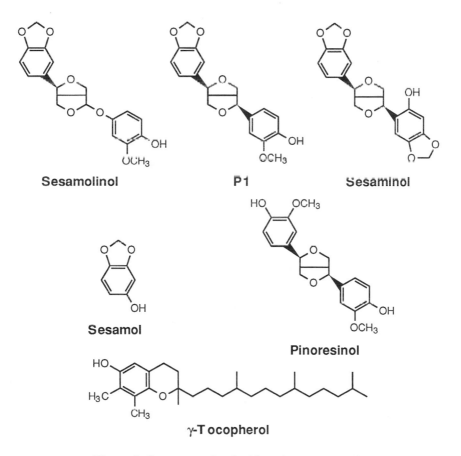

Figure 1. Structures of antioxidants in sesame seed.

The content of P1, sesamolinol and sesaminol is low as shown in Table I
(*15*), but content of their glycosides has not been determined. It may be possible
that these glycosides are decomposed to each lignanphenol and saccharides by
enzymes in the intestine (*16*) and then both types (free and bound) lignanphenols
act as antioxidants to inhibit lipid peroxidation *in vivo*. In order to determine their
in vitro antioxidative activity against lipid peroxidation, enzymatic lipid
peroxidation induced in the rat liver microsomes was used (*17*). All of the
compounds studied effectively inhibited lipid peroxidation (Table II).

**Table I. Contents of Sesamol, P1 and Sesaminol in
Different Strains of Sesame Seed in Japan (mg/100 g)**

Strain No.	Sesamol	P1 (a)	Sesamolinol (b)	Sesaminol (c)	Total (a)+(b)+(c)
48	2.0	1.6	1.0	1.4	4.0
611	2.5	1.3	1.0	1.0	3.3
630	2.5	2.3	0.9	0.3	3.5
638	n.d.	2.9	1.1	1.0	5.0
643	5.0	2.0	1.1	1.0	4.1
785	trace	2.0	0.9	0.3	3.2
673	2.5	1.8	1.5	1.1	4.4
675	trace	3.8	0.6	0.7	5.1
126	4.0	2.9	1.2	1.0	4.2
201	3.6	2.5	1.2	1.1	4.7
601	10.8	1.6	1.9	1.1	4.6
631	2.5	1.5	0.8	0.5	2.8
792	4.9	1.5	0.9	0.9	3.3
801	6.5	1.6	1.1	1.2	3.9
mean	3.4	2.1	1.1	0.9	4.0
S.D.	2.9	0.7	0.3	0.3	0.7

Source: Reprinted with permission from reference 15.
Copyright 1988 Nippon Shokuhin Kogyo Gakkai.

**Table II. Relative Antioxidative Activity of
Lignanphenols in Sesame Using Rat Liver Microsome**

	ADP-Fe^{3+} /NADPH	ADP-Fe^{3+}/EDTA-Fe^{3+} /NADPH
Control	100.0	100.0
P1	14.9	13.2
Sesamolinol	4.6	6.3
Sesaminol	8.6	10.3
Pinoresinol	17.2	14.4
Sesamol	24.1	19.0
α-Tocopherol	9.2	19.0

Antioxidants in sesame oil

Two different types of sesame oil, refined unroasted seed oil (RUSO) and roasted seed oil (RSO), are produced. The RUSO from unroasted seed is extracted with an expeller and refined by alkaline treatment, water washing, bleaching with acid clay and a deodorizing process as is done with other vegetable oils. On the other hand, the RSO from roasted seed is only filtered to remove contaminants. The RSO has a characteristic flavor and a brown color, and is widely used in Oriental countries for seasoning and cooking, while the RUSO is widely used in Europe and America for salad dressing and frying (*18*). Both sesame oils have been found to be more stable to oxidative deterioration than other vegetable oils. Figure 2 shows a comparison of autoxidation of commercial oils at 60°C. Soybean, rapeseed, safflower and corn oils begin to oxidize after 5 to 20 days incubation, but RUSO is oxidized only after 35 days, and RSO remains unaltered even after 50 days.

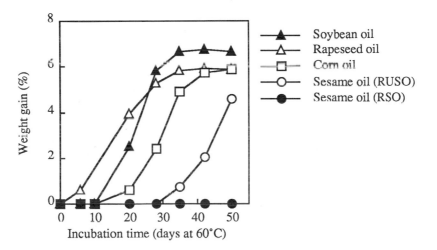

Figure 2. Antioxidative activity of commercial vegetable oils using the weighing method.

Antioxidants in RUSO

Sesaminol, found in the sesame seed, was also isolated from RUSO. Although sesaminol in the seed was a minor component, in RUSO a large amount of sesaminol was found (0.5–1.0 mg/g). It is clear by comparing the sesaminol content during the refining processes of the oil (Table III) (*19*) that at the bleaching step, drastic changes in the amounts of sesamin, episesamin, sesamolin, sesamol and sesaminol occur. Our investigation of this mechanism revealed that sesaminol was formed from sesamolin under anhydrous condition in the presence of an acid (acid clay) as catalyst with warming. As shown in Figure 3 sesamolin was first decomposed into sesamol by protonolysis to form oxonium ion and then the carbon-carbon bond (at C2) was formed, thus, the speculation is that sesaminol is formed from sesamolin by intermolecular transformation. In order to support the mechanism, the scrambling test was examined by addition of *m*-cresol as a competitor. The result suggested involvement of intermolecular processes in the formation of sesaminol from sesamolin *(20)*.

Table III. Amounts of Sesamolin, Sesamol and Sesaminol in RUSO During the Refining Process (mg/g oil)

Refining process	Sesamolin	Sesamol	Sesaminol
Crude oil	5.10	0.043	n.d.
Alkaline treatment	4.58	0.002	n.d.
Warm water treatment	4.24	0.007	n.d.
Bleaching	n.d.	0.463	0.919
Deodorizing	n.d.	0.017	0.727

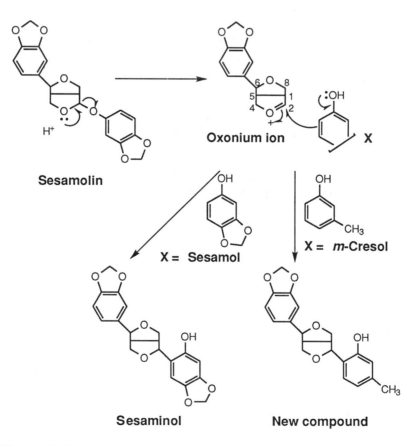

Figure 3. Scheme for the mechanism of formation of sesaminol from sesamolin.

The heat stability of sesaminol in cooking oils at frying temperature was examined. The amount of sesaminol and sesamol added to corn oil when heated at 180°C was quantified, and it was shown that 40.5% of the added sesaminol still survived even after being heated at 180°C for 6 hrs. On the other hand, sesamol was completely decomposed after 4 hrs at 180°C (*21*). Sesaminol also depressed the degradation of tocopherol in corn oil. When the amounts of tocopherol in corn oil with or without sesaminol was determined after 6 hrs at 180°C, 45.8% of total tocopherol was decomposed with sesaminol, 65.1% without sesaminol. The result suggested that sesaminol had a synergistic effect with the tocopherol during thermal oxidation and inhibited the formation of lipid peroxides in the oil (*22*).

Antioxidants in RSO

The roasting conditions of sesame seed are the most important factors determining the quality (flavor, color and oxidative stability) of RSO. The degree of browning and oxidative stability of the RSO increased with higher roasting temperature, but the amount of sesamol, which is produced by degradation of sesamolin during roasting of the seed did not increase significantly with roasting temperature (*23*). It is suggsted that the degree of browning affected the oxidative stability of the oil more than the content of sesamol produced from sesamolin. Studies on the browning components that possess antioxidative activity are under progress. In this connection, it was found that large amounts of sesamol were produced from sesamolin in RSO during the frying process (*24*) and contributed to the stability of fried food and oil as shown in Figures 4 and Figure 5 (*26*). In Figure 5, the amounts of tocopherol, sesaminol and sesamol in the oil in fried croutons were determined by HPLC (Table IV). Tocopherol remains in both sesame oils after 30 days at 60°C, but that in corn oil is completely decomposed. From this result, it is suggested that sesaminol and sesamol are used as radical scavengers with tocopherol.

**Table IV. Changes in Antioxidants in
Oils Separated From Fried Croutons During Storage (mg/100 g)**

| Days of storage | Sesame oil | | | | Corn oil |
| | Roasted | | Unroasted | | |
	Sesamol	Tocopherol[1]	Sesaminol	Tocopherol[1]	Tocopherol[1]
Fresh oil after frying	10.0	32.5	107.5	23.0	37.1
0	41.8	31.9	107.0	23.2	37.1
14	30.8	32.0	107.3	22.9	37.1
22	28.6	31.5	95.2	21.5	21.5
31	27.7	30.7	70.9	15.5	0
38	26.4	30.2	41.8	8.5	0

[1]Sum of α-, β-, γ-, and δ-tocopherol.
SOURCE: Reproduced with permission from reference 26. Copyright 1988 Nippon Shokuhin Kogyo Gakkai.

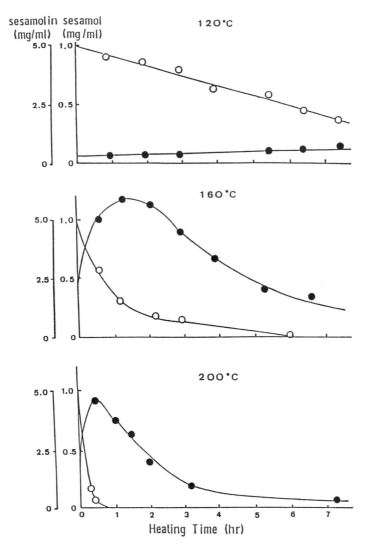

Figure 4. Degradation of sesamolin and formation of sesamol in roasted sesame seed oil at frying temperatures; sesamolin, —○—; sesamol, —●—. (Reproduced with permission from reference 25. Copyright 1987 Japanese Society of Science of Cookery.)

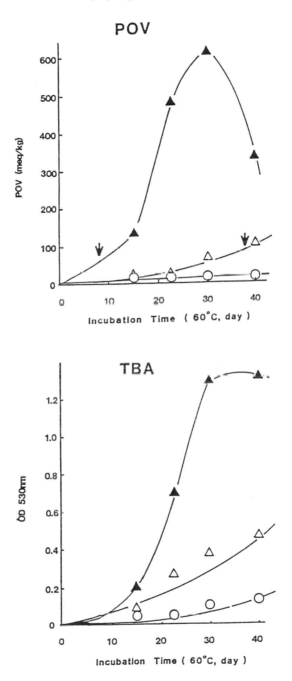

Figure 5. Deterioration of the oils in croutons fried with corn and sesame (RUSO and RSO) oils; corn oil, —▲—; sesame oil (RUSO), —△—; and sesame oil (RSO), —○—. Arrows indicate the point at which a rancid flavor was produced. (Reproduced with permission from reference 26. Copyright 1988 Nippon Shokuhin Kogyo Gakkai.)

It is clear that both sesame oils had one or more specific antioxidants other than tocopherol (Table V), and sesamolin was a very important precursor of the antioxidants in the oils (Figure 6). Sesamolin is one of the characteristic lignan compounds in sesame seed. The content of sesamolin in 14 domestic species was determined by HPLC. In general, white seed has more sesamolin than black seed, but ratio of sesamolin/sesamin is higher in black seed than white seed (15).

Recently sesamin, one of the lignans in sesame seed, has received notice as a regulator of tocopherol and cholesterol metabolism in rats (27,28) and dihomo-γ-linolenic acid biosynthesis (29).

In addition to our research on the antioxidative effects of sesame seed and its oils, we are undertaking studies on the antioxidants in black sesame seed (30).

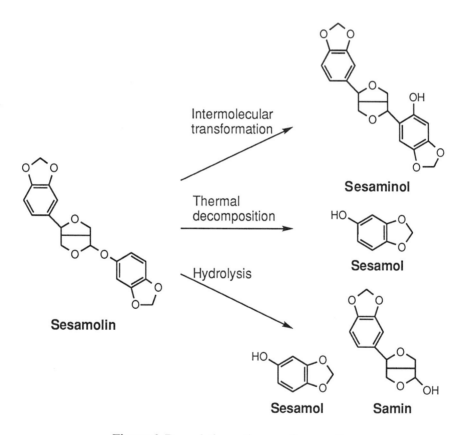

Figure 6. Degradation pathways of sesamolin.

Table V. Amounts of Antioxidants in
Fresh and Heated RUSO and RSO (mg/g in oil)

Antioxidants	Fresh		Heated (180°C for 2 hr.)	
	RUSO	RSO	RUSO	RSO
Sesamolin[1]	n.d.	3.5	n.d.	0.72
Sesaminol	0.95	n.d.	0.83	n.d.
Sesamol	trace	0.04	n.d.	0.76
Tocopherol	0.34	0.42	0.28	0.38
Browning color[2]	0.027	1.75		
Relative activity[3]	++	+++	++	+++

[1]Precursor of sesaminol and sesamol
[2]Absorbance at 470 nm (0.5 mg/ml in isooctane)
[3]Compared to the antioxidative activity of soybean oil (0.8 mg/g tocopherol) by the weighing method.

Literature Cited

1. Namiki, M.; Kobayashi, T., Eds. *Goma no Kagaku (Science of Sesame)*; Asakura Shoten: Tokyo, 1989; pp 1-2.
2. Rizitin, Honsokomoku, Syomuinsyokan
3. Hitomi, H., Ed. *Hontyoshokukan No. 1 Toyobunko* **1976**, *269*, 88–92
4. Budowski, P.; Markey K. S. *Chem. Rev.* **1951** *48*, 125–151.
5. Bhagwan Dash *Massage Therapy in Ayurveda*; Microtech Advance Printing Systems Ltd.: New Delhi, 1992.
6. Budowski, P.; Menezes F. G. T.; Dollear F. G. *J. Am. Oil Chem. Soc.* **1950**, *27*, 377–380.
7. Budowski, P. *J. Am. Oil Chem. Soc.* **1964**, *41*, 280–285.
8. Fukuda, Y.; Osawa, T.; Namiki, M. *Nippon Shokuhin Kogyo Gakkaishi* **1981** *28*, 461–464.
9. Yagi, K. *Chem. Phys. Lipids* **1987**, *45*, 337–341.
10. Pryor, W. A. *Free Radical Involvement in Chronic Diseases and Aging*; Fineley, J. W.; Schwass, D. E., Eds., American Chemical Society: Washington, D.C., 1985; pp 77–96.
11. Fukuda, Y.; Osawa, T.; Namiki, M.; Ozaki T. *Agric. Biol. Chem.* **1985**, *49*, 301–306.
12. Osawa, T.; Nagata, M.; Namiki, M.; Fukuda, Y. *Agric. Biol. Chem.* **1985**, *49*, 3351–3352.
13. Nagata, M.; Osawa, T.; Namiki, M.; Fukuda, Y. *Agric. Biol. Chem.* **1987**, *51*, 1285–1289.
14. Katsuzaki, H.; Kawasumi, M.; Kawakishi S.; Osawa T. *Biosci. Biotech. Biochem.* **1992**, *56*, 2087–2088.
15. Fukuda, Y.; Osawa, T.; Kawakishi, S.; Namiki, M. *Nippon Shokuhin Kogyo Gakkaishi* **1988**, *35*, 483-486.
16. Tamura, G.; Gold, C.; Luzzi, A. Z.; Ames, B. M. *Proc. Natl. Acad. Sci. USA* **1980**, *77*, 4961–4965.

17. Osawa, T.; Ide, A.; J. De Su; Namiki, M. J. *Agric. Food Chem.* **1987**, *35*, 808–812.
18. Namiki, M.; Kobayashi, T., Eds. *Goma no Kagaku (Science of Sesame)* Asakura Shoten: Tokyo, 1989, pp 119–126.
19. Fukuda, Y.; Nagata, M.; Osawa, T.; Namiki, M. *J. Am. Oil Chem. Soc.* **1986**, *63*, 1027–1031.
20. Fukuda, Y.; Isobe, M. Nagata, M.; Osawa, T.; Namiki, M. *Heterocycles* **1986**, *24*, 923–926.
21. Osawa, T; Kumon, H.; Namiki, M. Kawakishi, S.; Fukuda, Y. *Mutagens and Carcinogens in the Diet*; Pariza, M. W.; Aeschbacher, H. U.; Felton, J. S.; Sato, S., Eds.; Wiley–Liss Inc., 1990; pp 223–238.
22. Fukuda, Y.; Osawa, T.; Kawakishi, S.; Namiki, M. *Nippon Shokuhin Kogyo Gakkai Taikai* 1988, Tokyo.
23. Fukuda, Y.; Matumoto, K.; Koizumi, K.; Yanagida, T.; Namiki, M. *Nippon Shokuhin Kogyou Gakkai Taikai* 1992, Tokyo.
24. Fukuda, Y.; Nagata, M.; Osawa, T.; Namiki, M. *Agric. Biol. Chem.* **1986**, *50*, 857–862.
25. Fukuda, Y. *Science of Cookery* **1987**, *20*, 9–19.
26. Fukuda, Y.; Osawa, T.; Kawakishi, S.; Namiki, M. *Nippon Shokuhin Kogyo Gakkaishi* **1988**, *35*, 25–32.
27. Yamashita, K.; Nohara, Y.; Katayama, K.; Namiki, M. *J. Nutrition* **1992**, *122*, 2440–2446.
28. Sugano, M.; Inoue, T.; Koba, K.; Yoshida, K.; Hirose, N.; Shinmen, Y.; Akimoto, K. *Agric. Biol. Chem.* **1990**, *54*, 2669–2673.
29. Shimizu, S.; Akimoto, K.; Kawashima, S.; Shinmen, Y.; Yamada, H. *J. Am. Oil Chem. Soc.* **1989**, *66*, 237–241.
30. Fukuda, Y.; Osawa, T.; Kawakishi, S.; Namiki, M. *Nippon Shokuhin Kogyo Gakkaishi* **1991**, *38*, 915–919.

RECEIVED October 4, 1993

Chapter 28

Chemistry and Antioxidative Activity of Lignan Glucosides in Sesame Seed

Hirotaka Katsuzaki, Toshihiko Osawa, and Shunro Kawakishi

Department of Food Science and Technology, Nagoya University, Chikusa, Nagoya 464–01, Japan

Novel lipid-soluble lignans, sesamolinol, sesaminol and pinoresinol were isolated from sesame seed. While the quantity of these lignans was small, most of these lignans were found to exist as glucosides in sesame seed. Three novel lignan glucosides were isolated as the water-soluble antioxidative components from the 80% ethanol extracts of sesame seed. Their structures have been confirmed by instrumental analysis as pinoresinol 4'-O-β-D-glucopyranosyl(1→6)-β-D-glucopyranoside (KP1), pinoresinol 4' O-β-D-glucopyranosyl-(1→2)-β-D-glucopyranoside (KP2), and the lignan triglucoside pinoresinol 4'-O-β-D-glucopyranosyl(1→2)-O-[β-D-glucopyranosyl-(1→6)]-β-D-glucopyranoside (KP3). These lignan glucosides have unique glucosidic linkages, in particular KP3, which has branched (1→2)- and (1→6)-linkages. Chemistry and antioxidative activity of lignan glucosides using several *in vitro* systems are discussed.

Recently much attention has been focused on studies which suggest the involvement of active oxygen and free radicals in a variety of pathological events, cancer, and even the aging process. In particular, oxygen species such as hydrogen peroxide, superoxide anion radical, singlet oxygen and other radicals, are proposed as agents that attack polyunsaturated fatty acid in cell membranes, giving rise to lipid peroxidation. Lipid peroxidation may cause oxidative damage in the living cell, and finally, that damage leads to aging and susceptiblity to cancer. Normal cell membranes do not undergo lipid peroxidation, however, because of the efficient protective mechanism against damage caused by active oxygen and free radicals. On the other hand, dietary antioxidants may effectively protect from peroxidative damage in living systems. Many natural antioxidants have been found in numerous plant and food products. We have isolated lipid-soluble lignan antioxidants, such as sesamolinol (*1*) and sesaminol (*2*), from sesame seed (Figure 1). Presently, we describe the isolation of antioxidants from sesame seed which are novel lignan glucosides. These compounds were determined by instrumental analyses (including 1D and 2D NMR) and methylation analysis (*3*) using GC-MS (*4, 5*).

0097–6156/94/0547–0275$06.00/0

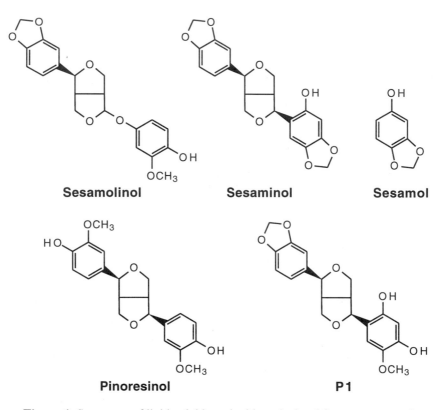

Figure 1. Structures of lipid-soluble antioxidants isolated from sesame seed

Isolation

Sesame seed (500 g) was ground, defatted with *n*-hexane, and extracted with 80% ethanol. The extract was charged onto an Amberlite XAD-2 column, and eluted with H_2O, 50% methanol, methanol and acetone. The 50% methanol fraction has antioxidative activity. This fraction was purified by preparative HPLC (ODS column). From two active fractions, two active compounds, temporarily named KP1 (10.3 mg) and KP2 (21.3 mg), were isolated. The third active fraction was further purified by preparative HPLC (phenyl column), and two compounds, temporarily named KP3 (26.1 mg) and KP4 (25.4 mg), were isolated.

Structural Determination

These compounds generated the aglycone pinoresinol and D-glucose with β-glucosidase digestion. FAB-MS of KP, KP2 and KP4 showed the same $[M+H]^+$ at m/z 683 and m/z 705 as a $[M+Na]^+$ peak, and KP3 showed $[M+H]^+$ at m/z 845 and m/z 867 as a $[M+Na]^+$ peak. The results of hydrolysis and FAB-MS spectroscopy indicate that KP1, KP2, and KP4 are constituted from one pinoresinol and two D-glucoses, and KP3 is constituted from one pinoresinol and three D-glucoses.

KP4 was identified as pinoresinol di-*O*-β-D-glucopyranoside by comparison of the UV, MS, IR, NMR and [α] $_D$ data with those of an authentic sample, which was isolated from *Eucommia ulmoides* by Deyama (6).

In the ^{1}H-NMR spectrum, KP1 and KP2 had two anomeric protons and KP3 had three anomeric protons. The anomeric configuration of D-glucosyl residues were deduced from the homonuclear vicinal coupling constants. Values obtained for the ^{1}H-homonuclear coupling constants of D-glucose moieties (all coupling constant showed ca. 7.5 Hz) were characteristic of the β-configurations. This configuration also supported the conclusion that these four compounds can be hydrolyzed to D-glucose and pinoresinol with β-glucosidase.

In the ^{13}C-NMR spectrum, each C-4′ signal of KP1, KP2, and KP3 showed downfield shifts in comparison with each C-4″ signal. These data indicate that, for each compound, the sugar residues were linked at the C-4′ positions of the aglycone.

Chemical shift of C-G6 of KP1 showed a downfield shift in comparison with C-G6′; therefore, the C-G6 position was assumed to be linked with D-glucosyl residues. The methylation analysis using GC-MS gave two peaks, the first peak was determined to be 1,5-di-*O*-acetyl-2,3,4,6-tetra-*O*-methyl-D-glucitol and second peak was determined to be 1,5,6-tri-*O*-acetyl-2,3,4-tri-*O*-methyl-D-glucitol from the mass fragments (7). The O-6 position of 1,5,6-tri-*O*-acetyl-2,3,4-tri-*O*-methyl-D-glucitol was not methylated, and indicating that KP1 had a (1→6)-linked D-glucosyl residue. From these results, the structure of KP1 was determined to be pinoresinol 4′-*O*-β-D-glucopyranosyl(1→6)-β-D-glucopyranoside (Figure 2).

Chemical shift of C-G2 of KP2 showed a downfield shift in comparison with C-G2′. From this results, glucosidic linkage of KP2 was assumed to be a (1→2)-linkage. Two peaks given by methylation analysis using GC-MS were characteristic for sugars, and determined to be 1,5-di-*O*-acetyl-2,3,4,6-tetra-*O*-methyl-D-glucitol and 1,2,5-tri-*O*-acetyl-3,4,6-tri-*O*-methyl-D-glucitol from the mass fragments (7). Since the O-2 position of 1,2,5-tri-*O*-acetyl-3,4,6-tri-*O*-methyl-D-glucitol was not methylated, a (1→2)-glucosidic linkage was confirmed. Thus KP2 was determined to be pinoresinol 4′-*O*-β-D-glucopyranosyl(1→2)-β-D-glucopyranoside (Figure 2).

For the ^{13}C-NMR spectrum of KP3, one of the 2 position carbon signals of D-glucosyl residues showed a downfield shift in comparison with the other 2 position carbon signals of D-glucosyl residues, and one of the 6 position carbon signal of D-glucosyl residues showed a downfield shift in comparison with the other 6 position carbon signals. These results suggest that KP3 has both (1→2)- and (1→6)-glucosidic linkages. Methylation analysis using GC-MS for the KP3 gave two peaks, which were determined to be 1,5-di-*O*-acetyl-2,3,4,6-tetra-*O*-methyl-D-glucitol and 1,2,5,6-tetra-*O*-acetyl-3,4-di-*O*-methyl-D-gluciotl from mass fragments (7). The O-2 and O-6 position of 1,2,5,6-tetra-*O*-acetyl-3,4-*O*-dimethyl-D-glucitol were not methylated; therefore, the pinoresinol linked D-glucosyl residue had branched (1→2)- and (1→6)-glucosidic linkages. From these results, KP3 was confirmed as pinoresinol 4′-*O*-β-D-glucopyranosyl(1→2)-*O*-[β-D-glucopyranosyl-(1→6)]-β-D-glucopyranoside (Figure 2).

Antioxidative Activity

KP1, KP2, and KP3 had strong antioxidative activity, but KP4 did not, using rabbit erythrocyte membrane ghost system (8) (Figure 3). Moreover, KP1, KP2, KP3, and

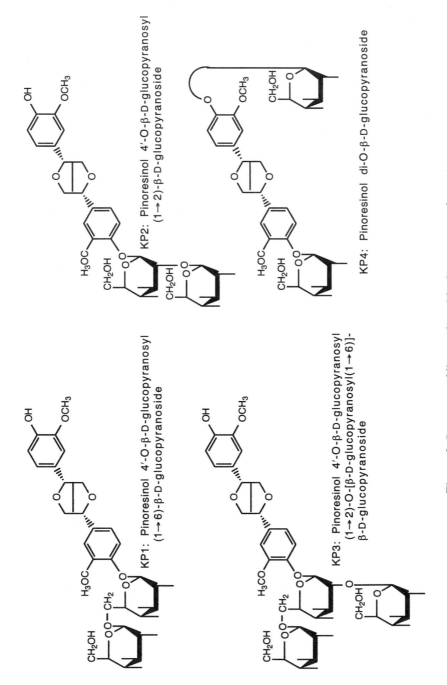

Figure 2. Structures of lignan glycosides in sesame seeds.

KP4 were classified in the category of pro-antioxidant, because they produce the lipid-soluble antioxidant pinoresinol by hydrolysis with β-glucosidase of intestinal bacteria after intake as food components (9). It is thought that when we eat sesame seed, these lignan glucosides produce the lipid-soluble antioxidative lignan pinoresinol, which protects against oxidative damage in the membrane lipids.

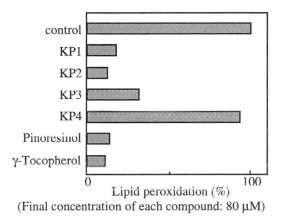

(Final concentration of each compound: 80 μM)

Figure 3. Antioxidative activity of lignan glucosides using the rabbit erythrocyte membrane ghost system

Conclusions

1. Three new lignan glucosides (KP1, KP2, and KP3) were isolated from sesame seed. The yield of new lignan glucoside and pinoresinol 4'-di-*O*-β-D-glucopyranoside (KP4) from 500 g sesame seed were: KP1, 10.3 mg; KP2, 21.3 mg; KP3, 26.1 mg; KP4, 25.5 mg.
2. Isolated lignan glucosides were easily hydrolyzed to pinoresinol and D-glucose by β-glucosidase.
3. The three new lignan glucosides were determined to be the follwing:
   KP1: pinoresinol 4'-*O*-β-D-glucopyranosyl(1→6)-β-D-glucopyranoside.
   KP2: pinoresinol 4'-*O*-β-D-glucopyranosyl(1→2)-β-D-glucopyranoside.
   KP3: pinoresinol 4'-*O*-β-D-glucopyranosyl(1→2)-*O*-[β-D-glucopyranosyl(1→6)]-β-D-glucopyranoside.
4. KP1, KP2, and KP3 had antioxidative activity using the rabbit erythrocyte membrane ghost system, and all of the lignan glucosides formed the lipid-soluble antioxidative aglycone, pinoresinol.

Literature Cited

1. Osawa, T.; Nagata, M.; Namiki, M.; Fukuda, Y. *Agric. Biol. Chem.* **1985**, *49*, 3351–3352.
2. Fukuda, Y.; Osawa, T.; Namiki, M.; Ozaki, T. *Agric. Biol. Chem.* **1985**, *49*, 301–306.
3. Hakomori, S. *J. Biochem.* **1964**, *55*, 205–207.

4. Katsuzaki, H.; Kawasumi, M.; Kawakishi, S.; Osawa, T. *Biosci. Biotech. Biochem.* **1992**, *56*, 2087–2088.
5. Katsuzaki, H.; Kawakishi, S.; Osawa, T. *Heterocycles*, in press.
6. Deyama, T.; *Chem. Pharm. Bull.* **1983**, *31*, 2993–2997.
7. Bjorndal, H.; Lindberg, B.; Svenssn, S. *Carbohyd. Res.* **1967**, *5*, 433–440.
8. Osawa, T.; Ide, A.; Su, J.-De; Namiki, M. *J. Agri . Food Chem.* **1987**, *35*, 808–812.
9. Tamura, G.; Gold, C.; Ferro-Luzzi, A.; Ames, B. N. *Proc. Natl. Acad. Sci. USA* **1980**, *77*, 4961–4965.

RECEIVED May 8, 1993

Chapter 29

Antioxidative and Anticancer Components Produced by Cell Culture of Sesame

A. Mimura[1], K. Takebayashi[1], M. Niwano[1], Y. Takahara[1], Toshihiko Osawa[2], and H. Tokuda[3]

[1]Biotechnology Research Laboratory, Kobe Steel Ltd., Tsukuba, Japan
[2]Department of Food Science and Technology, Nagoya University, Chikusa, Nagoya 464–01, Japan
[3]Department of Biochemistry, Kyoto Prefectural University of Medicine, Kyoto, Japan

Suspension cultured cells of sesame contained lignan compounds such as sesamin and sesamolin in higher quantity than sesame seeds or seedlings. They also accumulated several kinds of glycoside antioxidants with diverse solubilities in water. Among them, two compounds were identified as 3,4-dihydroxy-β-phenethyl-O-α-rhamnopyranosyl (1–3)-4-O-caffeoyl-β-glucopyranoside (acteoside), and a novel compound, 3,4-dihydroxy-β-phenyl O ethylcarboxyl O α L rhamnopyranosyl (1–3)-4-O-caffeoyl-β-glucopyranoside. The components of the ethanol extract of the cultured cells were found to have marked inhibitory activity on the induction of Epstein-Barr virus associated antigen. Furthermore, these components remarkably inhibited skin carcinogenesis in mice initiated by dimethylbenz[a]anthracene (DMBA) and promoted by phorbol ester (TPA). Antioxidants extracted from cultured cells of sesame also had protective activity against ultraviolet light-induced peroxidation of membrane lipids.

The balance of oxygen stress and oxygen tolerance is considered important for the maintenance of health in animals and plants. Overproduction of superoxide radicals causes oxygen stress, and consequently will induce toxic effects like inflammation, carcinogenesis, aging and geriatric disease in animals, or wither plants and flowers. Self defense systems against the toxic effects of oxygen free radicals may be supported by antioxidative compounds taken as foods, cosmetics or medicines.

Since sesame seeds and sesame oil have been familiar health foods for many years, their activities have been evaluated. Recently, it was found that the potent antioxidant activity of sesame seeds and sesame oil originates from lignan compounds such as sesamol, sesaminol, sesamolinol and pinoresinol, phenolic compounds such as syringic acid and ferulic acid, and tocopherols (*1–4*). From this research it has been concluded that the antioxidative activity of sesame seeds and sesame oil is one of the reasons why they have been considered health foods for a long time.

Callus cells of *Sesamum indicum* L., which were induced and established by us, grow rapidly in suspension culture and enable us to prepare a large biomass of cells in a short time (5,6). Some components isolated from the biomass, which have potent antioxidative activity, were found to possess anti-tumor promoting activity, chemopreventive activity of skin papillomas (7) and suppression of peroxidation of membrane lipid caused by ultraviolet light.

This report deals with the production of antioxidative compounds by plant cell culture technology, the elucidation of the chemical structures of the antioxidative compounds, and biological activities of the components extracted from biomass of sesame cultured cells.

Development of tissue culture from *Sesamum indicum* L.

The primary growing cultures of *S. indicum* L. were obtained from seedlings on Murashige-Skoog's medium supplemented with cytokinins and auxins. Subcultures of growing cultures were performed at intervals of 2 weeks for three months, and one of the cultures that showed stable growth on solid medium was established and designated ISC-1.

The methanol extract of ISC-1 cells has potent antioxidative activity. Sesamol and tocopherols, which have potent antioxidative activity, were not detected in the methanol extract. Lignan compounds such as sesamin and sesamolin were found to be present in larger quantity in ISC-1 cells than in sesame seeds or seedlings as shown in Table I. These compounds, however, do not have any antioxidative activity. These results indicated that cultured sesame cells must contain compounds other than sesamol and tocopherol with potent antioxidative activity.

Table I. Composition of Lignan Compounds and Tocopherol in Callus, Seeds, Seedings and Oil of Sesame

Sample	Lignan					Tocopherol[a]
	Sesamin	Sesamolin	Sesamol[a]	Sesaminol[a]	Sesamolinol[a]	
Callus (μg/g cell)	460	950	nd	nd	nd	nd
Seeds (μg/g seeds)	2,125	1,522	nd	—	—	205
Seedings (μg/g seeds)	330	260	nd	—	—	80
Oil (mg/g oil)	4–10	3–5	0.002	0.2	0.2	0.3–0.5

[a] Antioxidant
nd: not detected

Production of Antioxidants by Cell Culture of *Sesamum indicum* L.

The cell culture ISC-1-S, which can grow rapidly in suspension culture, has been developed from ISC-1, and the optimum cultivation conditions for ISC-1-S were established for the mass production of cells in liquid medium. It is an interesting finding that sesame cells grew abundantly at a relatively high temperature of 35–36°C as shown in Figure 1, although many plant cells proliferate moderately under 25–27°C. The sesame cells were found to acquire heat tolerance after subcultures of ISC-1-S at 35°C for more than 6 months. The heat tolerant culture, designated ISC-35, grew abundantly at 38–39°C. In this research, ISC-35 was used throughout all experiments. The time course of cultivation of ISC-35 at 35°C is shown in Figure 2.
    The antioxidative activity of ethanol extracts from biomass was evaluated by the method using rabbit erythrocyte membranes (8). This method is based on the idea that lipid peroxidation will lead to destabilization and disintegration of cell membrane. Ethanol extract of sesame cells harvested after 7 days cultivation had potent antioxidative activity; it inhibited about 60% of the peroxidation at a concentration of 250 µg/ml in the assay reaction mixture.

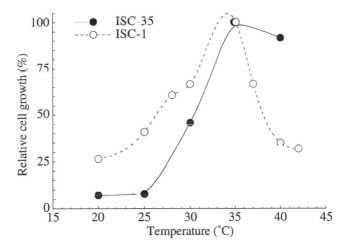

Figure 1. Effect of temperature on the growth of the cultured cells of *Sesamum indicum* L.

Purification and Identification of Antioxidants in the Cultured Cells of *Sesamum indicum* L.

The suspension cultured cells were extracted with 80% methanol, and the extract was subjected to column chromatography on Amberlite XAD-2 (Figure 3).
    The strongest antioxidative activities were concentrated in the fractions eluted by 40% (M-40) and 60% (M-60) methanol. These fractions were subsequently purified by HPLC, and one of the antioxidative compounds in 40% methanolic fraction (M-40-3) was isolated as single compound. By analyzing the

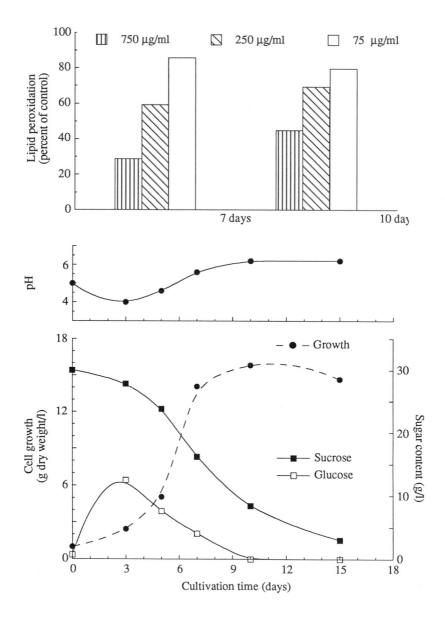

Figure 2. Production of antioxidative compounds by cultivation of *Sesamum indicum* L.

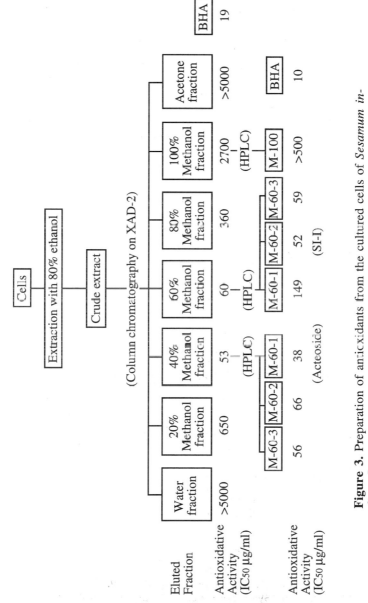

Figure 3. Preparation of antioxidants from the cultured cells of *Sesamum indicum* L.

NMR, MASS, UV, and IR spectrum data, the chemical structure was identified as 3,4-dihydroxy-β-phenethyl-O-α-rhamnopyranosyl (1–3)-4-O-caffeoyl-β-glucopyranoside (acteoside, Figure 4). Acteoside has been found in *Syringa vulgaris* (9), *Conandron ramoidioides* (10), *Rehmannia glutinosa* (11), and *Echinacea pallida* (12) as one of polyphenol glycosides in medicinal plants. Although it was reported to have antimicrobial, antiviral and cell growth inhibitory activities (11,12), the potent antioxidative activity—almost the same activity as BHA (butylated hydroxyanisol)—was the first observation by the erythrocyte membrane method.

In the M-60 fraction from Amberlite XAD-2 column chromatography, a novel polyphenol glycoside was been isolated and its structure was determined to be 3,4-dihydroxy-β-phenyl-O-ethylcarboxyl-O-α-L-rhamnopyranosyl (1–3)-4-O-caffeoyl-β-glucopyranoside (SI-1, Figure 4). The polyphenol glycoside SI-1 has strong antioxidative activity. The antioxidative activity of these polyphenol glycosides is due to the two caffeoyl moieties combined with glycoside.

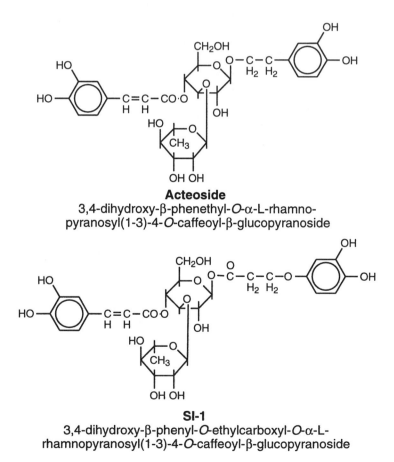

Acteoside
3,4-dihydroxy-β-phenethyl-O-α-L-rhamno-
pyranosyl(1-3)-4-O-caffeoyl-β-glucopyranoside

SI-1
3,4-dihydroxy-β-phenyl-O-ethylcarboxyl-O-α-L-
rhamnopyranosyl(1-3)-4-O-caffeoyl-β-glucopyranoside

Figure 4. Phenol glycosides with potent antioxidative activity found in cells of *Sesamum indicum* L.

Anti-tumor Promotion Activity of Components Extracted from Cultured Sesame Cells

Anti-tumor promotion activity was estimated using a short-term assay which measures induction by a tumor promoter such as 12-O-tetradecanoylphorbol 13-acetate (TPA) combined with n-butyric acid of Epstein-Barr virus (EBV) associated early antigen in EBV genome-carrying human lymphoblastoid Raji cells (*13,14*). This method has been used to detect tumor promoters or anti-tumor promoters in foods and environmental pollutants (*15,16*).

The most potent inhibition of EBV activation was observed in the fraction eluted by 100% methanol (M-100) from Amberlite XAD-2 chromatography. Moderate activity was found in fractions M-40 and M-60 (Table II).

Acteoside, which is contained in the M-40 fraction, has strong antioxidative activity, but it was less active in the anti-tumor promotion assay. On the other hand, SI-1, which was isolated from the M-60 fraction, showed marked antioxidative activity and also potent inhibition of EBV activation.

Antioxidative compounds such as β-carotene, α-tocopherol, BHA, BHT (butylated hydroxytoluene), curcumin and EGCG (epigallocatechin gallate) have been found to be potent anti-tumor promoters evaluated by the EBV activation method as shown in Table III. From these results, it can be considered that one of the mechanisms of anti-tumor promotion will be based on radical scavenging activity. It is interesting to notice that although fraction M-100 has low antioxidative activity, it shows strong inhibitory activity of EBV activation. The active compounds in the fraction M-100 are being purified.

Inhibition of Papilloma Formation on Mouse Skin

Anti-tumor promotion activity of topically and orally administered sesame components was tested in ICR mice using a two-stage skin tumorigenesis model (*7*).

Topical Administration. Female ICR mice (15 per group, 6 weeks old) were shaved, and 2 days later they were initiated with 390 nmol 7,12-dimethyl-benz[a]anthracene (DMBA). One week after initiation, promotion was begun by applying TPA (1.7 nmol) topically twice weekly for 20 weeks. The sesame components (85 nmol in 0.1 ml acetone) were applied topically one hour before each TPA application (pretreatment), or 30 minutes after each TPA application (post-treatment). The dose level of sesame components was set to 50 times that of TPA by assuming that the molecular weight of the active components in the M-60 and M-100 fractions was 620 and 500, respectively. Skin papilloma formation was recorded weekly in each experimental group.

Figure 5 shows the inhibition of papilloma formation by the components of M-100. They showed remarkable inhibition of papilloma formation especially in the post-treatment experiment. The formation of papillomas were observed to be delayed for 5 weeks, and resulted in a 60% reduction in the number of papillomas per mouse when compared to the TPA control group. The components of M-60 were also found to have considerably strong inhibitory activity in the post-treatment experiment.

It is interesting to notice that there are remarkably different effects between the pretreatment and post-treatment experiments. It can be considered that the sesame components may inhibit the actions of TPA in skin cells most effectively

**Table II. Inhibitory Effects of Sesame Cell Fractions Against
TPA-induced Epstein-Barr Virus Activation**

Fraction	Concentration (μg/ml)		
	500	100	10
Ethanol extract	24.4[a]	86.4	100.0
XAD-2 chromatography			
Water	57.5	91.7	100.0
Methanol 20%	55.2	91.4	100.0
Methanol 40%	16.5	48.3	88.1
Methanol 60%	11.7	51.7	71.3
Methanol 80%	0[b]	45.4	73.8
Methanol 100%	0[b]	24.1	58.4
Acetone	0[b]	79.1	100.0

TPA: 32 pmol (20 ng/ml)
[a] Percent of control
[b] Cytotoxicity

**Table III. Inhibitory Effects of Antioxidants Against
TPA-induced Epstein-Barr Virus Activation**

Fraction	Concentration (molar ratio of test compound to TPA)		
	100	10	1
BHA	42.1	63.5	88.6
BHT	58.4	89.7	100.0
α-tocopherol	34.6	57.6	79.4
β-Carotene	26.8	89.4	100.0
Curcumin	78.6	94.5	100.0
Ascorbic acid	76.7	92.6	100.0
Uric acid	61.3	90.6	100.0
EGCG	22.4	65.1	87.9
M-60-2	48.5	78.7	93.4
M-100	0[a]	58.3	70.6

TPA: 32 pmol (20 ng)
[a] Weak cytotoxicity

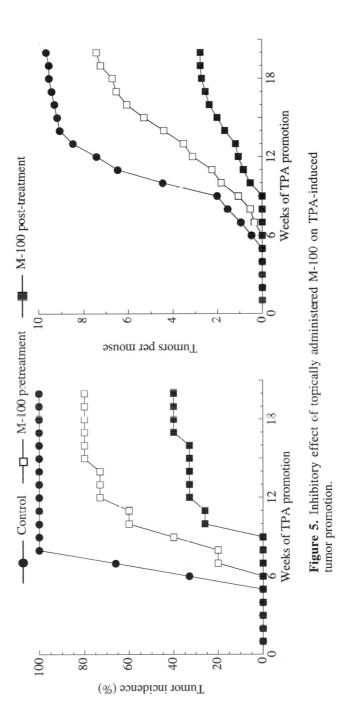

Figure 5. Inhibitory effect of topically administered M-100 on TPA-induced tumor promotion.

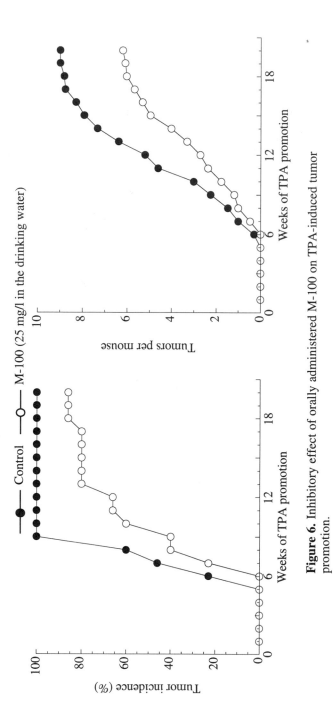

Figure 6. Inhibitory effect of orally administered M-100 on TPA-induced tumor promotion.

when TPA is generating its biological activity. Radical scavenging activity by the sesame components in the skin cells of active oxygen species or peroxidized molecules generated by TPA could be one of the possible scenario where skin papilloma formation was prevented by the sesame components. The experimental result that stronger inhibition of tumorigenesis was observed in the post-treatment experiment indicated that there may be metabolic inactivation of the sesame components in skin cells after topical administration on skin surface.

Oral Administration.   It is also important to examine oral administration to evaluate putative chemopreventive agents found in food so a two-stage tumorigenesis experiment was conducted using oral administration of the sesame components in M-100. DMBA and TPA were applied as in the topical experiment, and the M-100 fraction was given in the drinking water (25 mg/liter) starting with the first administration of TPA and continuing until the end of the experiment.

Figure 6 shows the result that the oral administration of M-100 also inhibited the formation of skin papilloma effectively, indicating that the components of M-100 can be absorbed and are active even after they are taken up through digestive organs. Chemopreventive effects by oral administration of sesame components on pulmonary cancer will be presented elsewhere (*17*).

Protection of Ultraviolet Light Damage by Sesame Antioxidants

Ultraviolet light induces the generation of active oxygen molecules in skin cells, and consequently it can cause skin cancer and aging of skin. For the preliminary evaluation of protective activity against ultraviolet B light (UVB), which induces peroxidation of membrane lipids, a simple assay method has been constructed by us using rabbit erythrocyte membranes (*18*). Rather high doses of UVB light have been found to induce peroxidation of the membrane lipid and peroxidized compounds such as malonaldehyde can be detected by the thiobarbituric acid (TBA) method. The general assay method is shown in Figure 7.

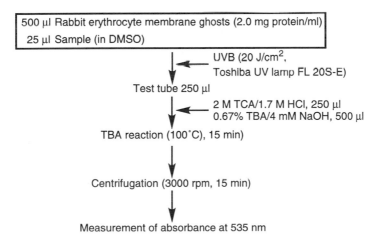

Figure 7. Procedure for evaluation of protective activity against UVB light-induced peroxidation of erythrocyte membrane lipid.

Lipid-soluble antioxidants such as BHA, α-tocopherol and curcumin were found to protect against the lipid peroxidation caused by UVB light as shown in Figure 8. Resorcinol, which does not have antioxidative activity but absorbs UVB light, was also tested for protective activity. The profiles of protection against lipid peroxidation are quite different between antioxidants and the UVB absorber resorcinol. The concentrations of BHA, α-tocopherol and curcumin that inhibited half of the UVB light-induced peroxidation of membrane lipid were 1.8, 60 and 4.0 μg/ml, respectively.

The fractions of sesame cell extract that have potent antioxidative activity also showed remarkable protection activity against UVB light-induced membrane lipid peroxidation. As shown in Figure 9, the concentrations of the M-40 and M-60 fractions that inhibited half of the UVB light-induced peroxidation were 14.2 and 20.8 μg/ml, respectively—3 to 4 times stronger than α-tocopherol.

Antioxidative activity, as measured by the method using rabbit erythrocyte membranes initiated by t-butylhydroxyperoxide, has found to be well correlated to protective activity against lipid peroxidation caused by UVB light. These results indicate the possibility that sesame antioxidants protect against skin damage induced by UVB light.

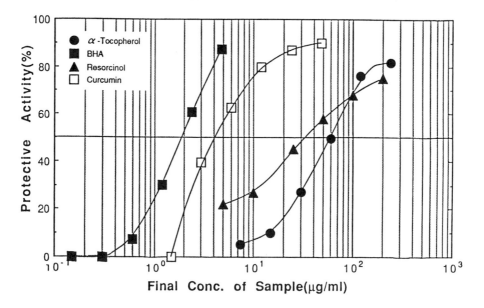

Figure 8. Protective activity of antioxidants against UVB light-induced peroxidation of erythrocyte membrane lipid.

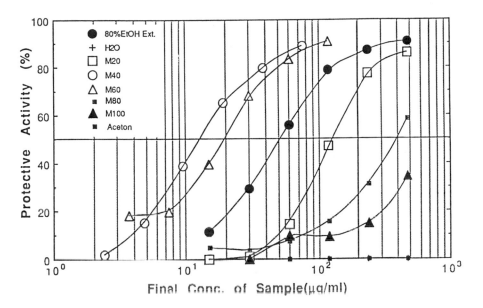

Figure 9. Protective activity of sesame extract fractions against UVB light-induced peroxidation of erythrocyte membrane lipid.

Literature Cited

1. Fukuda, Y.; Osawa, T.; Namiki, M.; Ozaki, T. *Agric. Biol. Chem.* **1985**, *49*, 301
2. Osawa, T.; Nagata, M.; Namiki, M.; Fukuda, Y. *Agric. Biol. Chem.* **1985**, *49*, 3351.
3. Fukuda, Y.; Isobe, M.; Nagata, M.; Osawa, T.; Namiki, M. *Heterocycles* **1986**, *24*, 923.
4. Osawa, T.; Namiki, M.; Kawakishi, S. In *Antimutagenesis and Anticarcinogenesis Mechanisms II*; Yuroda, Y.; Shankel, D. M.; Waters, M.D., Eds.; Plenum Press: New York, 1989; p 139.
5. Mimura, A.; Osawa, T. *Bio-industry* **1989**, *6*, 414.
6. Mimura, A. *Fragrance Journal* **1991**, *4*, 96
7. Mimura, A.; Takebayashi, K.; Niwano, M.; Tokuda, H.; Nishino, H.; Iwashima, A.; Takahara, Y. *Proc. Jpn. Cancer Assoc.* **1991**, 57.
8. Osawa, T.; Ide, A.; Su, J-De; Namiki, M. *J. Agric. Food Chem.*, **1987**, *35*, 808.
9. Birkofer, L.; Kaiser, C.; Thomas, U. *Z. Naturforschg.* **1968**, *23b*, 1051.
10. Nonaka, G.; Nishioka, I. *Phytochem.* **1967**, *16*, 1265.
11. Shoyama, Y.; Matsumoto, M.; Nishioka, I. *Phytochem.* **1986**, *25*, 1633.
12. Cheminat, A.; Zawatzky, R.; Becker, H.; Brouillard, R. *Phytochem.* **1988**, *27*, 2787.
13. Ito, Y.; Yanase, S.; Fujita, J.; Harayama, T.; Takashima, M.; Imanaka, H. *Cancer Lett.* **1981**, *13*, 29.

14. Ito, Y.; Kawanishi, M.; Harayama, T.; Takabayashi, S. *Cancer Lett.* **1981**, *12*, 175.
15. Yanase, S.; Ito, Y. *Cancer Lett.* **1984,** *22*, 183.
16. Koshimizu, K.; Ohigashi, H.; Tokuda, H.; Kondo, A.; Yamaguchi, K. *Cancer Lett.* **1988**, *39*, 247
17. Mimura, A.; Takebayashi, K.; Niwano, M.; Tokuda, H.; Nishino, H.; Iwashima, A. *Proc. Jpn. Cancer Assoc.* **1992**, 48.
18. Kato, T.; Murakami, H.; Tanimura, H.; Mimura, A.; Takahara, Y. *Proc. Ann. Mtg. Soc. Fermentat. Technol. Japan* **1992**, 213.

RECEIVED September 13, 1993

Chapter 30

Development of Stability-Indicating Analytical Methods for Flaxseed Lignans and Their Precursors

R. K. Harris, J. Greaves, D. Alexander, T. Wilson, and W. J. Haggerty

Midwest Research Institute, 425 Volker Boulevard, Kansas City, MO 64110

Analytical method development was undertaken to identify and quantify lignan and lignan precursors in flaxseed powder. Experiments indicated that the most successful modes for extracting the lignans from the flaxseed powder involved a hexane defatting process followed by extraction with either methanol or dioxane: ethanol (1:1, v:v). High performance liquid chromatography (HPLC) and thin layer chromatography (TLC) assays were performed on flaxseed powder extracts. Ultimately, six components from these extracts were separated and identified by HPLC versus phenolic standards. In descending order of abundance, the identified lignans were: gallic acid, ferulic acid, chlorogenic acid, coumaric acid, secoisolariciresinol diglucoside (SDG) and 4-hydroxybenzoic acid. The concentrations of these phenolics ranged from 0.1 to 12.0 µg/g, with only traces of 4-hydroxybenzoic acid. The flaxseed lignans identified remain stable when stored at -20°C for nine months.

Nutritional research has indicated a relationship between diet and chronic diseases (1–4). In addition, certain plant derived chemicals (phytochemicals) have shown inhibitory activity toward mutagens and carcinogens (5). These reports have increased interest in the anticarcinogenic activity of a wide variety of phytochemicals. Of particular interest are those found in vegetables and cereal grains. For example, there is a positive correlation in higher cancer incidence (colon, rectal, and mammary) when humans or other mammals consume large amounts of fat and red meat (6). The same study found this correlation is reversed, however, with diets containing large amounts of cereals.

Cancer incidence has been shown to be lower in countries where the diets are vegetarian or semivegetarian (7–10). Close review of these studies suggests these benefits are not totally due to fiber content alone, but may also be caused by other substances associated with the fiber. One source of these effects may be the presence of certain lignans and/or their precursors. Lignans demonstrate a broad spectrum of biological activity. They include antitumor, antimitotic, and antiviral

0097–6156/94/0547–0295$06.00/0

activities. They also inhibit enzymes, interfere with nucleic acid synthesis, and have shown other physiological effects. Two reviews on these effects are found in recent publications (11,12). They have also been investigated as potential anticancer agents (17–19).

Phenolic compounds such as caffeic, ferulic, coumaric, gallic, and syringic acids, commonly found in grains, have exhibited marked inhibitory effects in the presence of carcinogens (13). Plant phenolics can occur as monomers or dimers, or they can react with other chemical units to form esters or glycosides. Examples of dimeric phenolics are diferulic, ellagic, guaiaretic, and nordihydroguaiaretic acids. An example of a phenolic ester found in oats is chlorogenic acid formed from caffeic and quinic acids. Ellagic acid (a dimer of gallic acid) isolated from various green plants and diferulic acid from wheat are lignans that exhibit remarkable inhibitory activity. For example, studies with benzo[a]pyrene indicate that ellagic acid destroys the diol-epoxide of benzo[a]pyrene, making it incapable of attacking DNA in a cell (14). Caffeic acid and ferulic acid, both hydroxycinnamic acids, inhibit the formation of nitrosamines (15).

Lignans are "plant products of low molecular weight formed primarily by the oxidative coupling of p-hydroxyphenylpropene units" (16). They are believed to be chemical intermediates for lignin, a natural polymer and a major constituent of plants. The monomeric units, which form lignans, are primarily the hydroxy- and hydroxy-methoxy derivatives of cinnamic and benzoic acids. Cinnamic, caffeic, coumaric, ferulic, and sinapic acids represent the cinnamic acid group; while benzoic, hydroxybenzoic, protocatechuic, vanillic, and syringic acids compose the benzoic acid group. These phenolic acid monomers are commonly found in cereal grains (20) and are "biochemically...the building blocks for most of the other phenolics in plants" (13).

One lignan relevant to this project is 2,3-bis(3-methoxy-4-hydroxy-benzyl)butane-1,4-diol. Its common name is secoisolariciresinol (I). (Refer to Figure 1 for the structures of the lignans discussed in this paragraph.) Until a decade ago, lignans were thought to occur only in higher plants. In 1980, however, Setchell (21–22), Stitch (23), and their coworkers discovered the first mammalian lignans. These lignans are structurally related to I and are named enterodiol (II) and enterolactone (III). In 1982, Axelson (24) reported very convincing evidence that the likely precursor for both II and III was the diglucoside (Ia) of I. Later, Booriello et al. (25) reported that the in vivo conversion of I to II and III is caused by facultative anaerobes within the gut of the human or animal species. They also observed that matairesinol could also be a source for III.

The importance of large amounts of both II and III as human metabolites and their role in mitigating breast and other cancers in humans has been vigorously studied since their discovery (26–29). Recently, Thompson (30) reported using in vitro bacterial techniques to screen for the production of II and III from 68 common foods. These studies found that flaxseed flour produced greater amounts (from 75 to 800 times) of the mammalian lignans than the other food sources. This increasing evidence for using flaxseed in the diet as a method for preventing certain cancers has led to increased activity in evaluating flaxseed in human test diets.

Isolation Procedures for Secoisolariciresinol

Preventing the formation of artifacts during the extraction process is an important factor which needs to be considered when devising an extraction procedure. Ayres and Loike (*12a*) discuss some of the factors to be considered when milling, extracting, and purifying plant extracts. For example, high temperatures should be avoided. In addition, acids and bases can alter lignans through hydrolysis, rearrangement, condensations, and decomposition. Obviously, any alteration of the lignans or their chemical precursors during extraction would interfere in establishing the true stability of the lignans in the flaxseed powder.

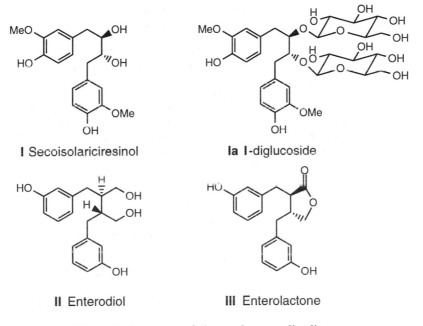

I Secoisolariciresinol

Ia I-diglucoside

II Enterodiol

III Enterolactone

Figure 1. Structures of plant and mammalian lignans.

   A literature review for solvent extraction methods for **I** or **Ia** (SDG) revealed only one citation for flaxseed. In 1956, Bakke and Klosterman (*33*) isolated **Ia** from flaxseed. Since then, no other isolation procedures for **I** or **Ia** from flaxseed have been reported. The review did reveal, however, that **I** or its derivatives have been isolated from the fruit, stems, leaves, needles, resin, and wood of 13 other plant species. Secoisolariciresinol (**I**) was found in 10 species. The other species contained a diferulic acid ester, a monoglycoside, and a bisrhamnoside of **I**, respectively. It is believed that plant lignans are formed biosynthetically by the union of two phenylpropane units or cinnamic acid residues (*12b*). This mechanism is known as the Shikimic Acid-Cinnamic Acid pathway. Thus, it is possible for chemicals occurring in this pathway to be found as intermediates or as decomposition products in flaxseed extracts.

Isolation Procedures for Flaxseed Lignans and Precursors

The flaxseed powder received by us from the National Cancer Institute was not defatted. We evaluated extraction of the flaxseed powder with various organic solvents such as methanol, ethanol, acetone, ethyl acetate, and acetonitrile. The initial extraction method, performed at room temperature, utilizing ethanol containing 1% hydrochloric acid (v:v) yielded chromatograms with the greatest number of peaks. It became apparent, however, that acidic solvents could (as previously discussed) and were altering the analytical profiles and interfering with the goals of the project. Therefore, acidic or basic solution extractions were not utilized for future extractions. Experimentation with extraction procedures revealed that the powder requires defatting prior to analysis. Without prior removal of the linseed oil, it was not possible to obtain reproducible analytical results.

The defatting of the flaxseed powder was achieved by overnight (16 hr) Soxhlet extraction using high purity hexane. Portions of bulk flaxseed powder (20 g each) were weighed into Whatman cellulose thimbles and extracted with 150 mL of solvent. After the defatted flaxseed powder (DFFSP) air dried for several hours, the samples were then suitable for further solvent extractions.

Extraction of DFFSP was performed with various solvents or mixed solvents. Solvent mixtures included ratios of methanol and chlorinated solvents which produced similar chromatographic profiles. The solvents which gave the best results were high purity methanol and 95% ethanol:1,4-Dioxane (1:1, v:v). All extractions were performed in duplicate.

A typical extraction procedure was performed by weighing out 10 g of DFFSP into 250-mL short-necked, round-bottom boiling flasks. The sample flasks were then charged with 100 mL extraction solvent and Teflon® boiling chips. After the samples refluxed for 8 hr, the cooled mixtures were filtered and collected. The filtered samples were reduced *in vacuo* at 65°C to a syrup, transferred to 5-mL volumetric flasks using small amounts of methanol, and then brought to volume with methanol. The samples were then transferred to 10-mL amber serum vials, capped, and stored at 4°C until the chromatographic analyses were performed.

Stability Indicating Analytical Methods for Flaxseed Lignans and Their Precursors

In order to assess the anticancer effects of flaxseed phytochemicals in diets, it is important that suitable analytical methods are available. The methods should be confirmatory and quantitative if measurement of the phytochemical's stability in food matrices is desired. Our approach for flaxseed chemicals was to develop chromatographic techniques capable of monitoring changes in the amounts of the phytochemicals at the parts-per-million levels and, when possible, to confirm the identity of the chemical(s) using mass spectrometry (MS).

We have developed thin layer chromatography (TLC) and high pressure liquid chromatography (HPLC) methods to monitor and quantitate the presence of lignans and their precursors in flaxseed. The TLC method was useful for monitoring the chemical profile of flaxseed when extracted by various solvents. The HPLC method also can be used to monitor and quantitate the lignans and their precursors. Table I lists the estimated lignan levels in fresh and aged (~9 months at ~-20°C) flaxseed powder. The instrumentation and conditions for the HPLC procedure used to quantitate the lignans and the TLC parameters used to verify the HPLC data are given below. Example chromatograms of flaxseed extract, spiked with phenolic standards and unspiked, are shown in Figures 2 and 3.

Thin Layer Chromatography Parameters

Plate: HP Keiselgel 60F (254 nm)
Solvent: chloroform/methanol/acetic acid (90/10/1, v/v)

Standards	R_f value
Chlorogenic acid	0.01
Gallic acid	0.17
4-Hydroxybenzoic acid	0.57
Coumaric acid	0.59
Ferulic acid	0.70
Secoisolariciresinol diglucoside	0.71

High Performance Liquid Chromatography Parameters

Varian 5000 pump, Waters 440 UV detector at 280 nm, Waters Wisp 710B
autosampler, integration by Nelson Analytical Interface and software.
Column:Zorbax C_8, 7 μm
Mobile phase: (A) Deionized water/acetic acid (99/1, v/v); (B) Methanol
Injection volume: 25 μL

Gradient:	Time (min)	%A	%B
	0	100	0
	5	90	10
	20	90	10
	70	0	100
	72	100	0
	85	100	0

Phenolic acid	Retention Time (min)
Gallic acid	4.4
4-Hydroxybenzoic acid	9.2
Chloroenic acid	17.2
Coumaric acid	20.2
Ferulic acid	29.4
Secoisolariciresinol diglucoside	34.2 & 35.7 (doublet)

Table I. Quantitation of Lignans and Their Precursors in Defatted Flaxseed Powder (μg/g) by HPLC

Lignan standard	Ethanol/dioxane		Methanol		
	Fresh (10 g)	Aged (10 g)	Fresh (10 g)	Fresh (5 g)	Aged (5 g)
Chlorogenic	15.1	14.9	12.5	7.5	6.0
Coumaric	—	—	2.0	—	—
Ferulic	—	—	7.6	10.9	11.4
Gallic	19.9	18.6	3.4	2.8	1.9
4-OH-Benzoic	—	—	trace	trace	—
SDG (**Ia**)	2.5	3.2	0.7	0.7	0.9

Note: The estimated levels for all of the lignans have been corrected for the defatted flaxseed powder. The estimated levels are for raw flaxseed powder.

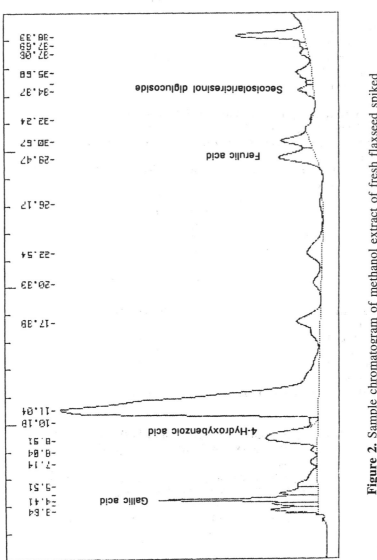

Figure 2. Sample chromatogram of methanol extract of fresh flaxseed spiked with gallic acid, 4-hydroxybenzoic acid, ferulic acid and SDG.

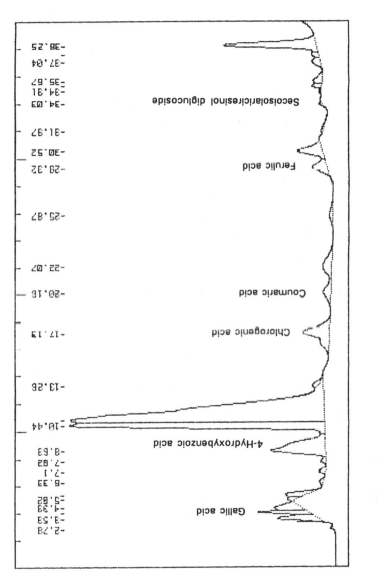

Figure 3. Sample chromatogram of methanol extract of unspiked fresh flaxseed.

The identity of the chemicals was based on retention time matching with pure standards. TLC indicated the presence of the same lignan standards. More definitive identification experiments, such as HPLC/MS, however, should be used to confirm these results. These data demonstrate that, for those lignans and lignan precursors present in the flaxseed, there is little difference in lignan concentrations between aged and fresh flaxseed powder. In addition, increasing the solvent to solute ratio by using only 5 g portions of flaxseed, instead of 10 g portions, did not significantly enhance the extraction efficiency for any of the lignans.

Bakke and Klosterman (*31*) utilized a rigorous extraction and degradation process to isolate the diglucoside of SDG from flaxseed. They noted that the SDG existed in a polymeric form. Fong and colleagues at the University of Illinois in Chicago have used a similar procedure to successfully isolate bulk quantities of secoisolariciresinol **I**. Therefore, it is not surprising that the quantitative estimates of this lignan in the extracts using less rigorous methods were lower than might be expected.

The HPLC method used to quantitate the lignans in the above table required a programmed gradient of water:acetic acid (99:1) to 100% methanol over 70 minutes. We have developed a Coupled Column Chromatography Particle Beam (CCC-PB) HPLC/mass spectrometry (MS) technique that is capable of identifying eluting peaks at very low levels. The optimum solvent for efficient transfer of an analyte through the PB/MS interface is methanol or acetonitrile with a volatile buffer sometimes added. Using water in the mobile phase can seriously degrade overall performance of PB-LC/MS. As the aqueous concentration is increased in the mobile phase, the sensitivity of the PB technique begins to deteriorate. Frequently, a 50:50 solution of water/methanol will reduce the overall sensitivity by a factor of 10 to 100. Coupled column chromatography allows the use of any mobile phase composition for the primary separation of an analyte. The primary separation can be performed on any type of LC column with any desired solvent (including water) flow rate or buffer necessary to separate an analyte from possible interferences. The primary separation LC column is coupled to a secondary LC column via a switching valve that is manually or electronically controlled. After proper calibration, a portion of the eluent from the primary column is "heart-cut" or diverted by a switching valve to a secondary column. The analyte of interest is then flushed off the secondary column and directed to the PB-MS system with more compatible solvents (i.e. methanol or acetonitrile) and flow rates (0.4 mL/min). The CCC techniques combined with LC/MS have greatly extended the range of applications successfully performed at our institute.

Several lignan standards that are found in flaxseed powder were used to develop a prototype method capable of identifying 1 µg on a column using this technique. With this technique, we are confident that the CCC-PB-HPLC/MS method will be useful as a confirmatory procedure. Initial experiments, while promising, indicated that a more robust "cleanup" technique is required before the technique is adopted as a routine method to identify the lignans.

Discussion and Considerations for Future Work

The analytical procedures employed to establish lignan concentrations for this program indicated lower than expected concentrations of the lignans and their precursors, the phenolic acids. These lower concentrations, however, are most likely due to the extraction procedures used, which were selected primarily to prevent degradation of the lignans to monomeric species. Comparison of concentrations of lignans in aged flaxseed powder with fresh powder indicated very little change in the characteristic lignan fingerprint during ~9 months storage at -20°C.

Because identification of the phenolic acid lignan precursors and the secoisolariciresinol derivatives was based on retention time matching with neat standards by both HPLC and TLC, further identification with an improved compound-specific method such as HPLC/MS analysis should be utilized for flaxseed powder extracts. In addition, the flaxseed powder extracts should be evaluated after hydrolysis experiments in order to establish the presence of higher concentrations of the polymeric lignans. Due to the success of the TLC experiments, methods employing high performance TLC (HPTLC) may allow for a larger number of samples to be analyzed in a shorter period of time. One of the major advantages that HPTLC could provide is the potential of analyzing the flaxseed powder extractions without employing a defatting procedure.

Supercritical fluid extraction (SFE) techniques could possibly provide improved defatting or extraction of lignans or lignan precursors, without contributing to hydrolysis problems. The SFE technique has been successfully applied to extraction of cinnamon from chewing gum; pesticides from alfalfa, onions, almonds, oranges, strawberries, and lettuce; caffeine from coffee; and cholesterol from eggs *(32)*. Modifications of similar techniques may be applicable to flaxseed powder and other physiologically functional foods.

Another follow-up series of experiments could encompass the characterization of flaxseed powder after its use in complex food matrices. Baked products with incorporated flaxseed should be analyzed before and after the cooking process to identify the presence or absence of the lignans or lignan precursors such as the phenolic acids. After the product has baked at elevated temperatures in the presence of other matrices, greater quantities of the phenolic monomeric species may be present. Additionally, depending upon the matrix of the food product, natural enzymatic activity or hydrolytic processes may increase the concentrations of lignans with greatest physiological interest (e.g. secoisolariciresinol diglucoside, enterolactone, and enterodiol).

Summary

Analytical procedures employed to establish lignan concentrations indicated lower than expected concentrations of the lignans and their precursors, the phenolic acids. The six lignans and phenolic acid components found in flaxseed extracts were: gallic acid, ferulic acid, chlorogenic acid, coumaric acid, secoisolariciresinol diglucoside (SDG) and 4-hydroxybenzoic acid. The concentrations of these phenolics ranged from 0.1 to 12.0 μg/g, with only traces of 4-hydroxybenzoic acid. These lower concentrations, however, are most likely due to the extraction procedures used, which were selected primarily to prevent degradation of the lignans and establish stability. The concentration of the lignans present in the aged flaxseed powder as compared to fresh flaxseed demonstrated very little change in the characteristic lignan fingerprint during ~9 months storage at -20°C. Future experimentation will focus on utilizing supercritical fluid extraction as an alternative method for isolating lignans from the flaxseed powder for analysis.

Acknowledgements

The authors wish to thank Ms. Jill Guthrie of Midwest Research Institute for her contributions to the mass spectrometry work discussed in this article. This project was funded by the National Cancer Institute, Division of Cancer Prevention and Control, Diet and Cancer Branch, under Contract NO1-CN-05281-01 and the Victor E. Speas Foundation of Kansas City, Missouri.

Literature Cited

1. Kritchevsky, D. *Cancer* **1986**, *58*, 1830.
2. Klurfeld, D. M.; Kritchevsky, D. *D. J. Am. Col. Nutr.* **1986** *5*, 477.
3. Slattery, M. L.; Sorenson, A. W.; Mahoney, A. W.; French, T. K.; Kritchevsky, D.; Stree, J. C. *J. Natl. Can. Inst.* **1988**, *80*, 1474.
4. Kritchevsky, D. *Cancer* **1988**, *62*, 1839.
5. Davis, D. L. *Environmental Research* **1989**, *50*, 322.
6. Correa, P. *Cancer Res.* **1981**, *41*, 3685.
7. Armstrong, B.; Doll, R. *Int. J. Cancer.* **1975**, *15*, 617.
8. Draser, S.; Irving, D. *Brit. J. Cancer.* **1973**, *27*, 167.
9. Phillips, R. L.; Garfinkel, K.; Kuzman, J. W.; Beeson, W. L.; *et al. J. Natl. Can. Inst.* **1980**, *75*, 1097.
10. Reddy, B. S.; Cohen, L. A.; McCory, G. D.; Hill, P.; Weisburger, J. H.; *et al. Adv. Cancer Res.* **1980**, *32*, 237.
11. MacRae, W. D.; Towers, G. H. N. *Phytochemistry* **1984**, *23*, 1207.
12. Ayres, D. C.; Loike, J. D. In *Lignans: Chemical, Biological and Clinical Properties*; Cambridge Univ. Press: Cambridge, 1990; 12a) Chapter 3, 12b) Chapter 5, p 139, 12c) Chapter 7, p 269.
13. Newmark, H. L. *Nutr. Cancer.* **1984**, *6*, 58.
14. Smart, R. C.; Huang, M. T.; Chang, R. C.; Sayer, J. M.; Jerina, D. M.; Wood, A. W.; Conney, A. H. *Carcinogenesis* **1986**, *7*, 1669.
15. Kuenzig, W.; Chau, H.; Norkins, E.; Holowaschenko, H.; Newmark, H.; Mergens, W.; Conney, A. H. *Carcinogenesis* **1984**, *5*, 309.
16. *The Merck Index, 11th Edition*, Budavari, S. Ed.; Merck & Company, Inc.: Rahway, NJ, 1989.
17. Hartwell, J. L. *Cancer Treat. Rep.* **1976**, *60*, 1031.
18. Barclay, A. S.; Perdue, R. E., Jr. *Cancer Treat. Rep.* **1976**, *60*, 1081.
19. Lim, C. K.; Ayres, D. C. *J. Chromatography* **1983**, *255*, 247.
20. Wetzel, D. L.; Pussayanawin, V.; Fulcher, G. *Frontiers of Flavor* Charalambous, Ed.; *Proceedings of the 5th International Flavor Conference*; 1987.
21. Setchell, K. D.; *et al. Nature* **1980**, *287*, 740.
22. Setchell, K. D.; *et al. Biochem. J.* **1981**, *197*, 447.
23. Stich, S. R.; *et al. Nature* **1980**, *287*, 738.
24. Axelson, M.; Sjovall, J.; Gustafsson, B. E.; Setchell, K. D. R. *Nature* **1982**, *298*, 859.
25. Borriello, S. P.; Setchell, K. D. R.; Axelson, M.; Lawson, A. M. *J. Appl. Bacteriol.* **1985**, *58*, 37.
26. Adlercreutz, H.; Fotsis, T.; Bannwart, C.; Wahala, K.; Malela, T.; Brunow, G.; Hase, T. *J. Steroid Biochem.* **1986**, *25*, 791.
27. Setchell, K. D. R.; *et al. Lancet* **1981**, 4.

28. Setchell, K. D. R.; Adlercreutz, H. in *The Role of the Gut Flora in Toxicity and Cancer*; 1988; Chapter 14, p 316.
29. Adlercreutz, H. *Front. Gastrointest. Res.* **1988**, *14*, 165.
30. Thompson, L. U.; Robb, P.; Serraino, M.; Cheung, F. *Nutr. and Cancer* **1991**, *16*, 43.
31. Bakke, J. E.; Klosterman, H. J. *Proceedings of the North Dakota Academy of Science* **1956**, *10*, 18.
32. Costelli, P. *Lab Tech '91: Industrial and Environmental Laboratory Technology Conference and Exhibitions*; Atlantic City, NJ; 1991.

RECEIVED May 28, 1993

LICORICE, GINSENG, AND OTHER MEDICINAL PLANTS

Chapter 31

Antitumor-Promoting and Anti-inflammatory Activities of Licorice Principles and Their Modified Compounds

Shoji Shibata

Shibata Laboratory of Natural Medicinal Materials, 3rd Tomizawa Building, 4th Floor, Yotsuya 3–2–7, Shinjuku-ku, Tokyo 160, Japan

Glycyrrhetinic acid, the aglycone of the licorice saponin glycyrrhizin, showed anti-tumor promoting effects *in vitro* and *in vivo* in the two-stage tumorigenic experiment using DMBA/TPA. Based on this finding, 18β- or 18α-olean-12-ene-3β,28-diol and 3β,23,28-triol and their hemiphthalates, prepared from oleanolic acid and hederagenin, were tested and found to possess potent antitumor promoting activities. Licochalcone A from Xin-jiang licorice showed remarkable inhibitory activities against DMBA/TPA-induced tumorigenesis and TPA-induced inflammatory edema of mice, which suggests an intimate correlation between antitumorigenic and antiinflammatory mechanisms.

Licorice is the sweet tasting root of various species of *Glycyrrhiza* (Leguminosae) which has been used in the East and West since ancient times as a medicine and a sweetening agent. In Dioscorides' De Materia Medica, licorice was described as an antiinflammatory drug for pharyngitis, and in old Chinese medical books it was described as being an antispasmodic agent as well as being a harmonizer to soften various drug actions. Therefore, licorice has frequently been combined with other herb drugs in the numerous prescriptions commonly used in traditional Chinese medicine. On the other hand, licorice is widely employed as a food sweetener.

Various species of licorice are sold on market under various trade names according to their habitat. Their botanical origins have been identified. Some species of *Glycyrrhiza* are not used for medicinal purpose or as food additives owing to their bitter taste (Table I).

Licorice's Saponin Component, Glycyrrhizin

All the licorice species used medicinally contain a sweet tasting triterpenoid saponin, glycyrrhizin (GL), at a fairly high level (average: 4–5% of dry wt.) as the main principle. Some satellite triterpenoid saponins and various types of flavonoids are also found in licorice at a much lower level.

0097–6156/94/0547–0308$06.00/0

Table I. Source Plants of Licorice for Medicinal Use

Plant	Location
Glycyrrhiza uralenesis Fisch.	Don-bei, Xi-bei, Xin-jiang (China)
Gl. glabra L. var. *glandulifera* Rgl. et Herd var. *pubescens* Litw.	Xin-jiang, Turkumenistan Azerbaijan, Afganistan Iran, Iraq, Turkey
Gl. inflata Batal *Gl. korshinskyi* G. Hrig. *Gl. aspera* Pall.	Xin-jiang " "

Gl. echinata L. and *Gl. pallidiflora* Maxim. have no medicinal use

GL is a $\beta(1\rightarrow2)$ linked glucuronyl glucuronide of glycyrrhetinic acid (GA, 18β-olean-11-one-12-ene-3β-ol-30-oic acid). The stereochemical β-linkage of glucuronyl moiety at the 3β-hydroxyl of GA has often been wrongly referred to as α, as noted in the Merck Index (*1*), based on the classical data of molecular rotation (*2*). In truth, it is β on the basis of ^{13}C NMR (103.3 ppm, C-1') and ^{1}H NMR (4.31 ppm ^{1}H d, J=6.9 Hz; 4.47 ppm 1H d, J=7.3 Hz) data obtained by Russian workers (*3*) and by us (Shibata, S.; Iwata, S., unpublished data, 1991) (Figure 1, Table II).

Kitagawa and his group have intensively investigated the minor satellite saponins in the roots of *Gl. uralensis* and *Gl. inflata* to determine their chemical structures (*4–6*) (Table III). All of these have β-linkages with their sugar moieties.

Biological Activities of Glycyrrhizin and Glycyrrhetinic Acid

The biochemical and pharmacological activities of GL and GA have been investigated by several workers. More distinct *in vivo* biological activities have been demonstrated by GL than by GA, owing probably to its water-solubility and higher absorbability; nevertheless, that does not exclude the basic biological activity of GA, the aglycone.

Clinical. The well-recognized pharmacological effects of licorice are the anti-inflammatory (*7*) and antiallergic effects (*8*) of the principal component, GL. Recently, interferon inducing activity (*9*) and an antiviral effect against HSV (*10*) by GL has been demonstrated clinically. Several preparations of GL and GA are now available for clinical use, but side effects, such as pseudo-aldosteronism causing edema, hypertension, and hypopotassemia in patients during long term and high dosage administration, should be noted.

Strong Neo Minophagen C (SNMC, a GL-preparation combined with L-cysteine and glycine) for i.v. injection and Glycyron Tablet (GL combined with L-methionine and glycine) for p.o. administration are available in Japan. The amino acids are added to reduce the pseudo-aldosteronism of GL. These GL-preparations, which have originally been used as an antiallergic drug, are clinically employed against viral chronic hepatitis (*11*) and recently, applied to HIV-carriers to prevent

Table II. ^{13}C-NMR Spectrum of Glycyrrhizin

Glycyrrhetinic acid moiety		Glucuronic acid moiety	
C-3	88.1 ppm	C-1′	103.3 ppm
C-11	198.8	C-2′	82.5
C-12	127.2	C-3′	75.0
C-13	169.5	C-4′	71.1
C-18	47.9	C-5′	75.7
C-30	177.5	C-6′	170.0
		C-1″	104.6
		C-2″	74.5
		C-3″	75.5
		C-4″	71.4
		C-5″	76.1
		C-6″	169.8

^{1}H-NMR signal of anomeric protons
4.31 ppm (^{1}H d, J = 6.9 Hz)
4.47 ppm (^{1}H d, J = 7.3 Hz)

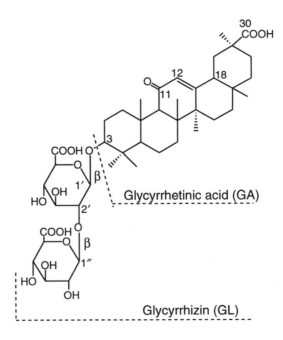

Figure 1. Structures of glycyrrhetinic acid (GA) and glycyrrhizin (GL).

Table III. Oleanane-type Triterpenoid Saponins of Licorice

Source	Saponins	Sapogenin	Sugar moiety
Gl. uralensis, Gl. inflata	glycyrrhizin (GL)	glycyrrhetinic acid (GA)	GlcA(β2→1)GlcA
Gl. uralensis, Gl. inflata	licorice saponin (LS)-A₃	GA	GlcA-GlcA(30)GlcA
Gl. uralensis	LS-B₂	deoxo-GA	GlcA-GlcA
Gl. uralensis	LS-C₂	deoxo-GA-11,13-diene	GlcA-GlcA
Gl. uralensis	LS-D₃	deoxo-GA	GlcA-GlcA(β2→1)Rha
Gl. uralensis	LS-E₂	GA-22-lactone	GlcA-GlcA
Gl. uralensis	LS-F₃	GA-22-lactone	GlcA-GlcA-Rha
Gl. uralensis, Gl. inflata	LS-G₂	GA-24-OH	GlcA-GlcA
Gl. uralensis, Gl. inflata	LS-H₂	liquiritic acid (30α-GA)	GlcA-GlcA
Gl. uralensis	LS-J₂	deoxo-GA-24-OH	GlcA-GlcA
Gl. uralensis	LS-K₂	deoxo-GA-11,13-diene-24-OH	GlcA-GlcA
Gl. uralensis (Xin-jiang)	LS-L₃	deoxo-GA-24-OH	GlcA(β2→1)Ara(β2→1)Rha
Gl. inflata	arabo-GL	GA	GlcA(β2→1)Ara
Gl. inflata	apio-GL	GA	GlcA(β2→1)Api
Gl. inflata	18α-GL	uralenic acid (18α-GA)	GlcA-GlcA

the progression of infection (Ishida, N.; Shigeta, S., personal communication, 1992). Carbenoxolone sodium, hemisuccinate of GA, has been developed in Britain for the treatment of peptic ulcer (*12*).

The pseudo-aldosteronism of GL and GA has been explained by the competitive binding of GA to the renal glucocorticoid receptor (*13*) and inhibition by GA of $\Delta^4 5\beta$-steroid reductase (*14,15*) and Na^+K^+-ATPase associated with the Na^+-extrusion pump (*16*). The inhibition by GA of 11β-hydroxysteroid dehydrogenase in skin (*17*) and lung tissues (Schleimer, R.P., personal communication, 1991) resulting in retention of endogenous cortisol may explain the clinical anti-inflammatory action of GL, GA, and their preparations.

In Vitro. Inhibition of several distinct biological and biochemical functions has been demonstrated *in vitro* by GA in such as intercellular gap juncture communication (*18,19*), phosphotransferase activity of protein kinase C (*20*), phospholipid metabolism promoted by 12-*O*-tetradecanoylphorbol-13-acetate (TPA) in HeLa cells (*21*), and 3-*O*-methylglucose transport stimulated by TPA (*22*).

Biological activities of GL on enzymatic actions and on cultured cells and organs are mostly demonstrated by the aglycone form, GA, but GA has fairly high cell-toxicity against host cells. Therefore, for *in vitro* experiments on antiviral action, GL is employed (*23–28*).

Antiviral activities of GL were shown *in vitro* against HSV-1, New Castle disease, Vescular stomatitis, and Polio-1 viruses (*23, 24*). Inhibition of proliferation of Varicella Zoster virus (*27*) and HIV-1 (*28*) by GL were also demonstrated. The antiviral effect against HIV-1 is attributed to the inhibition of adsorption of virus on the surface of lymphocytes, but not to the inhibition of viral RNA-reverse transcriptase.

Antitumor Promoting Effects of Glycyrrhetinic Acid

In the course of carcinogenesis, there are at least two stages, initiation and promotion (*29*). The promotion stage would be followed by progression establishing malignant state of cancer.

The best established experimental model is the two-stage mouse skin carcinogenesis system. The initiator, 7,12-dimethylbenz[*a*]anthracene (DMBA), is applied to the back skin of mice in a single dose that is under the threshold dosage for carcinogenesis. After some interval, the promoter TPA — an irritating principle of croton oil isolated from the seeds of *Croton tiglium* — is applied to the back by painting repeatedly for 18–20 weeks. TPA is effective at very low concentrations (10^{-9}–10^{-8} M), and the formation of papillomas has been demonstrated within a few weeks. Other irritant products, such as teleocidin, aplysiatoxin, and palytoxin have also been found to be effective cancer promoters for DMBA-initiated mice (*30*).

These two-stage carcinogenesis systems are useful for the survey of antitumor promoters and have been used to study cancers and organs other than skin papillomas induced by the DMBA/TPA system. But the *in vivo* experiments performed by either topical or oral administration of inhibitors require a long time to evaluate their potencies. More convenient, short-term, experimental systems have been developed as screening tests. For example, an *in vitro* experiment measuring the inhibitory potency of test compounds against TPA-promoted phospholipid metabolism in cultured cells is employed (*31,32*).

Since the tumor promoting agents are mostly skin irritating compounds, antiinflammatory agents have been tested for the antitumor promoting activities. As antiinflammatory agents occurring in nature, GL and GA were tested for inhibition of TPA-promoted phospholipid metabolism. Incorporation of $^{32}P_i$ into phospholipids of HeLa cells was increased 4-fold by 50 nM TPA. GA (100 μM) inhibited the increase by 50%. In this experiment, GL was not significantly effective (*32*).

The result of *in vitro* screening tests prompted us to investigate the effect of GA on *in vivo* two-stage carcinogenesis. The *in vivo* test was carried out as follows: The backs of 7-week-old ICR-mice (15 mice per group) were shaved with an electric clipper. After 2 days, a single application of 100 μg DMBA was applied to the shaved area. TPA (0.5 μg, 0.81 nmol) was applied by painting twice a week starting 1 week after initiation. GA (10 μM) was dissolved in 100 μl DMSO: acetone (1:99) and applied topically 40 min before each TPA application. The number and sizes of the skin tumors induced were determined once a week for 18 weeks. The percentage of tumor bearing mice in the group treated with GA was 40% at week 20 in contrast to 97% in the control. The average number of tumors per mouse in the DMBA/TPA alone group was 9.6 at week 20, whereas that in the DMBA/TPA plus 10 μM GA group was 1.8 (*32*).

Antitumor promoting activity has now been added to the multifunctional biological activities of GA, the sapogenin of GL. When tumor promoters do not follow initiation, they merely act as inflammatory agents. Therefore, the anti-inflammatory activities against TPA-induced mouse ear edema and those induced by other agents have been used as screens for antitumor promoting effects.

Destruction of α,β-unsaturated carbonyl at the 11-position and reduction of carboxyl attached at the 20-position of GA produced 18β-olean-12-ene-3β, 30-diol (deoxoglycyrrhetol), 18β-olean-9(11)12-diene-3β,30-diol (homoannular diene), and olean-11,13(18)-diene-3β,30-diol(heteroannular diene) (Figure 2). These compounds and their hemiphthalates, prepared by several chemical reaction steps, have inhibitory activity against lipoxygenase and cyclooxygenase, showing their anti-inflammatory actions. Actually, by topical and oral administration of the dihemiphthalates or the sodium salts of these GA-modified compounds showed dose-dependent inhibitory actions against TPA-induced and arachidonic acid (AA)-induced mouse ear edemata (*33–35*).

Antitumor Promoting Effects of Oleanane-type Triterpenoid Compounds Prepared from Oleanolic Acid and Hederagenin

The fact that GA showed antitumor promoting effects in the two-stage tumorigenesis system revealed a possibility that such a biological activity might be shared by other members of naturally occurring triterpenoids. Thus, oleanolic acid, hederagenin, ursolic acid, echinocystic acid, entagenic acid, and saikogenin A were tested in the same assay system.

Oleanolic acid and hederagenin, which are the most abundant triterpenes in natural sources except for glycyrrhetinic acid, were chemically modified in their functional groups, modulating their biological effects. The 28-COOH group was converted to CH_2OH, and 18β-H partly converted to 18α-H to yield 18β- or 18α-olean-12-ene-3β,28-diol from oleanolic acid and 18β- or 18α-olean-12-ene-3β,23,28-triol from hederagenin (Figure 3). The enhancement of biological activities by these functional and stereochemical conversions has been demonstrated in GA.

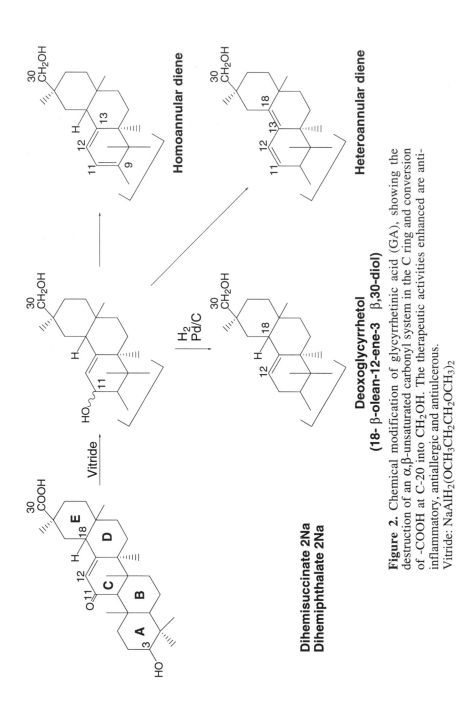

Figure 2. Chemical modification of glycyrrhetinic acid (GA), showing the destruction of an α,β–unsaturated carbonyl system in the C ring and conversion of -COOH at C-20 into CH₂OH. The therapeutic activities enhanced are anti-inflammatory, antiallergic and antiulcerous.
Vitride: NaAlH₂(OCH₃CH₂CH₂OCH₃)₂

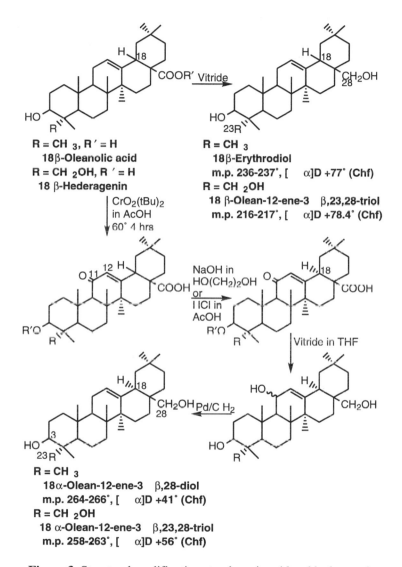

Figure 3. Structural modifications to oleanoic acid and hederagenin.

In Vitro **Tests.** Antitumor promoting activities of oleanane-type triterpenes were primarily screened by their inhibitory action against TPA-induced phospholipid metabolism by measuring $^{32}P_i$-incorporation into phospholipids of HeLa cells. The chemical and stereochemical conversion of functional structures of the triterpenoid compounds resulted in potentiation of their inhibitory effects as expected.

18α- and β-olean-12-ene-3β,23,28-triol also showed almost the same level of inhibition of $^{32}P_i$-incorporation using C3H10T 1/2 cells and Swiss 3T3 cells.

In Vivo **Tests.** Topical application of 18α- or 18β-olean-12-ene-3β,23,28-triol (81 nM) on the shaved back of mice treated with DMBA (100 μg) and TPA (0.5 μg, 0.81 nmol/mouse, twice a week) resulted in 80% or 60% inhibition, respectively, of tumor formation at week 18. The average number of tumors per mouse was 10.6 in the control group at week 18, whereas it was 0.6 (p<0.001) and 1.8 (p<0.001) in groups treated with 18α- and 18β-triol compounds, respectively (*36*).

The effect of oral administration of trihemiphthalate sodium of 18β-olean-12-ene-3β,23,28-triol (1.5–2.5 mg/day as 0.25 mg/ml in drinking water, mice drank 6–10 ml/day *ad libitum*) to mice treated with DMBA/TPA (100 μg, painted once/0.5 μg, painted twice a week) was also tested. The average number of tumors per mouse in the control group was 7.36 at week 18, whereas that of the triterpenoid-treated group was 1.00 (*37*).

The two-stage antitumor promoting experiment has been extended to another type of tumor. For a two-stage lung cancer model in mice, 4-nitroquinoline *N*-oxide (0.3 mg/mouse) was used as an initiator, and glycerol (5% in drinking water) as a promoter. The test compound was also dissolved in the drinking water (1.5–2.5 mg/day as 0.25 mg/ml in drinking water, mice drank 6–10 ml/day *ad libitum*). 18α-Olean-12-ene-3β,23,28-triol trihemiphthalate Na showed a significant inhibition of lung cancer formation in mice at week 18. The percent of tumor bearing mice in the control group was 85%, while that in the treated group 70%. Though that difference was not large, the mean number of tumors per mouse was 4.4 in the control group, and 1.3 in the triterpenoid-treated group (p<0.01) (*38*). It is noteworthy that the active triterpenoid compound is effective in different tumor promotion models using a different initiator and promoter.

Chalcones in Licorice

Origin. Apart from the main saponin, glycyrrhizin, and the minor homologous saponins, several flavonoid compounds have been isolated from various species of licorice. The distribution of most of these flavonoids, which consist of flavone, flavanone, flavonol, flavanonol, isoflavone, isoflavanone, isoflavan, chalcone, pterocarpan, coumestan, 3-arylcoumarin, 2-aryl-benzofuran, dibenzoylmethane, and dibenzyl groups, is species specific, contributing to the chemotaxonomical identification of plant species (*39*).

Licochalcone A, a flavonoid isolated from Xin-jiang licorice (the root of *Glycyrrhiza inflata* Betal.), is especially noted for its high content in the root (ca. 1%). The isolated compound is a yellow crystal with the following properties: m.p. 99–100°C; UV λ_{max} EtOH 254, 312, 379 nm; IR ν_{max} KBr 3236, 1602 cm^{-1}; EI mass m/z (%) M+ 338 (26), M+ (-OMe) 307 (100).

The novel chemical structure of licochalcone A differs from most naturally occurring normal chalcones in the absence of a hydroxy group at the 2′-position of the A-ring (Figure 4). According to the biosynthetical concept, this type of compound has been named retrochalcone, whose A-ring is of shikimate origin and B-ring is of polyketide origin, which is contrary to ordinary chalcones.

Other members of the retrochalcone group (Figure 4) have been found in much lower levels in the root of *Gl. inflata* and designated licochalcone B and echinatin — the latter has also been isolated from *Gl. echinata* L., a species of no medicinal value (*40*).

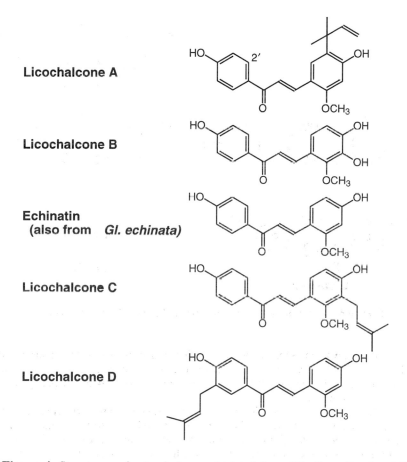

Licochalcone A

Licochalcone B

Echinatin
(also from   *Gl. echinata*)

Licochalcone C

Licochalcone D

Figure 4. Structures of retrochalcones isolated from Xin-jiang licorice (*Gl. inflata* root).

Recently, other minor retrochalcones and dibenzoylmethane derivatives, which might be intermediate metabolites during biosynthesis, have been isolated from the callus of *Gl. echinata* (*41,42*) and the root of *Gl. inflata* (*43*).

The biosynthetical experiment using ^{14}C and ^{3}H-labelling technique in callus culture of *Gl. echinata* demonstrated the occurrence of this type of 1–3 rearrangement of carbonyl, forming the retrochalcone echinatin from the normal chalcone isoliquiritigenin (*40*) (Figure 5).

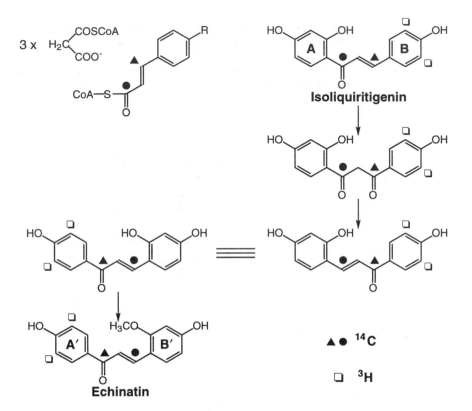

Figure 5. Biosynthesis of retrochalcones. Adapted from reference 40.

The pharmacological action of licorice is mainly due to the saponin, GL, and its sapogenin, GA, but the antispasmodic (44), antiulcerous (45), and anti-allergic effects of licorice (46) have been found to be attributed partly to the flavonoid components.

Recently, Okuda and his coworkers investigated some biological activites of retrochalcones in licorice. They reported antiviral activity against human immunodeficiency virus (HIV) (47), radical scavenging effect (48), inhibition of leukotriene synthesis in human polymorphonuclear neutrophiles (49), and inhibition of xanthine oxidase (50), which induces gout.

Antiinflammatory and Antitumor Promoting Effects. We have dealt with the *in vivo* antiinflammatory and antitumor promoting effects of licochalcone A, which were predicted by its inhibitory effect against TPA-enhanced $^{32}P_i$-incorporation into phospholipids of HeLa cells (51). The ID_{50} of licochalcone A was 5.3 μg/ml (50.5% inhibition with 5 μg/ml). Echinatin and isoliquiritigenin (5 μg/ml) showed 64.1 and 76.8% inhibition, respectively, under the same experimental conditions (Nishino, H., unpublished data).

Mouse ear edema induced by either AA (2 mg/ear) or TPA (2 µg/ear) was significantly inhibited by application of licochalcone A (500 µg/ear) 30 min before the irritant treatment.

Topical application of licochalcone A to the dorsal area of mice initiated by DMBA and promoted by TPA resulted in a remarkable suppression of tumor formation. The first tumor appeared at week 6 in the control group, while at week 11 in the licochalcone A-treated group. At week 20, 73.3% (11 out of 15) of the mice in the control group were bearing tumors in contrast to 26.7% (4 out of 15) of those in the licochalcone A-treated group (p<0.02). The control group had 10.8 tumors per mouse, while the licochalcone A-treated group 0.8 (p<0.001).

Based on the antiinflammatory and antitumor promoting effects of lico-chalcone A, several new chalcone derivatives are being synthesized to be tested for the biological activities in relation to their chemical structures.

Mechanism. As mentioned above, skin inflammation reaction seems partly involved in tumor promotion. We used AA and TPA to induce inflammatory ear edema of mice. AA induces ear edema instantly, whereas with TPA there is a time lag (5 h) in producing edema. AA induced acute inflammation has been found to be caused mainly by leukotrienes (LTs) and promoted by prostaglandin E_2 (PGE_2). 5-Lipoxygenase inhibitors such as AA 861, EN-105, and AC-5-1 (an Indonesian plant product) remarkably inhibited AA-induced ear edema, whereas aspirin, a cyclooxygenase inhibitor which suppresses PGE_2 formation, did not inhibit edema.

On the other hand, topical treatment of TPA increased local PGE_2 concentration by the activation of cyclooxygenase, whereas LTC_4 and LTB_4 were not increased at the time of maximum swelling (5 h after application of TPA). TPA-induced mouse ear edema was suppressed by cyclooxygenase inhibitors such as aspirin, naproxen, and indomethacin. Therefore, PGE_2 seems to be a mediator of TPA-induced inflammatory ear edema. TPA stimulated mast cells and released histamine which may also contribute to edema formation.

The TPA-induced edema was inhibited by protein biosynthesis inhibitors such as cycloheximide, and actinomycin D, whereas AA-induced inflammation was not affected by these reagents. This may suggest that TPA-induced inflammation involves protein synthesis in cells. It has been known that TPA activates either Ca^{2+} dependent or independent protein kinase C (PKC) group enzymes which stimulate phosphorylation of some target protein (52). It is likely that the protein synthesis promoted by the activated PKC might induce the edema response. It has been noted, however, that PKC inhibitors H-7 and H-8 did not inhibit TPA-induced mouse ear edema at the dose of 0.5 mg/ear when applied 30 min before the irritant application. Thus the PKC activation is one of the factors contributing to induction of inflammation, but it is not indispensable.

Almost all the tumor promoters so far known, whether TPA or non-TPA type, are strong skin irritants. They produce only inflammation without pretreatment with initiator, whereas with initiation they induce malignant tumors. Thus licochalcone A may have the same receptor in the course of antiinflammatory and antitumor promoting actions, and it may react as an antagonist of TPA at the site of the receptor, forming a ligand-receptor complex which ultimately inhibits inflammatory or tumorigenic protein synthesis, which might be a gene expression. Although the precise evidence has not yet been obtained, it may be correlated with nuclear type II estrogen binding site (53,54) which is associated with malignant cell growth and proliferation.

Acknowledgments

I wish to thank Dr. H. Nishino and his coworkers at Department of Biochemistry, Kyoto Prefectural Medical University, for their generous collaboration in biological and biochemical research. Thanks are also due to Dr. H. Inoue, Mr. S. Iwata, Mr. K. Hirabayashi, and Mr. N. Nagata, Research Laboratory of Minophagen Pharmaceutical Co., for their cooperation in pharmacological and chemical work.

Literature Cited

1. *The Merck Index*, 11th Ed.; Merck: Rahway, New Jersey, 1989, p 708.
2. Lythgoe, B.; Tripett, S. *J. Chem. Soc.* **1950**, 1983.
3. Khalilov *et al. Khim. Priv. Soedin* (in Russian) **1989**, 500. *Chem. Abstr.* **1990**, 217384-u.
4. Kitagawa, I.; Zhou, J. L.; Sakagami, M.; Taniyama, T.; Yoshikawa, M. *Chem. Pharm. Bull.* **1988**, *36*, 3710.
5. Kitagawa, I.; Sakagami, M.; Hashiuchi, F.; Zhou, J. L.; Yoshikawa, M.; Ren, J. *Chem. Pharm. Bull.* **1989**, *37*, 551.
6. Kitagawa, I.; Zhou, J. L.; Sakagami, M.; Uchida, E.; Yoshikawa, M. *Chem. Pharm. Bull.* **1991**, *39*, 244.
7. Finney, R.S.H.; Tarnoky, A.L. *J. Pharm. Pharmacol.* **1960**, *12*, 49.
8. Kuroyanagi, T.; Sato, M. *Allergy* **1966**, *15*, 67.
9. Abe, N.; Ebina, T.; Ishida, N. *Microbiol. Immunol.* **1982**, *26*, 535.
10. Patridge, M.; Poswilld, D. E. *Brit. J. Oral Maxillofacial Surgery* **1984**, *22*, 138.
11. Suzuki, H.; Ohta, Y.; Takino, T.; Fujisawa, R.; Hirayama, S.; Shimizu, N.; Aso, Y. *Igaku-no-Ayumi* (in Japanese) **1977**, *102*, 562.
12. Doll, R.; Hill, D.; Hutton, C.; Underwood, D. J. *Lancet* **1962**, *ii*, 793.
13. Ulmann, A.; Menard, J.; Corvol, P. *Endocrinol.* **1975**, *97*, 46.
14. Kumagai, A.; Yano, S.; Otomo, M.; Takeuchi, K. *Endocrinol. Jpn.* **1957**, *4*, 17
15. Atherden, L. M. *Biochem. J.* **1958**, *69*, 75.
16. Itoh, K.; Hara, T.; Shiraishi, T.; Taniguchi, K.; Morimoto, S.; Onishi, T. *Biochem. Intern.* **1989**, *18*, 81.
17. Teelucksingh, S.; Mackie, A. D. R.; Burt, D.; McIntype, M. A.; Brett, L.; Edwards, C. R. W. *Lancet* **1990**, *335*, 1060.
18. Davidson, J. S.; Barmgarten, I. M.; Harley, E. H. *Biochem. Biophys. Res. Comm.* **1986**, *134*, 29.
19. Davidson, J. S.; Barmgarten, I. M. *J. Pharmacol. Exp. Therap.* **1988**, *246*, 1104.
20. O'Brian, C.A.; Ward, N. E.; Vogel, V. G. *Cancer Lett.* **1990**, *49*, 9.
21. Nishino, H.; Kitagawa, K.; Iwashima, A. *Carcinogenesis* **1984**, 5, 1529.
22. Kitagawa, K.; Nishino, H.; Iwashima, A. *Cancer Lett.* **1984**, *24*, 157.
23. Pompei, R.; Flore, O.; Marciallis, M. A.; Pani, A.; Loddo, B. *Nature* **1979**, *281*, 689.
24. Pompei, R.; Marciallis, M. A. *L'Iigine Moderna* **1985**, *83*, 385.
25. Dargan, D. J.; Subak-sharp, J. H. *J. Gen. Virol.* **1985**, *66*, 1771.
26. Dargan, D. J.; Subak-sharp, J. H. *J. Antimicrob. Chemotherap.* **1986**, *18 Suppl. B*, 185.
27. Baba, M.; Shigeta, S. *Antiviral Res.* **1987**, *7*, 99.
28. Ito, M.; Nakashima, H.; Baba, M.; Pauwels, R.; De Clercq, E.; Shigeta, S.; Yamamoto, N. *Antiviral Res.* **1987**, *7*, 127.
29. Berenblum, I. *Cancer Res.* **1941**, *1*, 807.

30. Fujiki, H.; Suganuma, M.; Tahira, T.; Yoshioka, A.; Nakayasu, M.; Endo, Y.; Shudo, K.; Takayama, S.; Moore, R. E.; Sugimura, T. In *Cellular Interactions by Environmental Tumor Promoters*; Fujiki, H.; Hecker, E.; Moore, R. E.; Sugimura T., Eds; Japan Sci.Soc.Press: Tokyo, 1978, pp 37-45.
31. Nishino, H.; Fujiki, H.; Terada, M.; Sato, S. *Carcinogenesis* **1983**, *4*, 107.
32. Nishino, H.; Yoshioka, K.; Iwashima, A.; Takizawa, H.; Konishi, S.; Okamoto, H.; Okabe, H.; Shibata, S.; Fujiki, H.; Sugimura, T. *Jpn. J. Cancer Res.* **1986**, *77*, 33.
33. Shibata, S.; Takahashi, K.; Yano, S.; Harada, M.; Saito, H.;Tamura, Y.; Kumagai, A.; Hirabayashi, K.; Yamamoto, M.; Nagata, N. *Chem. Pharm. Bull.* **1987**, *35*, 1910.
34. Inoue, H.; Mori, T.; Shibata, S.; Koshihara, Y. *Brit. J. Pharmacol.* **1989**, *96*, 204.
35. Inoue, H.; Mori, T.; Shibata, S.; Koshihara, Y. *J. Pharm. Pharmacol.* **1988**, *40*, 272.
36. Nishino, H.; Nishino, A.; Takayasu, J.; Hasegawa, T.; Iwashima,A.; Hirabayashi, K.; Iwata, S.; Shibata, S. *Cancer Res.* **1989**, *48*, 5210.
37. Takayasu, J.; Nishino, H.; Hirabayashi, K.; Iwata, S.; Nagata, N.; Shibata, S. *J. Kyoto Pref. Univ. Med.* **1989**, *98*, 13.
38. Nishino, H.; Iwashima, A.; Hirabayashi, K.; Iwata, S.; Shibata, S. *Proc. Jpn. Cancer Ass.* **1992**, 45.
39. Shibata, S.; Saitoh, T. *J. Indian Chem. Soc.* **1978**, 55, 1184.
40. Saitoh, T.; Shibata, S.; Sankawa, U.; Furuya, T.; Ayabe, S. *Tetrahedron Lett.* **1975**, 4461.
41. Ayabe, S.; Kobayashi, M.; Hikichi, M.; Matsumoto, K.; Furuya, T. *Phytochem.* **1980**, *19*, 2179.
42. Ayabe, S.; Iida, K.; Furuya, T. *Phytochem.* **1986**, *25*, 2803.
43. Kajiyama, K.; Demizu, S.; Hiraga, Y.; Kinoshita, K.; Koyama,K.; Takahashi, K.; Tamura, Y.; Okada, K.; Kinoshita, T. *Phytochem.* **1992**, *31*, 3229.
44. Shibata, S.; Harada, M.; Budidarmo, W. *Yakugaku-Zasshi* 1960, 80, 620.
45. Takagi, K.; Ishii, Y. *Arzneim-Forsch.* **1967**, *17*, 1544 and Takagi, K.; Watanabe, K.; Ishii, Y. In *Pharmacology of Oriental Plant Drugs*; Chen, K. K.; Mukerji, B., Eds.; Oxford Pergamon Press: Oxford, 1969.
46. Sato, T.; Matsumoto, H. International Patent Application Number PCT/JP87/00143 (Japanese patent 61-49530), 1987.
47. Hatano, T.; Yasuhara, T.; Miyamoto, K.; Okuda, T. *Chem. Pharm. Bull.* **1988**, *36*, 2286.
48. Hatano, T.; Kagawa, H.; Yasuhara, T.; Okuda, T. *Chem. Pharm. Bull.* **1988**, *36*, 2090.
49. Kimura, Y.; Okuda, H.; Okuda, T.; Arichi, S. *Phytotherap. Res.* **1988**, *2*, 140.
50. Hatano, T.; Yasuhara, T.; Fukuda, T.; Noro, T.; Okuda, T. *Chem. Pharm. Bull.* **1989**, *37*, 3005.
51. Shibata, S.; Inoue, H.; Iwata, S.; Ma, R.; Yu, L.; Uyeyama, H.; Takayasu, J.; Hasegawa, T.; Tokuda, H.; Nishino, A.; Nishino, H.; Iwashima, A. *Planta Med.* **1991**, *57*, 221.
52. Nishizuka, Y. *Nature* **1984**, *308*, 693.
53. Markaverich, B. M.; Roberts, R. R.; Alejandro, M. A.; Clark, J. H. *Cancer Res.* **1984**, *44*, 1575.
54. Markaverich, B. M.; Gregory, R. R.; Alejandro, M. A.; Kittrell, F. S.; Medina, D.; Clark, J. H.; Varma, M.; Verma, R. S. *Cancer Res.* **1990**, *50*, 1470.

RECEIVED June 7, 1993

Chapter 32

Biological Activities, Production, and Use of Chemical Constituents of Licorice

Kenji Mizutani

Department of Research and Development, Maruzen Pharmaceuticals Company Ltd., 22-1 Sanada, Takanishi-cho, Fukuyama 729-01, Japan

To further expand the application of licorice, chemical and biological studies on glycyrrhizin derivatives as well as phenolic compounds were carried out. A new sweetener, monoglucuronide of glycyrrhetinic acid (MGGA) inhibited mouse skin tumor promotion more effectively than glycyrrhetinic acid. Our studies on antioxidative activity, antimicrobial activity, enzyme inhibition and other biological activities of phenolic compounds are summarized.

Licorice has been used as a crude drug in both the East and West since ancient times. Its main active principle is glycyrrhizin, which consists of a glycyrrhetinic acid and two molecules of glucuronic acid (Figure 1). Glycyrrhizin has been used not only as a medicine but also as a food additive because of its sweetness and flavoring property. Recently, the utilization of licorice extract and its active principles has been developed further.

Demand and Utilization of Licorice in Japan

Licorice grows widely throughout China, Russia, the Middle East and Southern Europe. Japan imports its entire demand of licorice, mainly from China, Russia, Afghanistan and Pakistan. The total amount of licorice imported into Japan averages about nine thousand tons a year. There are five companies extracting licorice in Japan. Our company extracts more than half of all the licorice in Japan, five thousand tons per year.

Present uses of licorice in Japan are shown in Tables I and II. Licorice root has been used as a Chinese medicine and as raw material to prepare extracts and active principles. The hydrophilic extract and its active principle, glycyrrhizin, are used in medicines, cosmetics, food additives, cigarette flavoring and so on. Glycyrrhetinic acid and its derivatives have been used in medicines and cosmetics. Recently the hydrophobic extract has been added to cosmetics and used as an antioxidant. Medicinal and cosmetic utilization is expanding — 60% of the licorice used by our company is extracted for use in medicines and cosmetics.

0097-6156/94/0547-0322$06.00/0
© 1994 American Chemical Society

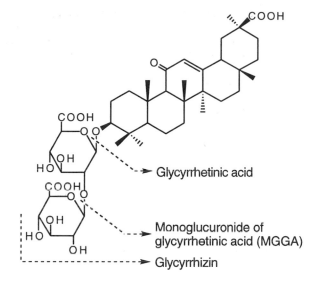

Figure 1. Structure of glycyrrhizin, MGGA and glycyrrhetinic acid

Table I. Uses of Licorice in Japan

	Licorice root	Hydrophilic extract	Glycyrrhizin and its salt
Medicine	Kampo (Chinese medicine)	<u>Oral</u> Cold remedy Antitussive Anti-inflammatory Antiulcer	<u>Oral Prep.</u> Anti-inflammatory Antiulcer Antidotic <u>Injections</u> Hepatitis remedy Antidotic Antiallergic <u>Eye Drops</u> Anti-inflammatory <u>External Prep.</u>
Cosmetic		Shampoo	Facial cream Lotion Tonic
Food	Sweetener Pigment	Sweetener Flavor	Sweetener Flavor
Other	Raw material (for extract) Compost (Extracted residue)	Cigarette flavor	Toothpaste

Table II. More Uses of Licorice in Japan

	Glycyrrhetinic acid and its derivatives	Hydrophobic extract
Medicine	Oral Prep. Antiulcer External Prep. Dermatitis	
Cosmetic	Facial cream Lotion	Facial cream
Food		Antioxidant

Structure-sweetness Relationship of Glycyrrhizin Derivatives and Production of a New Sweetener

In Japan, glycyrrhizin and stevioside preparation are mainly employed as natural low calorie sweeteners. As for stevioside sweeteners, enzymic transglucosylated products are commercially used as better sweeteners than stevioside. By means of this transglucosylation, its bitter taste is removed and its sweetness-intensity is increased. Tanaka *et al.* studied the sweetness and structure relationship of enzymic transglucosylated products of stevioside (*1*). They found that the number of sugar molecules and their positioning are important for sweetness-intensity.

As for glycyrrhizin-sweeteners, a similar transglucosylated product was developed as an improved sweetener. But in this case, an increase in sweetness was not observed, in fact, its sweetness decreased. Kamiya *et al.* studied the sweetness of glycyrrhizin (*2*). These results show that glycyrrhetinic acid-glycosides with fewer sugar molecules are sweeter and that sweetness-intensity varies depending on the type of bound sugars.

We synthesized glycyrrhetinic acid-glycosides containing various sugars — glucose, galactose, xylose, arabinose, glucuronic acid, cellobiose — and evaluated their sweetness. In the case of glycyrrhetinic acid-glycosides, monoglycosides were sweeter than diglycosides. Glycosides of glucose, xylose and glucuronic acid were found to be superior in sweetness to glycyrrhizin. Of these glycosides, monoglucuronide of glycyrrhetinic acid (MGGA) was the sweetest compound. MGGA is approximately 5 times sweeter than glycyrrhizin and 941 times sweeter than cane sugar (Table III).

We have already devised a production method for deriving

Table III. Relative Sweetness Of Various Glycosides of Glycyrrhetinic Acid

Glycosides	Relative sweetness (x sucrose)
β-Cellobioside	71
β-D-Glucoside	218
β-D-Galactoside	100–120
α-L-Arabinoside	31–34
β-D-Xyloside	544
β-D-Glucuronide (MGGA)	941
Glycyrrhizin	170

MGGA from glycyrrhizin using an enzymic reaction. Usually glycyrrhizin is hydrolyzed into glycyrrhetinic acid through MGGA using commercial β-glucuronidase. We screened enzymes from several hundred soil microorganisms which selectively hydrolyze glycyrrhizin to produce MGGA. A new enzyme from *Cryptococcus magnus* hydrolyzed glycyrrhizin to quantitatively produce MGGA. Our company is now producing MGGA using this method.

Screening Tests of Inhibition of Two-stage Carcinogenesis with Glycyrrhizin and Its Derivatives

It has been observed that glycyrrhetinic acid, an aglycone of glycyrrhizin effectively inhibits tumor promotion in mouse skin (*3,4*). As a primary screening test for anti tumor-promoting agents, we studied the inhibition capabilities of the aforementioned glycyrrhetinic acid-glycosides. The effects were evaluated by using a short-term *in vitro* assay of Epstein-Barr virus early antigen activation in Raji cells induced by the tumor promotor 12-*O*-tetradecanoylphorbol-13-acetate (TPA). Of the glycosides, MGGA showed the strongest inhibition ability, to our surprise, even superior to glycyrrhetinic acid (Table IV).

Table IV. Inhibitory Effects of Glycosides of Glycyrrhetinic Acid on Epstein-Barr Virus Activation

Glycosides	Concentration (Mol ratio/TPA)			
	1000	500	100	10
β-Cellobioside	19.6[a] (80)	53.8	81.6	100.0
β-D-Glucoside	11.3 (80)	51.5	75.4	95.7
β-D-Xyloside	10.4 (80)	42.6	71.8	93.4
α-L-Arabinoside	38.9 (80)	63.5	91.6	100.0
β-D-Glucuronide (MGGA)	0 (70)	41.6	74.2	93.1
Glycyrrhetinic acid	15.6 (70)	54.3	100.0	100.0
Glycyrrhizin	26.4 (70)	63.5	82.3	100.0

[a]Percent of control (percent viability)

Then we compared the inhibition capabilities of MGGA with those of glycyrrhizin and glycyrrhetinic acid in two-stage mouse skin carcinogenesis. Initiation of carcinogenesis was carried out by a single application of 7,12-dimethylbenz[*a*]anthracene (DMBA) on the back of the mice. The tumor promoter TPA (1.7 nmol) was applied twice a week starting one week after the initiation. The test compounds were applied topically 60 minutes before TPA application at a molar concentration 50 times that of TPA (85 nmol). Each experimental group used 15 mice, and the experiment was carried out for 20 weeks.

Figure 2 shows the time course of mouse skin tumor formation. In the control group, the first tumor appeared during 6th week and all mice had tumors by the 9th week. In groups treated with test compounds, the first tumor appeared at week 8. Of the compounds tested, MGGA possessed the most inhibitory activity against skin tumor promotion. Inhibition by MGGA of the number of tumor-bearing mice compared with control was 60% at week 10, 47% at week 15 and 13% at week 20. The average number of tumors per mouse in the control group was 9.5 at week 20, whereas that of those treated with MGGA was 4.7. From these results, MGGA is expected to be an excellent cancer preventive agent. We are currently furthering our experimentation into the prevention of lung cancer in mice.

As previously mentioned, MGGA is about 5 times sweeter than glycyrrhizin and is more effective than glycyrrhetinic acid at inhibiting tumor promotion. While glycyrrhetinic acid is difficult to dissolve in water, MGGA is easily dissolved for use in medicines or foods. As shown in Table V, MGGA has no problem with acute toxicity, mutagenicity, or metabolism.

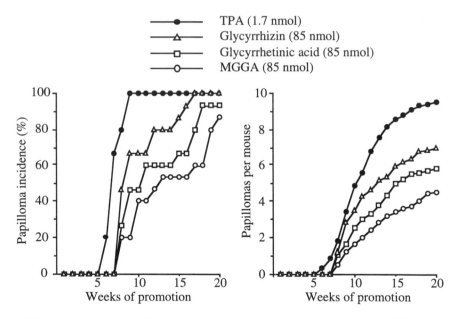

Figure 2. Inhibitory effects of glycyrrhizin, glycyrrhetinic acid and MGGA on DMBA-TPA tumor promotion.

Table V. Properties of MGGA

Formula	$C_{36}H_{54}O_{10}$
State	White powder
Sweetness	Glycyrrhizin x 5 (sucrose x 941)
Solubility	Soluble in water
Acute toxicity	$LD_{50} > 5000$ mg/kg (p.o. in mice)
Mutagenicity	Negative (by umu-test)
Metabolism	Glycyrrhizin-like (by human intestinal flora)

Studies on Antioxidative Activity, Antimicrobial Activity, Enzyme Inhibition and Other Biological Activities of Phenolic Compounds

Recently, chemical and biological studies of licorice have expanded to include components of hydrophobic extracts. In order to find new uses for the hydrophobic extracts of licorice, our research group studied the biologically active principles of commercially available licorice. Because the licorice comes from a number of different origins, the hydrophobic components varied.

We have previously reported the antioxidant and antimicrobial principles in Xinjiang licorice (5). The Xinjiang licorice was collected in Xinjiang Province of China and identified as *Glycyrrhiza inflata*. Recently, our group investigated its hydrophobic extracts in more detail (6-8). We isolated a total of 20 phenolic compounds. Of these compounds, licochalcone-B and -D showed more potent antioxidant capabilities than other compounds. As for antimicrobial activities, the main phenolic constituent, licochalcone-A inhibited the growth of gram-positive bacteria most effectively. In *in vitro* tests of enzyme inhibition, glycydione-C and licochalcone-A exhibited the most effective inhibitory abilities against glucosyl-transferase and tyrosinase, respectively. The anti-tyrosinase capability of licochalcone-A was very weak, however, when compared to glabridin, which was isolated from Russian-licorice. The anti-tyrosinase ability of glabridin will be described later.

We have identified the antioxidant and antimicrobial components licocoumarone and glycycoumarin in Xibei licorice (9). Xibei licorice was imported from northwestern China and assigned the name *Glycyrrhiza uralensis*. Licocoumarone exhibited a potent antioxidant capacity. As for antimicrobial capability, both licocoumarone and glycycoumarin inhibited the growth of gram-positive bacteria and yeasts.

From the Russian licorice botanically named *Glycyrrhiza glabra*, we isolated two main phenolic constituents, glabridin and glabrene (5). Glabrene showed the most potent antioxidant capacity. As for antimicrobial abilities, both glabridin and glabrene inhibited the growth of gram-positive bacteria, yeasts and fungi. Glabridin showed more significant growth-inhibition than glabrene. The most important discovery was the inhibition of tyrosinase by glabridin. In an *in vitro* anti-tyrosinase assay, the preparation containing 40% glabridin is 60 times and 270 times more effective than kojic acid and ascorbic acid, respectively. We have already observed that this preparation inhibits melanin synthesis and shows a more effective whitening activity for human skin than kojic acid and albutin which are now being used as whitening agents in Japan. Based on these results, we are producing some preparations of hydrophobic extracts from licorice (Table VI).

Based on our ongoing studies, we expect further developments in the use of licorice extracts and identification their biologically active principles.

Table VI. Products of hydrophobic extracts from licorice

Product	Origin	Utilization
SANKANON-30	*G. uralensis*	Antioxidant in vegetable oils
SANKANON-FC	*G. glabra*	Antioxidants in animal oils
LICORICE EXTRACT-PU	*G. inflata*	UV-absorbent in cosmetics
LICORICE EXTRACT-PT	*G.glabra*	Whitening agent in cosmetics

Acknowledgments

I wish to express my thanks to Professor Mutsuo Kozuka, Kyoto Pharmaceutical University and to Dr. Harukuni Tokuda, Kyoto Prefectural University of Medicine for inhibitory experiments on Epstein-Barr virus activation and mouse skin carcinogenesis. I would also like to thank my many co-researchers.

Literature Cited

1. Fukunaga, Y.; Miyata, T.; Nakayasu, N.; Mizutani, K.; Kasai, R.; Tanaka, O. *Agric. Biol. Chem.* **1989**, *53*, 1603.
2. Esaki, E.; Konishi, F.; Kamiya, S.*Agric. Biol. Chem.* **1978**, *42*, 1599.
3. Nishino, H.; Yoshioka, K.; Iwashima, A.; Takizawa, H.; Konishi, S.; Okamoto, H.; Okabe, H.; Shibata, S.; Fujiki,H.; Sugimura, T.*Jpn. J. Cancer Res. (Gann)* **1986**, *77*, 33.
4. Yasukawa, K.; Takido, M.; Matsumoto, T.; Takeuchi, M.; Nakagawa, S. *Oncology* **1991**, *48*, 72.
5. Okada, K.; Tamura, Y.; Yamamoto, M.; Inoue, Y.; Takagaki, R.; Takahashi, K.; Demizu, S.; Kajiyama, K.; Hiraga, Y.; Kinoshita, T. *Chem. Pharm. Bull.* **1989**, *37*, 528.
6. *Proc. 38th Annual Meeting Japanese Soc. Pharmacognosy, Kobe, Japan* **1991**, p 95 (Part 1), p 96 (Part 2).
7. *Proc. 112th Annual Meeting Pharmaceu. Soc. Japan, Fukuoka, Japan* **1992**, p 202.
8. Demizu, S.; Kajiyama, K.; Hiraga, Y.; Kinoshita, K.; Koyama, K.; Takahashi, K.; Tamura, Y.; Okada, K.; Kinoshita, T. *Chem. Pharm. Bull.* **1992**, *40*, 392.
9. Demizu, S.; Kajiyama, K.; Takahashi, K.; Hiraga, Y.; Yamamoto, S.; Tamura, Y.; Okada, K.; Kinoshita, T. *Chem. Pharm. Bull.* **1988**, *36*, 3474.

RECEIVED May 4, 1993

Chapter 33

Anticarcinogenesis of Licorice and Its Major Triterpenoid Constituents

Zhi Yuan Wang

Laboratory for Cancer Research, College of Pharmacy, Rutgers, The State University of New Jersey, Piscataway, NJ 08855-0789

Licorice (*Glycyrrhiza glabra* L.) is a traditional herb widely used as a food sweetening and flavoring agent. Oral administration of a water extract of licorice (1% in drinking water) to female A/J mice throughout the experimental period afforded significant protection against benzo[*a*]pyrene (BP) or *N*-nitrosodiethylamine (NDEA)-induced lung and forestomach tumorigenesis. The main water soluble constituent of licorice is glycyrrhizin (GL), a pentacyclic triterpene. Feeding 0.05% GL in the drinking water to female Sencar mice gave substantial protection against 7,12-dimethylbenz[*a*]-anthracene (DMBA)-induced skin tumor initiation. Glycyrrhetinic acid (GA) is an aglycone of GL which exists in 18α-GA and 18β-GA stereoisomeric forms. Topical treatment of female Sencar mice with both isomers significantly inhibited DMBA-induced skin tumor initiation as well as 12-*O*-tetradecanoylphorbol-13-acetate (TPA)-induced skin tumor promotion. Possible mechanisms were studied. Our results indicate that triterpenoids (e.g., glycyrrhetinic acids) may be useful cancer chemopreventive agents.

Licorice (*Glycyrrhiza glabra* L.) has been used as an antidote, demulcent and elixir folk medicine in China for thousands of years. In addition, licorice is also widely used in the food industry as a food sweetening and flavoring agent in candy, chewing gum, chocolate, cigarettes, liquors and beer. The main water soluble constituent of licorice is glycyrrhizin (GL), a saponin of pentacyclic triterpene derivative of β-amyrin type. GL has been shown to possess several important pharmacological activities, such as an antiinflammatory effect. GL has been shown to be partly hydrolyzed by glucuronidase to its aglycone, glycyrrhetinic acid (GA), which exists in 18α-GA and 18β-GA stereoisomeric forms. Our prior studies have shown that oral administration of GL inhibited DMBA-induced skin tumorigenesis in Sencar mice (*1*) and topical application of 18α-GA and 18β-GA inhibited skin tumor initiation and promotion in Sencar mice (*2*). We also demonstrated that oral administration of water extracts of licorice inhibited BP and DMBA-induced lung and forestomach tumorigenesis in A/J mice (*3*). In this review the anticarcinogenic effects of licorice and its major triterpenoid constituents are summarized.

0097–6156/94/0547–0329$06.00/0

Anticarcinogenesis of Licorice

Female A/J mice (6–8 weeks old, Jackson Laboratories, Bar Harbor, ME) were fed Purina Chow 5001 diet (Ralston-Purina Co., St. Louis, MO). Ten g licorice root (cut into small pieces) was extracted with 500 ml boiling water for 30 min and then filtered. The licorice root was extracted a second time with another 500 boiling water and filtered. The combined extract was referred as 1% water extract of licorice.

The protective effects of licorice against BP-induced lung and forestomach tumorigenesis in female A/J mice are summarized in Table I. The mice were treated with BP (100 mg/kg administered p.o. at 2-week intervals for 8 weeks) then sacrificed 33 weeks after the last BP treatment. Lung tumors were found in all the animals and the average number of tumors per mouse was 8.7 ± 0.7. Oral feeding of 1.0% water extract of licorice as the sole source of drinking water during the BP treatment (initiation) period caused a 20% reduction in lung tumor incidence, and a 60% reduction in lung tumor multiplicity. Treatment of A/J mice with BP also induced forestomach tumors in 96% of the mice and an average of 2.1 ± 0.3 tumors per mouse. Oral administration of 1.0% water extract of licorice caused a 33% reduction in lung tumor incidence, and 24% reduction in lung tumor multiplicity.

The inhibitory effect of licorice on NDEA-induced lung and forestomach tumorigenesis is also shown in Table I. The mice were treated with NDEA (20 mg/kg administered p.o. once weekly for 8 weeks) then sacrificed 33 weeks after the last NDEA treatment. Lung tumors were found in 92% of the animals and the average number of tumors per mouse was 2.9 ± 0.5. Oral administration of 1.0% water extract of licorice as the sole source of drinking water during the NDEA treatment period caused a 26% reduction in lung tumor incidence, and a 55% reduction in lung tumor multiplicity. Treatment of A/J mice with NDEA induced forestomach tumors in 80% of the mice and an average of 2.5 ± 0.7 tumors per mouse. Oral administration of 1.0% licorice water extract caused a 45% reduction in lung tumor incidence, and a 68% reduction in lung tumor multiplicity. In both experiments all forestomach tumors were papillomas and all lung tumors were adenomas.

Anticarcinogenesis of Glycyrrhizin

Female Sencar mice (6–8 weeks old, obtained from the National Cancer Institute Frederick Cancer Research Facility) were used in a DMBA/TPA-induced two-stage skin tumorigenesis protocol. A single topical application of 40 nmol DMBA followed after one week by application of 4 nmol TPA twice weekly for 16 weeks resulted in a 100% skin tumor incidence and 51 ± 4 skin tumors per mouse. Oral administration of 0.05% GL (as ammonium salt, purchased from Sigma Chemical) as the sole source of drinking water for 50 days prior to DMBA application (group GL-1) or 112 days post-DMBA application (group GL-2) both afforded protection against skin tumor initiation (Table II). The latent period prior to the onset of tumor development was prolonged in GL-fed animals compared with the control groups. The time at which 100% of the animals exhibited tumors in the control, GL-1 and GL-2 groups was after 8, 14 and 16 weeks of TPA application, respectively. After

Table I. Effect of Feeding Water Extract of Licorice on BP- and NDEA-induced Lung and Forestomach Tumorigenesis in Female A/J Mice

Group	Number of mice	Weight gain[a] (g)	Lung tumors		Forestomach tumors	
			Tumor incidence[b]	Tumors per mouse[c]	Tumor incidence[b]	Tumors per mouse[c]
Control	10	12.1	0	0	0	0
BP-induced tumorigenesis						
BP alone	25	10.7	100	8.7 ± 0.7	96	2.1 ± 0.3
+ 1% licorice water extract	25	9.5	30 (20%)	$3.4 \pm 0.8^*$	64**	1.6 ± 0.5 (24%)
NDEA-induced tumorigenesis						
NDEA alone	25	9.2	92	2.9 ± 0.5	80	2.5 ± 0.7
+ 1% licorice water extract	25	10.3	68** (26%)	$1.3 \pm 0.3^*$ (55%)	44** (45%)	$0.8 \pm 0.3^*$ (68%)

[a]Gain from 8 to 42 weeks of age.
[b]Percent of mice with tumors. The number in parentheses is the percent protection. **Significantly different from positive control (χ^2 test, $p<0.05$).
[c]Average number of tumors per mouse ± SE. The number in parentheses is the percent protection. *Significantly different from positive control (Newman-Keuls multiple comparison, $p<0.05$).

eight weeks of TPA application, when 100% of the animals in the control group had developed tumors, the number of tumors per mouse in the control group was 16 ± 2 compared with 5 ± 1 and 10 ± 2 in the GL-1 and GL-2 groups, respectively.

Table II. Effect of Oral Administration of Glycyrrhizin on DMBA/TPA-induced Skin Tumorigenesis

	8 weeks		12 weeks		16 weeks	
Treatment	Incidence[a]	Number[b]	Incidence	Number	Incidence	Number
H$_2$O	100	16	100	43	100	51
GL-1 (pre-DMBA)	80	10	90	29	100	37
GL-2 (post-DMBA)	35	5	90	30	100	36

[a]Percent of mice with tumors.
[b]Number of tumors per mouse.

Anticarcinogenesis of Glycyrrhetinic Acid

The comparative effect of 18α- and 18β-GA on DMBA/TPA skin tumorigenesis was evaluated. Female Sencar mice (6–8 weeks old, obtained from the National Cancer Institute Frederick Cancer Research Facility), were used in a DMBA/TPA-induced two-stage skin tumorigenesis protocol.

Inhibition of Skin Tumor Initiation. Animals received a single topical application of 0.2 ml DMSO:acetone (1:2) daily for 7 days (as control group), or 10 μmol 18α-GA or 18β-GA in the above solvent for 7 days (test groups). On the last day, 1 h following topical application of test compounds, 40 nmol DMBA was applied topically as an initiation agent. One week later 4 nmol TPA was applied topically twice weekly for 16 weeks. The data in Table III indicate that both 18α-GA and 18β-GA inhibited DMBA-induced skin tumor initiating activity. Between isomers, 18β-GA exerted stronger inhibitory effects than 18α-GA. For example, after 9 weeks of TPA application, the tumor incidences in the control group, 18α-GA and 18β-GA groups were 100, 65 and 45%, respectively. The average number of tumors per mouse at the termination of the experiment was inhibited 20 and 50% by 18α-GA and 18β-GA, respectively.

Inhibition of Skin Tumor Promotion. Animals received a single topical application of 40 nmol DMBA as initiating agent. One week later, the animals received 4 nmol TPA twice weekly for another 16 weeks. Thirty min prior to TPA application the animals were treated topically with 18α-GA or 18β-GA. Both compounds inhibited skin tumor promotion induced by TPA (Table III). After 9 weeks of TPA application, the tumor incidences in the control group, 18α-GA and 18β-GA were 100, 10 and 40%, respectively. The average number of tumors per mouse at the termination of the experiment was inhibited 80 and 60% by 18α-GA and 18β-GA, respectively. These data suggest that 18α-GA and 18β-GA possess comparable activity as inhibitors of TPA-induced skin tumor promotion.

Table III. Effect of Topical Administration of Glycyrrhetinic Acid on
DMBA/TPA-induced Skin Tumorigenesis

Treatment	8 weeks		12 weeks		16 weeks	
	Incidence[a]	Number[b]	Incidence	Number	Incidence	Number
H_2O	78	5	100	38	100	48
Initiation stage						
α-GA	53	5	100	23	100	38
β-GA	42	4	78	12	85	24
Promotion stage						
α-GA	5	1	68	6	79	10
β-GA	5	1	72	6	82	19

[a]Percent of mice with tumors.
[b]Number of tumors per mouse.

Discussion

GL has been reported to protect against saponin, alkaloid, urethan, benzene and CCl_4-induced toxicity (4–7). We found that oral feeding of 0.05% GL as the sole source of drinking water to Sencar mice resulted in 36% and 41% inhibition on *in vivo* binding of [³H] BP and [³H] DMBA, respectively, to epidermal DNA (1). It was also found that topical application of 18α-GA or 18β-GA resulted in 41% or 59% inhibition, respectively, of *in vivo* binding of [³H]-BP to epidermal DNA and 30% or 48% inhibition, respectively, of *in vivo* binding of [³H]-DMBA to epidermal DNA. These results suggest that protection against DNA damage induced by tumor initiation agents may be one of the mechanisms of their anti-tumor initiation activity.

It is widely accepted that most tumor promoters, such as TPA, cause skin inflammation and induce ornithine decarboxylase, which, when inhibited by antiinflammatory agents, results in prevention of skin tumor promotion (8,9). GA was reported to be an effective antiinflammatory drug (10,11). We observed that both 18α-GA and 18β-GA inhibited lipoxygenase activity and TPA-induced ODC induction (2). This evidence may be useful to understand the mechanism of their anti-promotion activity.

Comparative studies of the stereoisomers 18α-GA and 18β-GA on anti-tumor initiation and anti-promotion show some interesting findings. For anti-tumor initiating activity, 18β-GA was more potent than 18α-GA, whereas in the case of anti-promoting activity, the effects were approximately equal for 18α-GA and 18β-GA. Prior studies have also assessed the differences in biological functions of these two stereoisomers. Kiso *et al.* (12) indicated that in the case of CCl_4-induced hepatotoxicity, 18β-GA afforded higher protection than 18α-GA. The order of potency was exactly the same as observed by us in antimutagenesis and anti-tumor initiation assays. We also found that 18β-GA caused a remarkable type 1 spectra interaction with oxidized cytochrome P-450, while 18α-GA exerted almost no interaction. 18β-GA was found to be more effective than 18α-GA in the inhibition of monooxygenase activity (2). The explanation may involve the difference in the stereostructure between 18α-GA (*trans* conformation of the *D/E*-ring) and 18β-GA (*cis* conformation of the *D/E*-ring).

In summary, the results of this study suggest that oral administration of water extract of licorice to A/J mice protected against lung and forestomach tumorigenesis induced by BP or NDEA. Oral administration of the major water soluble component GL was found to inhibit DMBA-induced skin tumor initiation in Sencar mice. It was also shown that topical application of GA, the aglycone of GL, inhibited DMBA-induced skin tumor initiation and TPA-induced skin tumor promotion in Sencar mice.

Licorice is an ancient and traditional food additive agent and herb worldwide. Since Kubuke *et al.* (*13*) reported that the long-term oral administration of the sodium salt of GL in drinking water (up to 0.5%) to mice did not yield any evidence of chronic toxicity, licorice may be one of the attractive and practical chemopreventive agents for human. The relationship between licorice consumption and cancer warrants more laboratory and epidemiological studies. Terpenoids may be an important class of cancer chemopreventive agents, not only the mono-terpenoids (e.g., D-limonene), diterpenoids (e.g., Vitamin A), and tetraterpenoids (e.g., β-carotene), but triterpenoids (e.g., glycyrrhetinic acids) as well.

Literature Cited

1. Wang, Z. Y.; Agarwal. R.; Zhou, Z. C.; Bickers, D. R.; Mukhtar, H. *Carcinogenesis* **1991**, *12*, 187.
2. Agarwal, R.; Wang, Z. Y.; Mukhtar, H. *Nutrition and Cancer* **1991**, *15*, 187.
3. Wang, Z. Y.; Agarwal, R.; Khan, W. A.; Mukhtar, H. *Carcinogenesis* **1992**, *13*, 1491.
4. Segal, R.; Milo-Goldzweig, I.; Kaplan, G.; Weisenberg, E. *Biochem. Pharmacol.* **1971**, *26*, 643.
5. Mizoguchi, Y.; Katho, H.; Tsutsui, H.; Yamanoto, S.; Morisawa, S., *Gastroenterol. Jpn.* **1985**, *20*, 99.
6. Akimoto, M.; Kimura, M.; Sawano, A.; Iwasaki, H.; Nakajima, Y.; Matano; S.; Kasai, M. *Gan No Rinsho.* **1986**, *32*, 869.
7. Ko, M. Q., Ed. *The Active Components of Chinese Medicine: Chemical, Physical, and Pharmacological Characters* Hunan Scentific and Technological Publishing House: Chansa, China, 1982; pp. 254.
8. Gschwendt, M.; Kittstein, W.; Furstenberger, G.; Marks, F. *Cancer Lett.* **1984**, *25*, 77.
9. Weeks, C. E.; Slaga, T. J.; Hennings, H.; Gleason, G. L.; Bracken, W. M., *J. Natl. Cancer Inst.* **1979**, *63*, 401.
10. Sugishita, E.; Amagaya, S.; Ogihara, Y. *J. Pharmacol. Dynam.* **1982**, *5*, 379.
11. Amagaya, S.; Sugishita, E.; Ogihara, Y. *J. Pharmacol. Dynam.* **1984**, *7*, 923.
12. Kiso, Y.; Tohkin, M.; Hikino, H.; Hattori, M.; Sakamoto, T.; Namba, T. *Planta. Med.* **1984**, *50*, 298.
13. Kobuke, T.; Inai, K.; Nambu, S.; Ohe, K.; Takemoto, T.; Matsuki, K.; Nishina, H.; Huang, I. B.; Tokuoka, S. *Food Chem. Toxicol.* **1985**, *23*, 979.

RECEIVED August 20, 1993

Chapter 34

Ginseng and Its Congeners
Traditional Oriental Food Drugs

Osamu Tanaka

Department of Food and Nutrition, Suzugamine Women's College, Inokuchi, Nishi-ku, Hiroshima 733, Japan

Ginseng has been used not only as a traditional Oriental medicine but also as a health-giving vegetable in Oriental dietary therapy. Recent reports related to cancer prevention of ginseng and its congeners are summarized in this chapter. They include an anti-carcinogenic effect of ginseng on mice exposed to chemical carcinogens, an inhibitory effect of ginseng saponins on TPA-induced inflammation, a case-control study on correlation of ginseng intake to cancer, inhibitory effects of Sanchi-Ginseng on the Epstein-Barr virus early antigen activation in Raji cells induced by TPA and n-butylic acid, and inhibition of pulmonary tumorigenesis induced in mice by 4-nitroquinoline-N-oxide and glycerol.

Ginseng has been used as a well known traditional medicine in Oriental countries for more than 5,000 years. The source plant of this drug is *Panax ginseng* (Araliaceae), a herb which is now extensively cultivated in Korea, China and Japan. After four to six years cultivation, the carrot-like roots are steamed and dried to prepare "Red-Ginseng," while roots dried without steaming (sometimes after peeling) are named "White-Ginseng." Red-Ginseng is more commonly used in China, while White-Ginseng is used in Kampo medicine and health foods. In China and Korea, ginseng is taken not only as a medicine but in dietary therapy as a health-giving vegetable. For example, it is used in chicken-ginseng soup and a slice of fresh ginseng with honey is sometimes served to a special guest as a precious hors d'oeuvre.

Chemistry and Pharmacology of Ginseng

More than 500 scientific papers on ginseng are in the literature, and at present, ginseng is one of the most deeply and widely investigated herb drug among those used in the traditional medicines. The chemical constituents which are characteristic of ginseng have been investigated extensively since 1955.

Our group, organized by Dr. Shibata, isolated a number of saponins from ginseng. The major saponins of ginseng are not pentacyclic oleanane oligoglycosides which are very common in nature, but tetracyclic dammarane oligoglycosides (1–3). This was the first example of dammarane saponins in nature. More than twenty saponins of this type have been isolated from the roots and leaves of ginseng; these characteristic dammarane saponins, named ginsenosides, are classified into two groups: the glycosides of 20[S]-protopanaxadiol (20[S]-dammar-24-ene-3β,12β,20-triol) and those of 20[S]-protopanaxatriol (6α-hydroxy-20[S]-protopanaxadiol) (Figure 1).

In addition to these dammarane saponins, an oleanolic acid glucuronide saponin named ginsenoside Ro, which is identical with chikusetsusaponin V from rhizomes of *Panax japonicus*, has also been identified in ginseng.

Aglycone: 20[S]-protopanaxadiol (R$_1$=R$_2$=H)

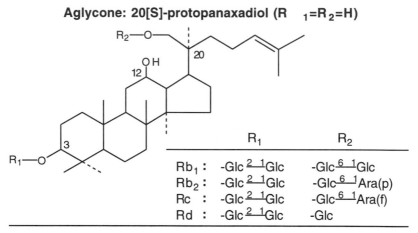

	R$_1$	R$_2$
Rb$_1$:	-Glc$\frac{2\quad1}{}$Glc	-Glc$\frac{6\quad1}{}$Glc
Rb$_2$:	-Glc$\frac{2\quad1}{}$Glc	-Glc$\frac{6\quad1}{}$Ara(p)
Rc :	-Glc$\frac{2\quad1}{}$Glc	-Glc$\frac{6\quad1}{}$Ara(f)
Rd :	-Glc$\frac{2\quad1}{}$Glc	-Glc

Glc: Glucose, Ara(p): α-L-arabinopyranose, Ara(f): α-L-arabinofuranose

Aglycone: 20[S]-protopanaxatriol (R$_1$=R$_2$=H)

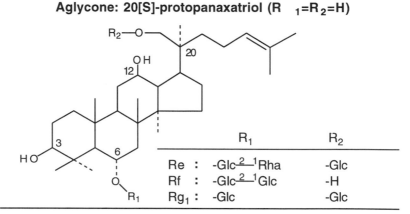

	R$_1$	R$_2$
Re :	-Glc$\frac{2\quad1}{}$Rha	-Glc
Rf :	-Glc$\frac{2\quad1}{}$Glc	-H
Rg$_1$:	-Glc	-Glc

Glc: Glucose, Rha: Rhamnose

Figure 1. Major dammarane-saponins of ginseng.

Traditionally, people have used ginseng for treatment of many kinds of diseases concerned with anemic feebleness. Since 1965, several pharmacological activities have been reported for ginseng extracts or ginseng dammarane saponins, including effects on the central nervous system, antipsychotic action, tranquilizing effects, protection from stress ulcers, increased gastrointestinal motility, antifatigue action, endocrinological effects (2–4), enhancement of sex behavior, and acceleration of metabolism or synthesis of carbohydrates, lipid, RNA and protein, etc.

Kitagawa *et al.* isolated a small amount of 3-*O*-β-glucoside of 20(S)-protopanaxadiol named ginsenoside-Rh2 from Red-Ginseng (5). Odazima *et al.* reported that ginsenoside-Rh2 demonstrated growth inhibition against several cultured tumor cells — Lewis lung cancer, HeLa human uterus adenocarcinoma, Morris hepatoma, and B16 melanoma (6). In the case of Morris hepatoma cells, reverse transformation of the tumor cells was also observed, and in the case of melanoma cells, morphological alteration and stimulation of melanogenesis took place in the presence of this glucoside.

Besides the saponins, several acetylene alcohols were isolated from the *n*-hexane soluble fraction of methanolic extract of ginseng (Figure 2). Of these unsaturated alcohols, panaxytriol was isolated not from White-Ginseng but from Red-Ginseng, so that this compound seems to be an artifact which is derived from the epoxide panaxydol during the process of steaming (5). Panaxytriol is reported to exhibit cytotoxic activities against cultured tumor cells — MK-1 human gastric adenocarcinoma, M14 human melanoma, SW620 human colonic cancer cells, HeLa human uterus adenocarcinoma, K562 human erythroleukemia, B16 mouse melanoma, and MRK-5 human embryo-derived fibrosarcoma (7).

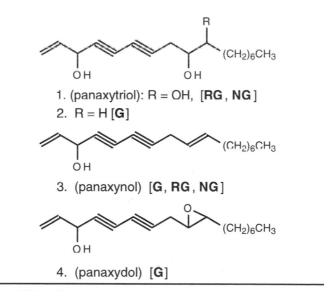

1. (panaxytriol): R = OH, [**RG** , **NG**]
2. R = H [**G**]

3. (panaxynol) [**G**, **RG** , **NG**]

4. (panaxydol) [**G**]

***

G : White-Ginseng
RG: Red-Ginseng (steamed ginseng)
NG: Sanchi-Ginseng (Konoshima, Kozuka *et al.*, 1991, ref. 21)

Figure 2. Acetylene alcohols from *Panax spp.*

In 1969, Breckman proposed the presence of "adaptogen" in ginseng (*8*). Adaptogen is a term for a prophylaxis against aging and diseases which recovers homeostasis. In other words, an adaptogen is a compound which increases non-specific resistance against toxic environmental agents as well as against physical stress such as hypothermia, hyperthermia and radiation.

In relation to this suggestion, Yonezawa *et al.* reported the repair of X-ray radiation injury by ginseng extract, increase of survival ratios of X-ray irradiated mice, rats and guinea pigs, and acceleration of recovery of blood cell counts, especially thrombocyte cells. They also observed enhancement of recovery of pluripotent blood forming stem cells in bone marrow and prevention of occult blood appearance in feces. They mentioned that this efficacy would be due to the enhancement of thrombopoietic hematogenesis and diminishing hemorrhage (*9*).

The concept of adaptogen suggests the possibility of using ginseng as a prophylactic or delaying agent against carcinogenesis rather than as therapeutic agent.

Cancer Prevention of Ginseng and Its Congeners

Anticarcinogenic Effect Against Chemical Carcinogens.   T. K. Yun and colleagues at the Cancer Research Hospital of the Korean Atomic Energy Research Institute have extensively investigated the anticarcinogenic effect of long term oral administration of ginseng on newborn mice exposed to various chemical carcinogens (*10*). The administration of ginseng extract decreased the incidence and average number of lung adenomas induced by urethane and reversed the decrease of natural killer cell activity induced by urethane (*11*). This evidence strongly suggests that the anticarcinogenic activity of ginseng extract against urethane-induced carcinogenesis may be related to the augmentation of natural killer cells.

In the case of lung adenoma induced by dimethylbenz[*a*]anthracene (DMBA), significant inhibition in the incidence of adenoma by ginseng extract was not observed. Administration of ginseng extract did, however, decrease the average diameter of the largest lung adenomas by 23% and the incidence of diffuse pulmonary infiltration by 63%. In the case of cancer induced by aflatoxin B_1, prolonged administration of ginseng extract decreased the incidence of lung adenoma by 29% and hepatoma by 75%.

Inhibitory Effect of Ginseng Saponins on TPA-Induced Inflammation. Yasukawa, Takito and their coworkers reported an inhibitory effect by ginseng saponins on TPA-induced inflammation in their extensive studies on the antitumor promoting activities of triterpene derivatives from traditional crude drugs and foods using a two stage carcinogenesis model (*12*). Significant inhibitory effects were observed for the saponins of 20(S)-protopanaxadiol such as ginsenosides Rb_1, Rb_2, Rc, and Rd, though the activities were lower than glycyrrhetic acid from licorice. The effects of saponins of 20(S)-protopanaxatriol, such as ginsenosides Re and Rg_1 as well as glucuronide saponins of oleanolic acid such as chikusetsusaponins IV and V, were less significant than those of 20(S)-protopanaxadiol.

Case-Control Study on the Relationship Between Cancer and Ginseng Intake. Yun and his colleagues reported a case-control study of ginseng and cancer in humans (*13*). The cases and controls were carefully selected from patients admitted to the Korea Cancer Center Hospital in Seoul, one of the major cancer hospitals in

Korea. In order to collect ginseng intake information, a dietary recall method commonly employed in epidemiological studies of diet and disease was used. The results obtained after statistical treatment show a potent preventive effect of ginseng against carcinogenesis. The odds ratio of cancer in relation to ginseng intake was 0.56. This means that the risk of cancer was reduced to almost one half by the intake of ginseng. In general, ginseng powder and extracts seem to be more effective than fresh sliced ginseng, juice or tea in reducing odds ratios. The lowest odds ratio was 0.14 for consumers of fresh ginseng extract. Significant preventive effects were observed for the combination of the different forms of ginseng; in the case of fresh ginseng, the odds ratio for the combination of the fresh slice and the extract was 0.18 and that for the powder and the extract was 0.34. No statistical significant reduction was shown in the case of the slice and juice of fresh ginseng and White-Ginseng tea. Data on Red-Ginseng intake was too sparse for reliable statistical analysis.

A dose response relationship between the consumption of all types of ginseng and cancer prevention was also demonstrated. The odds ratios decreased with increasing ginseng intake. In this study, the significance of the dose response relationship was much higher in males than in females.

Inhibitory Effect of Sanchi-Ginseng on Two Stage Carcinogenesis. A variety of ginseng congeners grow in the northern hemisphere from Central Nepal onward through China, Japan to North America. *Panax notoginseng* is cultivated in Yunnan, China. The root of this plant, called Sanchi-Ginseng or Tien-chi, is a famous Chinese medicine used as a tonic and a hemostatic.

Cooperative studies by the Kunming Institute of Botany with our group isolated a number of dammarane saponins (*1,3,14,15*), which are related to the ginseng saponins, as well as a reticuloendothelial system-activating arabinogalactane, from this crude drug.

A test of the inhibitory effects on Epstein-Barr virus early antigen activation in Raji cells induced by a combination of 12-*O*-tetradecanoylphorbol-13-acetate (TPA) and *n*-butylic acid (EBV-EA test) has been employed as a simple primary screening for antitumor promoting agents. Many compounds which are active in this test also show positive results in the conventional skin tumor promotion test (*16-19*).

Konoshima, Kozuka and their coworkers at Kyoto Pharmaceutical University with the cooperation of Tokuda's group at the Kyoto Prefectural University of Medicine observed the inhibitory effects of methanol extracts of ginseng and its congeners on the EBV-EA test. Of these extracts, Sanchi-Ginseng exhibited the strongest inhibitory effect (*20,21*). It was also demonstrated that the extract of Sanchi-Ginseng strongly inhibited the skin tumor promoting effect of TPA in mice initiated with DMBA.

Furthermore, the extract of Sanchi-Ginseng inhibited pulmonary tumorigenesis in mice initiated with 4-nitroquinoline-*N*-oxide (4-NQO) and promoted with glycerol. Daily administration of the extract reduced the incidence of tumors by nearly 50%, and average number of tumors per mouse decreased from 3.0 to 0.67.

The methanol extract of Sanchi-Ginseng was extracted to give hexane, chloroform, ethyl acetate and butanol (saponin-fraction) fractions. Each fraction was subjected to the EBV-EA test. Strong inhibiting activity was observed for the nonpolar hexane and chloroform fractions. From the hexane fraction, two acetylene alcohols — panaxytriol and panaxynol — were isolated, both of which, especially panaxytriol, exhibited potent inhibition in the EBV-EA test and cytotoxicity against Raji cells. As already mentioned, both these alcohols have already proved to have strong cytotoxicities against a variety of cultured cancer cells.

On behalf of the medicinal chemists examining ginseng, I have reviewed some of the recent studies on the cancer prevention of ginseng and its congeners. In order to elucidate whether or not ginseng and its congeners are really effective for the prevention of cancer and what type of compounds participates in this efficacy, much more extensive chemical, biochemical and clinical studies must be conducted.

Acknowledgements

I am very much grateful to Drs. T. Konoshima and H. Kozuka at Kyoto Pharmaceutical University and Dr. H. Tokuda at the Kyoto Prefectural University of Medicine for their permission to report on their studies on the anticarcinogenic activity of Sanchi-Ginseng before publication. My thanks are also due to Dr. T.-K. Yun of the Korea Cancer Hospital, Korea Atomic Energy Research Institute, Seoul, and Professor M. Takido and Dr. K. Yasukawa at the College of Pharmacy of Nihon University, Chiba, for detailed information about their recent studies.

Literature Cited

1. Tanaka, O.; Kasai, R. In *Saponins of Ginseng and Related Plants* Herz, W.; Griesebach, H.; Kirby, G.W.; Tamm, C., Eds.; Progress in the Chemistry of Organic Natural Products; Springer-Verlag: New York, 1984; Vol.46, pp 1–76.
2. Shibata, S.; Tanaka, O.; Shoji, J.; Saito, H. In *Chemistry and Pharmacology of Panax* Wagner, H.; Hikino, H.; Farnsworth, N.R., Eds.; Economic and Medicinal Plant Research; Academic Press: Tokyo, 1985; Vol. 1, pp 218–284.
3. Tanaka, O. In *Saponin-Composition of Panax Species* Shibata, S.; Ohtsuka, T.; Saito, H., Eds.; Recent Advances in Ginseng Studies; Hirokawa Publishing Company: Tokyo, 1990; pp 43–48.
4. Korea Ginseng and Tobacco Research Institute *Proceedings of the 1st-4th International Ginseng Symposia* Daejeon, Korea, **1976, 1978, 1980, 1984.**
5. Kitagawa, I.; Taniyama, T.; Shibuya, H.; Noda, T.; Yoshikawa, M. *Yakugaku Zasshi* **1987**, *107*, 495–505.
6. Odashima, S.; Nakayabu; Honjo, N.; Abe, H.; Arichi, S. *Europ. J. Cancer* **1979**, *15*, 885–892 and references cited therein.
7. Matsunaga, H.; Katano, M.; Yamamoto, H. *Chem. Pharm. Bull.* **1989**, *37*, 1279–1281 and references cited therein.
8. Breckman, I.I.; Dardymov, I.V. *Ann. Rev. Pharmacol.* **1969**, *9*, 419–430.
9. Takeda, A.; Katoh, N.; Yonezawa, M. *J. Radiat. Res.* **1982**, *23*, 150–167.
10. Yun, T.-K.; Yun, Y.-S.; Han, I.-W. *Cancer Detect. Prev.* **1983**, *6*, 515–525.
11. Yun, Y.-S.; Jo, S.-K.; Moon, H.-S.; Kim, Y.-J.; Oh, Y.-R.; Yun, T.-K. *Cancer Detect. Prev.* **1987**, *Suppl. 1*, 301–309.
12. Yasukawa, K.; Takido, M.; Matsumot, T.; Takeuchi, N.; Nakagawa, S. *Oncology* **1991**, *48*, 72–76.
13. Yun, T.-K.; Choi, S.-Y. *Internat. J. Epidemiol.* **1990**, *19*, 871–876.
14. Tanaka, O. *Pure Appl. Chem.* **1990**, *62*, 1281–1284.
15. Tanaka, O. *Gakujutu Geppo* **1991**, *44*, 590–596.

16. Ito, Y.; Kawanishi, M.; Harayama, T.; Takabayashi, S. *Cancer Lett.* **1981**, *12*, 175–180.
17. Yamamoto, N.; Hausen, H.Z. *Nature* **1979**, *280*, 244–245.
18. Hausen, H. Z.; Bornkamm, G. W.; Schmidt, R.; Hecker, E. *Proc. Natl. Acad. Sci. USA* **1979**, *76*, 782–785.
19. Konoshima, T.; Kokumai, M.; Kozuka, M.; Iinuma, M.; Mizuno, M.; Tanaka, T.; Tokuda, H.; Nishino, H.; Iwashima, A. *Chem. Pharm. Bull.* **1992**, *40*, 531–533.
20. Konoshima, T.; Takasaki, M.; Kozuka, M.; Tokuda, H.; Nishino, H.; Iwashima, A.; Tabe, M. *Proc. 37th Annual Meeting of the Japanese Society of Pharmacognosy (Chiba)* **1990**, 135.
21. Konoshima, T.; Kokumai, M.; Kozuka, M.; Haruna, M.; Ito, K.; Tokuda, H.; Nishino, H.; Iwashima, A.; Tabe, M. *Proc. 38th Annual Meeting of the Japanese Society of Pharmacognosy (Kobe)* **1991**, 125.

RECEIVED June 7, 1993

Chapter 35

Natural Products and Biological Activities of the Chinese Medicinal Fungus *Ganoderma lucidum*

Ming-Shi Shiao[1], Kuan Rong Lee[2], Lee-Juian Lin[1], and Cheng-Teh Wang[2]

[1]Department of Medical Research, Veteran's General Hospital—Taipei, Taipei, Taiwan 11217, Republic of China
[2]Institute of Life Science, National Tsing Hua University, Hsinchu, Taiwan 33034, Republic of China

Ganoderma lucidum (Fr.) Karst, a fungus (Polyporaceae) used in traditional Chinese medicine, has attracted great attention recently. It produces polysaccharides with antitumor and hypoglycemic activities and many highly oxygenated lanostanoid triterpenes including multiple pairs of C-3 stereoisomers and C-3/C-15 positional isomers. Separation by reverse phase HPLC and ^{13}C NMR correlation of stereochemistry for these triterpenes are described. Biosynthetic studies indicate that oxygenated triterpenes in the 3α series are derived from those in the 3β series through an oxidation-reduction pathway. Biological activities reported for the oxygenated triterpenes include bitterness, cytotoxicity to hepatoma cells, inhibition of histamine release, angiotensin converting enzyme and cholesterol biosynthesis, and stimulation as well as inhibition of thrombin-induced platelet aggregation.

Traditional Chinese medicine has been practiced in a clinical manner. The chemical and pharmacological bases, however, are not well understood in most cases. Chinese medicine emphasizes whole-body homeostasis instead of disease oriented therapy. There is no strict borderline between drug treatment and dietary manipulation. Food and herbal drugs are utilized together for the prevention and cure of illnesses. Ginseng (*Panax genseng*) and *Ganoderma lucidum* (Fr.) Karst are two famous medical herbs in this regard.

G. *lucidum*, a mushroom-like fungus, is used as a tonic and sedative drug. The fruiting bodies and cultured mycelia are also prescribed to treat chronic hepatopathy, hypertension, bronchitis, arthritis, neurasthenia and neoplasia in China and other Asian countries (*1*). This fungus has attracted great attention recently due to its production of polysaccharides with antitumor (*2–3*) and hypoglycemic activities (*4–5*). Most interestingly, it also produces over one hundred species of oxygenated triterpenes (*1,6–30*) with various biological functions (Table I). The natural product chemistry and biological activities of fungus G. *lucidum* are discussed.

0097–6156/94/0547–0342$06.00/0

Table I. Biological Activities Reported for the Natural Products of the Fungus
Ganoderma lucidum

Biological activity	Type of compound
Antitumor activity	polysaccharides
Hypoglycemic activity	polysaccharides
Inhibition of platelet aggregation	adenosine
Bitterness	triterpenes
Cytotoxicity to hepatoma cells	triterpenes
Inhibition of histamine release	triterpenes
Inhibition of ACE	triterpenes
Inhibition of cholesterol absorption	triterpenes
Inhibition of cholesterol synthesis	triterpenes
Stimulation of platelet aggregation	triterpenes
Inhibition of thrombin-induced platelet aggregation	triterpenes

Biological Activities of *G. lucidum* Components

Inhibitors of Platelet Aggregation. The water-soluble fraction of *G. lucidum* suppresses platelet aggregation and the inhibitory substance has been identified as adenosine (*31*). The content of adenosine varies greatly among different strains of *G. lucidum*. Several well known fungi in Chinese medicine, such as *Lentinus edodes*, *Auricularia polytricha* and *Cordyceps sinensis* all contain considerable amounts of adenosine. It has been reported that consumption of the fruiting bodies of *A. polytricha* in a large meal of Oriental foods induces transient inhibition of platelet aggregation and mild hemorrhagic diathesis (*32*). The levels of adenosine in *G. lucidum* and *A. polytricha* are about the same.

Antitumor Polysaccharides. Several polysaccharides isolated from *G. lucidum* were reported to possess antitumor properties (*3*). The bioassays are based on the inhibition of growth of Sarcoma 180 solid tumor and many cultured tumor cell lines. The antitumor polysaccharides contain branched glycan core in (1-3)-β-, (1-4)-β- and (1-6)-β-linkages (*2*). Polysaccharides from *G. lucidum* also enhance the production of interleukin-2 activity in concanavalin A treated splenocytes, which are isolated from the BALB/C mice previously treated with the antitumor polysaccharides (*33*). The production of antitumor polysaccharides does not seem limited to the species *G. lucidum*. Similar activity is also found in *G. applanatum* (*34*). The antitumor polysaccharides are reported to act as biological response modifiers in host defense mechanisms against cancer.

Hypoglycemic Polysaccharides. Hypoglycemic glycans A, B, and C have been isolated from the fruiting bodies of *G. lucidum* (*5,35*). The potencies of these hypoglycemic polysaccharides are not in parallel with their antitumor effects.

Ganoderan B is the major hypoglycemic glycan in *Ganoderma*. It contains (1-3)-β- and (1-6)-β-linked D-glucopyranose moieties (*4*). Ganoderan B increases plasma insulin levels in normal and glucose-loaded mice. Administration of ganoderan B (100 mg/kg, intraperitoneal injection) increases the activities of hepatic glucokinase, phosphofructokinase and glucose-6-phosphate dehydrogenase. Meanwhile, hepatic glucose-6-phosphatase and glycogen synthetase activities are decreased. Ganoderan B also reduces hepatic glycogen content (*4*).

Natural Product Chemistry of Oxygenated Triterpenes in *Ganoderma lucidum*

The structures, separation and biosynthesis of oxygenated triterpenes in *G. lucidum* are summarized briefly.

Structural Characteristics. The triterpenoid natural products produced by *G. lucidum* are primarily the lanostanoid type containing a very broad spectrum of oxygenated functionalities at C-3, 7, 11, 12, 15, 22, 23, 24, 25, 26 and 27 (Figure 1). Up to now, over one hundred species of triterpenes have been isolated from this fungus (*1,6–30*). Many of these triterpenes have been identified only in this species. Highly oxygenated triterpenoids with side chain cleavage at C-20 and C-24 (lucidenic acid series) are also encountered (*36*) (Figure 1). This fungus produces multiple pairs of oxygenated triterpenes which are C-3 stereoisomers and C-3/C-15 positional isomers (*37*) (Figure 2).

Separation and Isolation. Separation of numerous oxygenated triterpenes in *G. lucidum* by normal phase high performance liquid chromatography (HPLC) and silica gel TLC is extremely difficult (*38*). Reverse phase HPLC (RP-HPLC) has been chosen as the method for separation (*36,38–42*). This tool is particularly useful for the separation of a mixture of triterpenoid compounds containing stereo- and positional isomers (Figure 3). The capacity factors (k′) of several representative triterpenes have been measured with methanol/water and acetonitrile/water as the mobile phase in RP-HPLC (*41*), making it possible to identify the molecular polarity due to the contribution of an individual functional group. The molecular hydrophobic nature of these oxygenated triterpenes was compared in an empirical manner by measuring their partition coefficients (P_{oct}) between 1-octanol and water. The P_{oct} values correlate well with the k′ values in RP-HPLC.

Empirical rules were developed to correlate the molecular polarity due to the presence of hydroxyl, acetoxyl, oxo and carboxyl groups. Number, position, and stereochemistry of the functional groups each plays a characteristic role in governing the molecular polarity. The weights of polarity factors are in the following order: 3β-OH>3α-OH>3α-OAc>3β-OAc. The contributions to molecular polarity due to 15α-OAc and 22β-OAc are very similar (*40*). The trend of elution sequences for these oxygenated triterpenes provides valuable information to predict the stereochemistry of polar functional groups in a previously unknown triterpene. Many C-26 carboxyl containing triterpenes including several pairs of stereoisomers and positional isomers have been successfully purified by semi-preparative RP-HPLC using this approach (Figure 2).

Oxygenated Triterpenes for Chemical Taxonomy. Many oxygenated triterpenes have been identified only in the genus *Ganoderma*. The triterpenoid patterns vary during the formation of fruiting bodies (*39*). Strain specificity in the species *G. lucidum* was also reported (*36*). Evidently, the triterpenoid patterns give clues to distinguish this medicinal fungus from other taxonomically related species. They can serve as supporting evidence for classification (*42*). The determination of

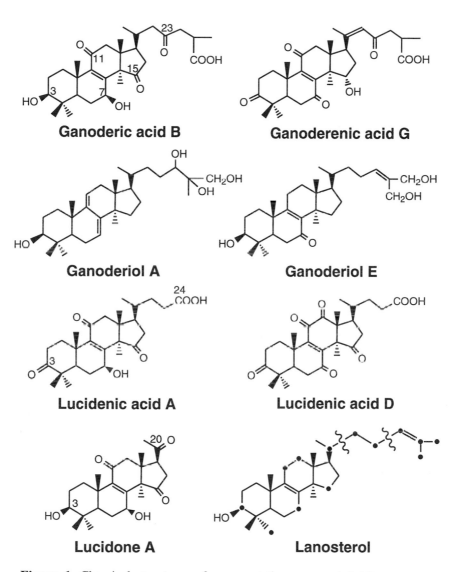

Figure 1. Chemical structures of representative oxygenated triterpenes to illustrate the post-lanosterol modifications in *Ganoderma lucidum* (Fr.) Karst. The positions of oxidative modifications and cleavage sites in the lanosterol skeleton are pointed out.

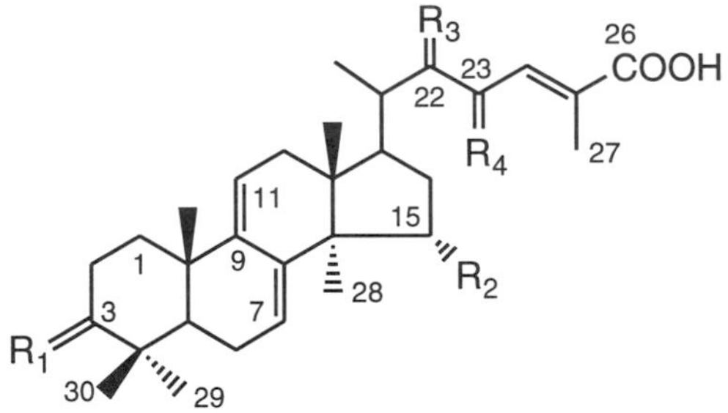

R_3 26
23 COOH
22
R_4 27

11
13
15
1 9
28
7 R_2
3
R_1
30 29

	R_1	R_2	R_3	R_4		R_1	R_2	R_3	R_4
1	₍ᵐ₎OAc / H	OAc	H_2	H_2	14	OH / H	OAc	H_2	O
2	H / OAc	OAc	H_2	H_2	15	OAc / H	OAc	H_2	O
3	OH / H	OAc	H_2	H_2	16	OAc / H	OH	H_2	O
4	OAc / H	OH	H_2	H_2	17	OAc / H	OH	OH / H	H_2
5	H / OH	OAc	H_2	H_2	18	OAc / H	OAc	OH / H	H_2
6	H / OAc	OH	H_2	H_2	19	H / OAc	OAc	OH / H	H_2
7	=O	OAc	H_2	H_2	20	H / OH	OH	H / OAc	H_2
8	OH / H	OH	H_2	H_2	21	OH / H	OH	H / OAc	H_2
9	H / OH	OH	H_2	H_2	22	H / OH	OH	H / OH	H_2
10	OAc / H	OH	H / OAc	H_2	23	OH / H	OH	OH / H	H_2
11	H / OH	OAc	H / OAc	H_2	24	OH / H	H	H_2	H_2
12	OAc / H	OAc	H / OAc	H_2	25	OAc / H	H	H_2	H_2
13	H / OAc	OAc	H / OAc	H_2					

Figure 2. Several C-26 carboxyl-containing triterpenes isolated from cultured mycelia of *Ganoderma lucidum* (Strain TP-1). Multiple pairs of C-3 stereo-isomers (compounds 1/2, 3/5, 4/6, 8/9, 12/13, 18/19 and 20/21) and C-3/C-15 positional isomers (compounds 3/4, 5/6 and 14/16) have been identified.

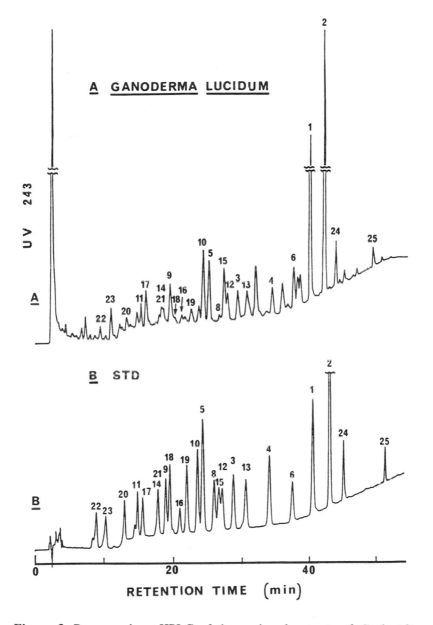

Figure 3. Reverse phase HPLC of the methanol extracts of *G. lucidium* (chromatogram A) and authentic triterpenoid compound mixture (chromatogram B).

triterpenoid profiles by RP-HPLC is useful to screen for bioactive compounds from various species of *Ganoderma* at different culture stages (*42*).

Structural Elucidation and Stereochemical Correlation by ^{13}C NMR Spectroscopy.

Accumulation of ^{13}C NMR and 2-D NMR correlational spectral information on these triterpenes supports the structural assignments based on other spectroscopic methods (*37,43–44*). The configuration of hydroxyl and acetoxyl substituents at C-3 affects ^{13}C chemical shifts of adjacent carbons in a consistent and predictable manner. C-3 carbons bearing equatorial substituents are more deshielded than those with axial substituents. The difference in ^{13}C chemical shifts due to this configurational difference is in the range of 2-3 ppm (*37*). For triterpenes with 3β-substituents, their C-30 carbons resonate at δ 15.40-16.90 ppm. They exhibit a large upfield shift of C-30 carbon signals (by 5–7 ppm) relative to the 3α-substituted counterparts which resonate at δ 22.14–22.81 ppm. The C-29 signals of all C-3 epimers (at δ 27–28 ppm) remain for the most part undisturbed. Epimerization at C-3 also affects other carbons in the vicinity. The C-1 and C-5 signals are shifted in upfield direction by about 5 ppm in the 3α-substituted series relative to the corresponding 3β series. This γ-effect of C-3 substituents on C-1 and C-5 signals provides the most useful evidence to support the assignment of configuration at C-3 (*37*).

A similar trend was also observed in compounds with varied configuration at C-22. The C-22 signals of the 22α-hydroxylated series have δ values about 5 ppm smaller than the corresponding 22β-hydroxylated epimers. Their C-22 signals appear in a narrow, characteristic range (at δ 74.36–74.87 ppm for compounds **10–13**, **20**, and **21** in Figure 2) further confirming that these C-22 acetylated triterpenes are all β-substituted. The substitution parameters on C-22 chemical shifts due to hydroxylation alone and further acetylation are additive.

Functionalization of C-23 as an oxo group results in a characteristic alternation of the signals corresponding to the carbons of the side chain (compounds **14–16**, Figure 2). The 23-oxo carbons resonate at δ 201.44–201.75 ppm and cause a concomitant upfield shift of C-20, C-24, and C-26 signals. C-22, C-25, and C-27 carbons, however, are deshielded. This effect confirms that the 23-oxo group has a configuration *cis* to the C-27 methyl groups.

Biosynthesis. The isoprenoid origin of triterpenes in *G. lucidum* has been demonstrated by an experiment using both incorporation of a stable isotope and enhancement of the intensity of the ^{13}C NMR (*45*) (Figure 4). The incorporation of [5-^{13}C]mevalonate into two oxygenated triterpenes (compounds **1** and **2** in Figure 2) by a liquid culture of *G. lucidum* has shown that the ^{13}C signal intensities of C-2, C-6, C-11, C-12, C-16, C-23 of compound **1** and compound **2** are significantly increased (*45*). The biosynthetic study by ^{13}C NMR also provides information for complete assignment of the ^{13}C chemical shifts of several olefinic carbons (*37,43–44*).

Many pairs of C-3 stereoisomers and C-3/C-15 positional isomers occur in this fungus. Further biosynthetic study has indicated that triterpenes in the 3α series are derived from those in the 3β series through an oxidation-reduction pathway (Figure 4) (*39,45*). Evidence that such a mechanism is involved in the formation of 3α stereoisomers is provided by the following observations. Multiple pairs of 3α stereoisomers are produced in the mycelial culture during the production period of 3β triterpenes. 3-Oxo triterpenes occur in the mycelia of the culture (e.g. compound

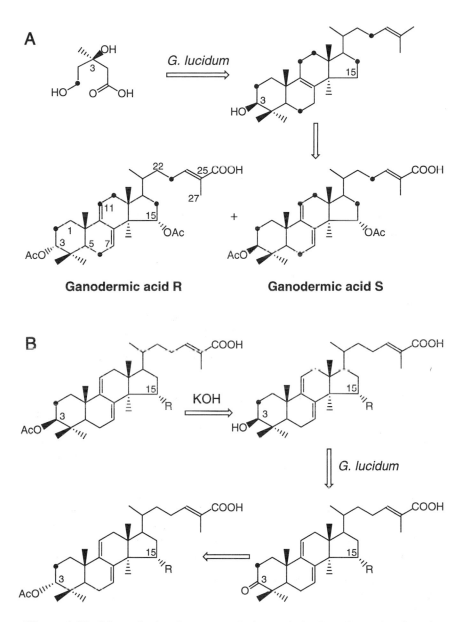

Figure 4. The biosynthesis of oxygenated triterpenes in *Ganoderma lucidum*. A: The isoprenoid origin confirmed by the incorporation of [5-^{13}C]mevalonate (•) into ganodermic acid R and ganodermic acid S. B: Proposed pathway for the conversion of 3β triterpenes into 3α series.

7 in Figure 2) (*18*). ^{3}H-labeled lanosterol is converted to triterpenes in both the 3α and 3β series with comparable efficiency. When added to the liquid culture at an appropriate time, labeled 3β triterpenes are converted to the 3α series metabolites (Figure 4).

Various oxidative modifications at the lanostanoid ring skeleton and oxidative cleavage of the side chain were found in *G. lucidum* (Figure 1). This fungus is a remarkable system to look for the biotransformation of terpenoid and steroid compounds.

Biological and Pharmacological Activities of Oxygenated Triterpenes

Bitterness. Oxygenated triterpenes exhibiting intense bitterness have been isolated from *G. lucidum*. They include a series of derivatives of ganoderic acids, lucidenic acids and a furanoganoderic acid (*6–8*). Relative bitterness is compared by the threshold values with equal concentrations to naringin and quinine sulfate. The intensity of bitterness is lucidenic acid D > ganoderic acid C > lucidone A > lucidenic acid A > ganoderic acid B > lucidenic acid B, lucidenic acid C and lucidenic acid E (*7*). Bitterness and structure relationship studies conclude that the spatial relationship of the hydrophobic methyl groups to the three functional oxygen atoms play an important role in generating bitterness (*46*).

Hepatoprotection and Cytotoxicity to Hepatoma Cells. The cultured mycelium of *G. lucidum* and a few purified triterpenoid compounds, namely ganodosterone and ganoderic acids (S and R, and trideacetyl ganoderic acid T) protect against D-galactosamine-induced hepatic injuries (*47*). At least six cytotoxic oxygenated triterpenes (ganoderic acid U–Z) were isolated from the cultured mycelium of this fungus by a cytotoxicity test on hepatoma cells grown *in vitro* (*12*).

Inhibition of Histamine Release. Two histamine release inhibitory triterpenes, namely ganoderic acid C (3β,7β,15α-trihydroxy-11,23-dioxo-5α-lansta-8-en-26-oic acid) and ganoderic acid D (7β-hydroxy-3,11,15,23-tetraoxo-5α-lanost-8-en-26-oic acid) were isolated from the methanolic extract of *G. lucidum* (*10*). Quantitative analysis of these triterpenes has been performed to evaluate the inhibitory action on histamine release from rat peritoneal mast cells.

Inhibition of Angiotensin Converting Enzyme (ACE). Several highly oxygenated triterpenes from aqueous methanol extracts have been identified to be mild inhibitors of ACE (*11*). At least five triterpenes (ganoderal A, ganoderol A and B, and ganoderic acids K and S) have been structurally determined. The IC_{50} values fall within 10^{-4} to 5×10^{-6} M. Several other triterpenes which contain oxygenated functionalities at C-3, C-15, C-23 and a carboxyl group at C-26 are active in an *in vitro* assay by using an artificial substrate of ACE. None of these triterpenes, however, are as active as the positive control captopril.

Hypolipidemic Activities. Several oxygenated triterpenes from *G. lucidum* are structurally similar to the post-lanosterol intermediates in the mammalian cholesterol biosynthetic pathway. Hence, to elucidate the hypocholesterolemic potentials of these oxygenated triterpenes seems worthwhile (*48–49*). Screening for cholesterol lowering activities of *G. lucidum* has been performed by using *in vivo* and *in vitro* assays. The potential to inhibit cholesterol absorption and key enzymes in cholesterol biosynthesis have been examined.

Pertaining to blockage of cholesterol absorption, several oxygenated tri-terpenes (compounds **1**, **8** and **9** in Figure 2) have been demonstrated to reduce the absorption of dietary cholesterol in experimental animals fed 2% cholesterol. These oxygenated triterpenes all contain a C-26 carboxyl group. They are poorly absorbed in the gastrointestinal tract of rats and are more potent than β-sitosterol.

Inhibition of the rate-limiting enzyme 3-hydroxy-3-methylglutaryl co-enzyme A reductase (HMG-CoA reductase) in the cholesterol biosynthetic pathway by oxygenated triterpenes has also been elucidated. Several oxygenated triterpenes inhibit cholesterol synthesis by inhibiting this key enzyme in an *in vivo* assay with mevinolin as a positive control. By using rat liver slices from animals initially fed cholestyramine, the incorporation of [2-^{3}H]acetate and [2-^{14}C]mevalonate into cholesterol in the presence of the test compounds were reduced. A few triterpenes, which contain oxygenated functionalities at C-15 and a hydroxyl group at C-26, have been demonstrated to inhibit cholesterol biosynthesis.

Post-mevalonate inhibition was also demonstrated for a few oxygenated triterpenes from *G. lucidum*. The target site is lanosterol 14α-methyl demethylase (*49*). Some oxygenated triterpenes containing a 15α hydroxyl substituent inhibit this enzyme (*50*).

Stimulation of Platelet Aggregation by High Dose Ganodermic Acid S.

Ganodermic acid S (GAS, compound **2** in Figure 2) is a major oxygenated triterpene in *G. lucidum* (Figure 3). The biochemical effects of this compound on human platelet aggregation have been investigated in great detail (*29–30*). Incubation of gel filtered human platelets in ganodermic acid S shows that within a minute, 80% of the agent is absorbed by cells. The process of uptake is a simple diffusion and the partition coefficient is about 10^5 (*29*).

GAS causes platelet aggregation at a concentration above 20 mM. Above the threshold, the extent of cell aggregation is linearly proportional to the agent concentration. The percentage of cell aggregation is comparable to the elevation of cytosolic free Ca^{+2} concentration, protein phosphorylation and serotonin release; it is correlated with altered interconversion of phosphoinositides. Platelets at various concentrations of GAS show distinct temporal profiles in alteration of [^{32}P]phospho-inositides and [^{32}P]phosphatidic acid (PA). Upon addition of the agent, platelets show an initial increase in all [^{32}P]phosphoinositides. Subsequently, the level of phosphoinositide of each kind decreases sequentially in phosphatidylinositol 4,5-*bis*-phosphate (PIP$_2$), phosphatidylinositol 4-phosphate (PIP) and phosphatidyl-inositol (PI). Below the aggregation threshold, platelets exhibit neither the resynthesis of [^{32}P]PIP$_2$ and [^{32}P]PIP nor the accumulation of [^{32}P]PA. At 25 and 50 μM, however, platelets show not only the resynthesis of [^{32}P]PIP$_2$ and [^{32}P]PIP but also the accumulation of [^{32}P]PA. At 100 mM GAS, platelets show no resynthesis of [^{32}P]PIP$_2$ and [^{32}P]PIP. In this case, the levels of accumulation of [^{32}P]PA and decrease of [^{32}P]PI are less than those found in platelets at 50 μM GAS. The results indicate that GAS causes activation of PIP$_2$ hydrolysis (*29*).

Scanning electron microscopy shows that the morphology of platelets below the aggregation threshold is in the shape of spiculate discoid. Above the threshold, the cells become rounded to spiculate irregular forms, which show elongation of filopodia after prolonged incubation. In addition, platelets at above 50 μM GAS display membrane vesiculation (*29*).

Inhibition of Thrombin-Induced Platelet Aggregation by Low Dose Ganodermic Acid S. GAS displays both concentration and time dependent inhibition of thrombin-induced platelet aggregation, in which the agent potency in response to inducers was ADP-fibrinogen > collagen > thrombin (*29*). GAS caused a biphasic time dependent effect on platelet phosphoinositide metabolism. Platelets initially become spiculate discs, then swelled to a potato-like morphology.

The incorporation of this triterpene into the platelet membrane results in altered membrane morphology (*29*). The distinct effects of GAS at high and low concentrations indicate that insertion of this agent into platelet membrane brings about multiple effects on phosphoinositide metabolism. The responses of platelets to GAS depend on the differential accumulation of this amphipathic agent in the two membrane leaflets.

Structure-Activity Relationship in the Induction of Platelet Aggregation by Oxygenated Triterpenes. As diverse structural variations appear among the oxygenated triterpenes of *G. lucidum*, it is possible to explore the roles of functional groups, molecular polarity and stereochemistry of oxygenated triterpenes on human platelet aggregation induced by various effectors. Eight oxygenated triterpenes, which are four pairs of C-3 stereoisomers (compounds 1/2, 3/5, 4/6 and 8/9 in Figure 2) and two pairs of C-3/C-15 positional isomers containing a common C-26 carboxyl group (compounds 3/4 and 5/6 in Figure 2), were elucidated. These lanostanoid triterpenes activated platelet phospholipase C (PLC) in a structurally related manner. The trend of activating PLC is triterpenes with two acetoxy substituents > one acetoxy and one hydroxyl substituent > two hydroxyl substituents. The triterpene containing a substituent at C-3β is more effective than its epimer.

Acknowledgments

The authors wish to thank Professors S. F. Yeh and Dr. T. C. Tseng for valuable advice. We thank C. S. Chou, K. J. Lee, J. S. Chen, C. Y. Lien, N. Y. Tsai, T. N. Wang and R. Chyr for technical assistance. The work carried out in this laboratory is supported by the National Science Council (NSC77-0412-B-075-03) and Veterans General Hospital-Taipei, Taiwan, Republic of China.

Literature Cited

1. Arisawa, M.; Fujita, A.; Saga, M.; Fukumura, H.; Hayashi, T.; Shimizu, M.; Morita, N. *J. Natl. Prod.* **1986**, *49*, 621.
2. Sone, Y.; Okuda, R.; Wada, N.; Kishida, E.; Misaki, A. *Agric. Biol. Chem.* **1985**, *49*, 2641.
3. Miyazaki, T.; Nishijima, M. *Chem. Pharm. Bull.* **1981**, *29*, 3611.
4. Hikino, H.; Ishiyama, M.; Suzuki, Y.; Konno, C. *Planta Medica* **1989**, *55*, 423.
5. Hikino, H.; Konno, C.; Mirin, Y.; Hayashi, T. *Planta Medica* **1985**, *51*, 339.
6. Kubota, T.; Asaka, Y.; Miura, I.; Mori, H. *Hel. Chim. Acta* **1982**, *65*, 611.
7. Nishitoba, T.; Sato, H.; Sakamura, S. *Agric. Biol. Chem.* **1985**, *49*, 1547.
8. Nishitoba, T.; Sato, H.; Kasai, T.; Kawagishi, H.; Sakamura, S. *Agric. Biol. Chem.* **1985**, *49*, 1793.
9. Nishitoba, T.; Goto, S.; Sato, H.; Sakamura, S. *Phytochemistry* **1989**, *28*, 193.

10. Kohda, H.; Tokumoto, W.; Sakamoto, K.; Fujii, M.; Hirai, Y.; Yamasaki, K.; Komoda, Y.; Nakamura, H.; Ishihara, S.; Uchida, M. *Chem. Pharm. Bull.* **1985**, *33*, 1367.

11. Morigiwa, A.; Kitabatake, K.; Fujimoto, Y.; Ikekawa, H. *Chem. Pharm. Bull.* **1986**, *34*, 3025.

12. Toth, J. O.; Luu, B.; Ourisson, G. *Tetrahedron Lett.* **1983**, *24*, 1081.

13. Kikuchi, T.; Matsuda, S.; Kadota, S.; Murai, Y.; Ogita, Z. *Chem. Pharm. Bull.* **1985**, *33*, 2624.

14. Nishitoba, T.; Sato, H.; Shirasu, S.; Sakamura, S. *Agric. Biol. Chem.* **1987**, *51*, 619.

15. Nishitoba, T.; Sato, H.; Sakamura, S. *Agric. Biol. Chem.* **1987**, *51*, 1149.

16. Nishitoba, T.; Oda, K.; Sato, H.; Sakamura, S. *Agric. Biol. Chem.* **1988**, *52*, 367.

17. Shiao, M.-S.; Lin, L.-J.; Yeh, S.-F. *Phytochemistry* **1988**, *27*, 873.

18. Lin, L.-J.; Shiao, M.-S.; Yeh, S.-F. *Phytochemistry* **1988**, *27*, 2269.

19. Shiao, M.-S.; Lin, L.-J.; Yeh, S.-F. *Phytochemistry* **1988**, *27*, 2911.

20. Lin, L.-J.; Shiao, M.-S.; Yeh, S.-F. *J. Natl. Prod.* **1988**, *51*, 918.

21. Yeh, S. F.; Lee, K. C.; Shiao, M.-S. *Proc. Natl. Sci. Counc. ROC(A)* **1987**, *11*, 129.

22. Shiao, M.-S.; Lin, L.-J.; Yeh, S. F.; Chou, C.-S. *J. Natl. Prod.* **1987**, *50*, 891.

23. Kikuchi, T.; Kanomi, S.; Kadota, S.; Murai, Y.; Tsubono, K.; Ogita, Z.-I. *Chem. Pharm. Bull.* **1986**, *34*, 3695.

24. Kikuchi, T.; Kanomi, S.; Murai, Y.; Kadota, S.; Tsubono, K.; Ogita, Z.-I. *Chem. Pharm. Bull.* **1986**, *34*, 4018.

25. Hiratani, M.; Ino, C.; Furuya, T.; Shiro, M. *Chem. Pharm. Bull.* **1986**, *34*, 2282.

26. Hiratani, M.; Asaka, I.; Ino, C.; Furuya, T.; Shiro, M. *Phytochemistry* **1987**, *26*, 2797.

27. Sato, H.; Nishitoba, T.; Shirasu, S.; Oda, K.; Sakamura, S. *Agric. Biol. Chem.* **1986**, *50*, 2887.

28. Wang, T.-N.; Chen, J.-C.; Shiao, M.-S.; Wang, C.-T. *Biochim. Biophys. Acta* **1989**, *986*, 151.

29. Wang, C.-N.; Chen, J.-C.; Shiao, M.-S.; Wang, C.-T. *Biochem. J.* **1991**, *277*, 189.

30. Fujita, A.; Arisawa, M.; Saga, M.; Hayashi, T.; Morita, N *J. Natl. Prod.* **1986**, *49*, 1122.

31. Shimizu, A.; Yano, T.; Saito, Y.; Inada, Y.; *Chem. Pharm. Bull.* **1986**, *33*, 3012.

32. Hammerschmidt, D. E. *New Engl. J. Med.* **1980**, *302*, 1191.

33. Cheng, H.-H.; Hsieh, K.-H.; Tung, Y.-C.; Tung, T.-C. *J. Chin. Oncol. Soc.* **1988**, *4*, 13.

34. Usui, T.; Iwasaki, Y.; Hayashi, K.; Mizuno, T.; Tanaka, M.; Shinkai, K.; Arakawa, M. *Agric. Biol. Chem.* **1981**, *45*, 323.

35. Tomoda, M.; Gonda, R.; Kasahara, Y.; Hikino, H. *Phytochemistry* **1986**, *25*, 2817.

36. Nishitoba, T.; Sato, H.; Shirasu, S.; Sakamura, S. *Agric. Biol. Chem.* **1986**, *50*, 2151.

37. Lin, L.-J.; Lee, K. R.; Shiao, M.-S. *J. Natl. Prod.* **1989**, *52*, 595.

38. Lin, L.-J; Shiao, M.-S., *J. Chromatogr.* **1987**, *410*, 195.

39. Hirotani, M.; Furuya, T. *Phytochemistry* **1990**, *29*, 3767.

40. Shiao, M.-S.; Lin, L.-J.; Chen, C.-S. *J. Lipid Res.* **1989**, *30*, 287.

41. Shiao, M.-S.; Lin, L.-J. *J. Chin. Chem. Soc.* **1989**, *36*, 429.
42. Chyr, R.; Shiao, M.-S. *J. Chromatogr.* **1991**, *542*, 327.
43. Shiao, M.-S.; Lin, L.-J.; Yeh, S.-F.; Chou, C.-S. *Proc. Natl. Sci. Counc. ROC(A)* **1988**, *12*, 10.
44. Arisawa, M.; Fujita, A.; Hayashi, T.; Shimizu, M.; Morita, N. *J. Natl Prod.*, **1988**, *51*, 54.
45. Yeh, S.-F.; Chou, C.-S.; Lin, L.-J.; Shiao, M.-S. *Proc. Natl. Sci. Counc. ROC(B)* **1989**, *13*, 119.
46. Nishitoba, T.; Sato, H.; Sakamura, S. *Agric. Biol. Chem.* **1988**, *52*, 1791.
47. Furuya, T.; Tsuda, Y.; Koga, N. *Jpn Kokai Tokkyo Koho* JP 62/270595 A2.
48. Shiao, M.-S.; Tseng, T.-C.; Hao, Y.-Y.; Shieh, Y. S. *Bot. Bull. Academia Sinica* **1986**, *27*, 139.
49. Komoda, Y.; Shimizu, M.; Sonoda, Y.; Sato, Y. *Chem. Pharm. Bull.* **1989**, *37*, 531.
50. Fisher, R. T.; Trzaskos, J. M.; Magolda, R. L.; Ko, S. S.; Brosz, C. S.; Larsen, B. *J. Biol. Chem.* **1991**, *266*, 6124.

RECEIVED July 6, 1993

INDEXES

Author Index

Affiliation Index

Subject Index

Production: Charlotte McNaughton
Indexing: Deborah H. Steiner
Acquisition: Rhonda Bitterli
Cover design: Amy Hayes

Printed and bound by Maple Press, York, PA

Bestsellers from ACS Books

The ACS Style Guide: A Manual for Authors and Editors
Edited by Janet S. Dodd
264 pp; clothbound ISBN 0–8412–0917–0; paperback ISBN 0–8412–0943–X

The Basics of Technical Communicating
By B. Edward Cain
ACS Professional Reference Book; 198 pp;
clothbound ISBN 0–8412–1451–4; paperback ISBN 0–8412–1452–2

Chemical Activities (student and teacher editions)
By Christie L. Borgford and Lee R. Summerlin
330 pp; spiralbound ISBN 0–8412–1417–4; teacher ed. ISBN 0–8412–1416–6

Chemical Demonstrations: A Sourcebook for Teachers,
Volumes 1 and 2, Second Edition
Volume 1 by Lee R. Summerlin and James L. Ealy, Jr.;
Vol. 1, 198 pp; spiralbound ISBN 0–8412–1481–6;
Volume 2 by Lee R. Summerlin, Christie L. Borgford, and Julie B. Ealy
Vol. 2, 234 pp; spiralbound ISBN 0–8412–1535–9

Chemistry and Crime: From Sherlock Holmes to Today's Courtroom
Edited by Samuel M. Gerber
135 pp; clothbound ISBN 0–8412–0784–4; paperback ISBN 0–8412–0785–2

Writing the Laboratory Notebook
By Howard M. Kanare
145 pp; clothbound ISBN 0–8412–0906–5; paperback ISBN 0–8412–0933–2

Developing a Chemical Hygiene Plan
By Jay A. Young, Warren K. Kingsley, and George H. Wahl, Jr.
paperback ISBN 0–8412–1876–5

Introduction to Microwave Sample Preparation: Theory and Practice
Edited by H. M. Kingston and Lois B. Jassie
263 pp; clothbound ISBN 0–8412–1450–6

Principles of Environmental Sampling
Edited by Lawrence H. Keith
ACS Professional Reference Book; 458 pp;
clothbound ISBN 0–8412–1173–6; paperback ISBN 0–8412–1437–9

Biotechnology and Materials Science: Chemistry for the Future
Edited by Mary L. Good (Jacqueline K. Barton, Associate Editor)
135 pp; clothbound ISBN 0–8412–1472–7; paperback ISBN 0–8412–1473–5

For further information and a free catalog of ACS books, contact:
American Chemical Society
Distribution Office, Department 225
1155 16th Street, NW, Washington, DC 20036
Telephone 800–227–5558